THE SYMBIOTIC PHENOMENON

ASTROPHYSICS AND
SPACE SCIENCE LIBRARY

A SERIES OF BOOKS ON THE RECENT DEVELOPMENTS
OF SPACE SCIENCE AND OF GENERAL GEOPHYSICS AND ASTROPHYSICS
PUBLISHED IN CONNECTION WITH THE JOURNAL
SPACE SCIENCE REVIEWS

VOLUME 145
PROCEEDINGS

THE SYMBIOTIC PHENOMENON

PROCEEDINGS OF THE 103RD COLLOQUIUM OF THE
INTERNATIONAL ASTRONOMICAL UNION,
HELD IN TORUN, POLAND, AUGUST 18-20, 1987

Edited by

JOANNA MIKOLAJEWSKA

Institute of Astronomy,
Nicolaus Copernicus University, Torun, Poland

MICHAEL FRIEDJUNG

Institut d'Astrophysique (CNRS), Paris, France

SCOTT J. KENYON

Harvard-Smithsonian Center for Astrophysics,
Cambridge, Massachusetts, U.S.A.

and

ROBERTO VIOTTI

Istituto Astrofisica Spaziale (CNR), Frascati, Italy

KLUWER ACADEMIC PUBLISHERS
DORDRECHT / BOSTON / LONDON

Library of Congress Cataloging in Publication Data

International Astronomical Union. Colloquium (103rd : 1987 : Torún,
Poland)
   The symbiotic phenomenon : proceedings of the 103rd Colloquium of
the International Astronomical Union, held in Torún, Poland, August
18-20, 1987 / edited by Joanna Mikolajewska ... [et al.].
      p.   cm. -- (Astrophysics and space science library ; v. 145)
   Includes index.
   ISBN-13: 978-94-010-7833-7   e-ISBN-13: 978-94-009-2969-2
   DOI: 10.1007/ 978-94-009-2969-2
   1. Stars, Symbiotic--Congresses.   I. Mikolajewska, Joanna.
II. Title. III. Series.
QB843.S96I58 1987
523.8'41--dc19   88-4392
                                                         CIP
ISBN-13: 978-94-010-7833-7

Published by Kluwer Academic Publishers,
P.O. Box 17, 3300 AA Dordrecht, The Netherlands.

Kluwer Academic Publishers incorporates
the publishing programmes of
D. Reidel, Martinus Nijhoff, Dr W. Junk and MTP Press.

Sold and distributed in the U.S.A. and Canada
by Kluwer Academic Publishers,
101 Philip Drive, Norwell, MA 02061, U.S.A.

In all other countries, sold and distributed
by Kluwer Academic Publishers Group,
P.O. Box 322, 3300 AH Dordrecht, The Netherlands.

TABLE OF CONTENTS

PREFACE

Symbiotic stars were identified spectroscopically as M giants with a very strong He II 4686 emission line. After five decades of study by many astronomers, the first internatioinal meetings devoted to symbiotics were held at the University of Colorado (Boulder) and at the Haute Provence Observatory during the Summer of 1981. These conferences emphasized exciting new results obtained by modern satellite (EINSTEIN, IUE) and ground-based observatories. Although the vast majority of the participants were already fairly sure that symbiotics are almost certainly interacting binary systems, and not extremely peculiar single stars, it was not clear exactly which types of physical processes were needed to be invoked to explain their observed behaviour. Many were even worried that it might not be possible to clearly define a class of "symbiotic stars", and thus establish a unique model applicable to any system.

Since the publication of the Haute-Provence proceedings, our understanding of the physical processes occuring in symbiotic stars (and in related objects such as cataclysmic variables and compact planetary nebulae) has greatly improved. We now speak confidently of a "symbiotic phenomenon", in which an evolved red giant and a hot companion object (usually thought to be an accreting main sequence star or a luminous white dwarf star) happily coexist. Given this basic advance in our field, it seemed appropriate to bring together symbiotic afficianados from around the world to summarize our basic understanding of these binaries, to delineate the physical problems we have yet to solve, and thus to plan new observational and theoretical attacks on this important group of interacting binary system.

Our apparently improved understanding of symbiotic objects motivated a presentation that was significantly different from those of Boulder and Haute Provence. Rather than describe observations in distinct wavelength regions or discuss physical peculiarities of individual objects, the organization of this colloquium emphasized a multifrequency approach to our study of symbiotic stars and a description of the basic physical components (and their interaction). Thus, the meeting began with introduction to the observations and the working models constructed for symbiotic stars, continued with discussions of techniques designed to probe the physical structure of a symbiotic binary (e.g., orbital solutions, IR photometry and spectroscopy of the cool mass-losing component, radio/optical imagery, and polarization), and then turned to a description of physical models developed to interpret the multifrequency observations. With this physical background in mind, we then confronted cherished ideas with observational reality to test our grasp of the symbiotic phenomenon. We hoped that this interaction would lead to a better understanding of our successes and failures, which could be used on the final day of the conference to place our field in the context of binary stellar evolution, to identify aspects of the symbiotic phenomenon in other stellar objects, and to discuss projects that remain to be accomplished.

The large attendance of this colloquium (91 participants from 23 countries) speaks highly of the growth of our field since Haute Provence. For this reason alone, we may consider the meeting successful, because many researchers entered the lively discussions and exchanged ideas. It was not possible to reproduce the full discussion in this volume, but summaries of the three general discussions are included. We are happy to see that many young scientists have begun to study the symbiotic phenomenon, so we can expect an exciting crop of new results and revolutionary ideas for the next symbiotic colloquium !

It was a happy circumstance that the Colloquium was held in the beautiful old town of Torun which actually was the birthplace of Nicolas Copernicus the founder of Modern Astronomy. The first director of the Torun Astronomical Observatory at Piwnice, the late W. Dziewulski was very active in the field of variable stars, and a postumus work is included in these proceedings.

We are grateful to the other members of the scientific organizing committee, D. Allen, A. Boyarchuk, E. Brandi, M. Hack, S. Kwok, B. Paczynski, R. Stencel, and A. Woszczyk, for their assistance in preparing the scientific program. We especially acknowledge D. Allen, H. Nussbaumer and R. Stencel for their extra efforts, particularly in planning the conference agenda. Light curves supplied by J. Mattei of the AAVSO highlighted the long-term behaviour of symbiotic stars, and emphasized the vital role that amateur astronomers play in our attempts to understand these objects. We finally thank the presidents of IAU Commissions 27, 29, 42 and 44 for sponsoring the Colloquium.

We extend our heartfelt thanks to the local organizing committee, R. Biernikow-icz, C. Iwaniszewska, M. Mikolajewski, B. Ridak, A. Woszczyk, and J. Ziolkowski. We are especially grateful to the chairman of this committee, Prof . A. Woszczyk, fot his careful attention to the many fine details and extracurricular activities that made this meeting so enjoyable. Additional thanks are due to many anonymous people for taking care of the day activities of a scientific meeting.

The Editors

xiii

D. A. ALLEN, Anglo-Australian Observatory, Epping, Australia
A. ALTAMORE, Universita La Sapienza, Rome, Italy
B. G. ANANDARAO, Physical Research Laboratory, Ahmedabad, India
N. E. ANDERSON, Princeton University Observatory, Princeton, USA
T. BELYAKINA, Crimean Astronomical Observatory, Nauchny, USSR
S. BENSAMMAR, Observatoire de Paris, Meudon, France
R. BIERNIKOWICZ, Nicolaus Copernicus University, Torun, Poland
M. F. BODE, Lancashire Polytechnic, Preston, United Kingdom
Yu. V. BORISOV, Institute of Astrophysics TSSR Acad. of Sciences, Dushanbe, USSR
E. W. BRUGEL, University of Colorado, Boulder, USA
A. CASSATELLA, European Space Agency, Madrid, Spain
D. CHOCHOL, Astronomical Institute, Tatranska Lomnica, Czechoslovakia
J. P. DE GREVE, Vrije Universiteit Brussel, Brussel, Belgium
E. DROBYSHEVSKY, A.F. Ioffe Institute of Physics and Technology, Leningrad, USSR
W. J. DUSCHL, M.P.I. fur Physik und Astrophysik, Garching, FRG
J. E. EATON, Indiana University, Bloomington, USA
T. FERNANDEZ-CASTRO, Planetario de Madrid, Madrid, Spain
J. FRANTSMAN, R.A.O. A.N. L.S.S.R., Riga, USSR
M. FRIEDJUNG, Institute d'Astrophysique, Paris, France
M. GARCIA, Smithsonian Astrophysical Observatory, Cambridge, USA
K. GESICKI, N. Copernicus Astronomical Center, Torun, Poland
A. GIMENEZ, Instituto de Astrofisica de Andalucia, Granada, Spain
L. HRIC, Astronomical Institute, Tatranska Lomnica, Czechoslovakia
CHANG-CHUN HUANG, Purple Mountain Observatory, Nanking, China
A.P. IPATOV, Astrosoviet, Moscow, USSR
C. IWANISZEWSKA, Nicolaus Copernicus University, Torun, Poland
I. V. Il'in, Crimean Astrophysical Observatory, Nauchny, USSR
E. JANASZAK, Planetarium and Astronomical Observatory, Olsztyn, Poland
SHI-YANG JIANG, Beijing Astronomical Observatory, Beijing, China
A. JORISSEN, Universite Libre de Bruxelles, Brussel, Belgium
J. KALUZNY, Warsaw University, Warsaw, Poland
S. J. KENYON, Smithsonian Astrophysical Observatory, Cambridge, USA
T. KHUDYAKOVA, Astronomical Observatory, Leningrad, USSR
S. KRAWCZYK, Nicolaus Copernicus University, Torun, Poland
J. KRELOWSKI, Nicolaus Copernicus University, Torun, Poland
M.I. KUMSIASHVILI, Abastumani Astrophysical Observatory, Abastumani, USSR
S. KWOK, University of Calgary, Calgary, Canada
P. G. LASKARIDES, University of Athens, Athens, Greece
E. M. LEIBOWITZ, Tel Aviv University, Tel Aviv, Israel
M. LIVIO, University of Illinois, Urbana, USA
R. LUTHARDT, Zentralinstitut fur Astrophysik der Adw. DDR, Sonneberg, DDR
J. LUTZ, Washington State University, Pullman, USA
L.S. LUUD, I.A.F A. / A.N. E.S.S.R., Tartu, USSR
A. M. MAGALHAES, Instituto Astronomico e Geofisico USP, Sao Paulo, Brasil
O.Yu. MALKOV, Astrosoviet, Moscow, USSR

E.V. MENCHENKOVA, Odessa Astronomical Observatory, Odessa, USSR
A. MICHALITSIANOS, NASA-Goddard Space Flight Center, Greenbelt, USA
J. MIKOLAJEWSKA, Nicolaus Copernicus University, Torun, Poland
M. MIKOLAJEWSKI, Nicolaus Copernicus University, Torun, Poland
M. MUCIEK, Nicolaus Copernicus University, Torun, Poland
H. NUSSBAUMER, Institute of Astronomy ETH Zentrum, Zurich, Switzerland
M. ORIO, Technion-Israel Institute of Technology, Haifa, Israel
J. PAPAJ, Nicolaus Copernicus University, Torun, Poland
H. PAUL, Zentral Institut fuer Astrophysik der Adw. DDR, Sonneberg, DDR
E.N. POPOVA, Astrosoviet, Moscow, USSR
A.F. PUGACH, Main Astronomical Observatory, Kiev, USSR
D. RAIKOVA, Dept. of Astronomy of the Bulgarian Acad. of Sci., Sofia, Bulgaria
O. REGEV, Technion-Israel Institute of Technology, Haifa, Israel
M. H. RODRIGUEZ, Main Astronomical Observatory, Kiev, USSR
B. RUDAK, N. Copernicus Astronomical Center, Torun, Poland
Yu.S. RUSTAMOV, Institute of Physics, Baku, USSR
H. M. SCHMID, Institute of Astronomy ETH Zentrum, Zurich, Switzerland
H. E. SCHWARZ, European Southern Observatory, Santiago, Chile
PARAG SEAL, Indian Institute of Astrophysics, Bangalore, India
E. R. SEAQUIST, University of Toronto, Toronto, Canada
P. L. SELVELLI, Osservatorio Astronomico di Triest, Trieste, Italy
N.M. SHAKHOVSKOY, Crimean Astronomical Observatory, Nauchny, USSR
A. SKOPAL, Astronomical Institute, Tatranska Lomnica, Czechoslovakia
M. H. SLOVAK, University of Wisconsin, Madison, USA
J. SMOLINSKI, N. Copernicus Astronomical Center, Torun, Poland
J. SOLF, Max-Planck-Institute fur Astronomie, Heidelberg, FRG
R. E. STENCEL, University of Colorado, Boulder, USA
A. STROBEL, Nicolaus Copernicus University, Torun, Poland
R. SZCZERBA, N. Copernicus Astronomical Center, Torun, Poland
SHIN'ICHI TAMURA, Tohuku University, Sendai, Japan
O. G. TARANOVA, Sternberg Astronomical Institute, Moscow, USSR
A. TARASOV, Crimean Astrophysical Observatory, Nauchny, USSR
A. R. TAYLOR, Nuffield Radio Astronomy Laboratories, Jordrell Bank, United Kingdom
T. TOMOV, National Astron. Observatory Rozhen, Smoljan, Bulgaria
R. TYLENDA, N. Copernicus Astronomical Center, Torun, Poland
V.A. URPIN, A. F. Ioffe Institute of Physics and Technology, Leningrad, USSR
R. VIOTTI, CNR/Instituto di Astrofisica Spaziale, Frascati, Italy
M. VOGEL, Institute of Astronomy ETH Zentrum, Zurich, Switzerland
R. E. WEBBINK, University of Illinois, Urbana, USA
P. A. WHITELOCK, South African Astron. Observatory, Cape, South Africa
A. WOSZCZYK, Nicolaus Copernicus University, Torun, Poland
B. F. YUDIN, Sternberg Astronomical Institute, Moscow, USSR
L.R. YUNGELSON, Astrosoviet, Mosco, USSR
D. ZAREMBA, Nicolaus Copernicus University, Torun, Poland
J. ZIOLKOWSKI, N. Copernicus Astronomical Center, Warsaw, Poland
S. ZOLA, Jagiellonian University, Cracow, Poland

SESSION 1. THE BASIC DATA

"Symbiotic stars are like platypus"

David Allen

A PERSPECTIVE ON THE SYMBIOTIC STARS

David Allen
Anglo-Australian Observatory
PO Box 296, Epping
NSW 2121
Australia

ABSTRACT. I give a very brief summary of the state of our knowledge of the symbiotic stars, together with some of my hopes for how the field will develop.

1. PREAMBLE

The task of the introductory speaker at a conference is a challenging one. The more so since the words one might choose for the verbal presentation differ from what the reader will seek in the final publication. So, I have decided to cheat: the text you are now considering reading is not what you would have heard if you attended the conference ... well, only partly so. I have taken the liberty of using different titles for the two presentations, to reflect their distinct emphases. But in one important way I have not cheated. This paper was written before the conference, and has not subsequently been modified. It may contain errors that are corrected by later papers; that is the risk I take. On the other hand, it is as fair a summary as I can give, as useful an introduction as I can conceive, to the view of symbiotic stars prevalent early in August 1987. I sincerely hope that the papers which follow will so overthrow the contents of this introduction that you will have no interest in reading it twice.

What I will present in the limited space available here cannot be regarded as a review, but only as a perspective. I eschew references (subsequent papers surely contain ample) save to draw attention to the proceedings of the 1981 conferences on the subject (Stencel 1981; Friedjung & Viotti 1982), and the only book published to date on these stars (Kenyon 1986).

Although not a review, it is appropriate to illustrate this paper with one optical spectrum of a classical symbiotic star, to show just what it is that characterises these objects. Because it is in the optical that they were first recognised, and still are classified, I have not broadened the waveband. Features to note in the spectrum, reproduced on the next page, include the TiO bands of the cool giant in the red; the Balmer jump in emission that shows the blue continuum to be gaseous rather than stellar; the high-excitation emission lines ($\lambda 4686$ of He II; $\lambda 6087$ of [Fe VII]); and the unidentified bands at 6830, 7088 Å, which are markedly broader than atomic emission lines of comparable intensity.

J. Mikolajewska et al. (eds.), The Symbiotic Phenomenon, 3–9.

4

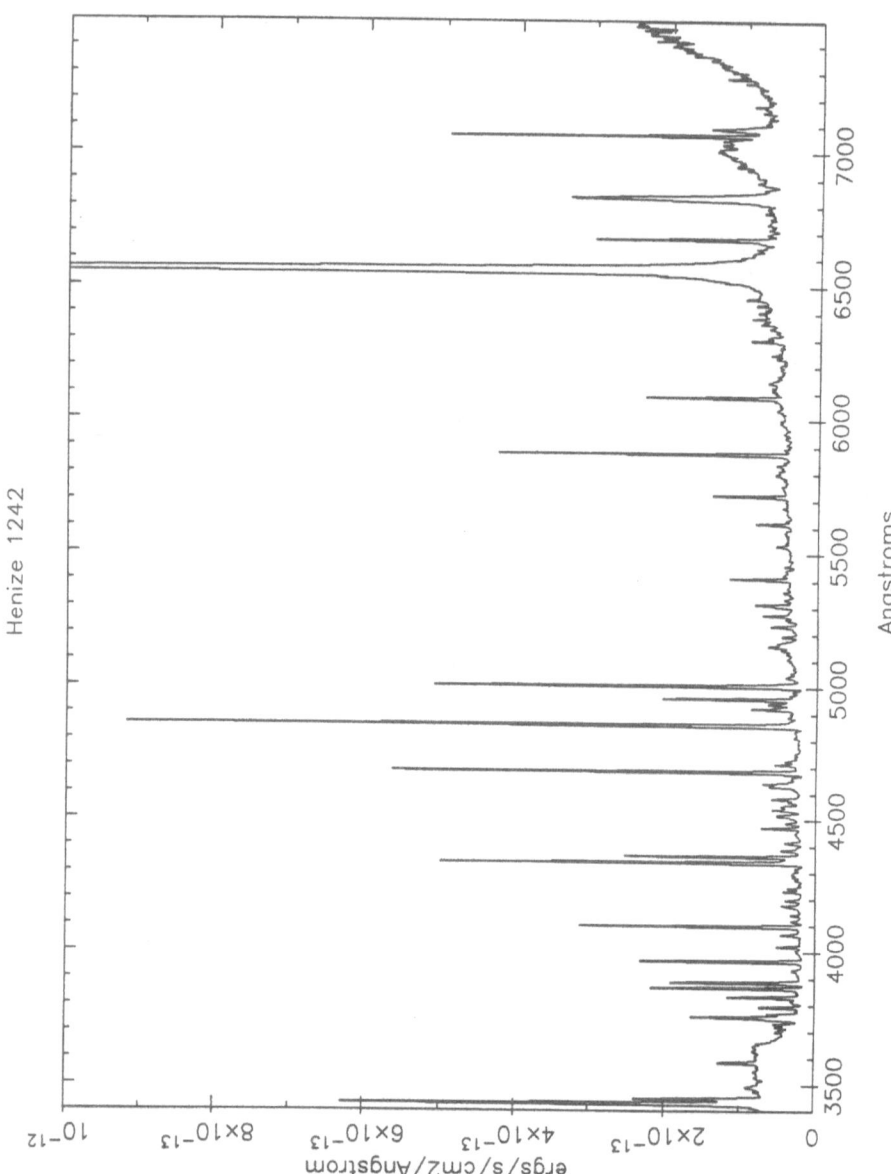

2. SYMBIOTICS AS BINARIES

When first found, in the 1920's, the symbiotic stars were mere curiosities: the platypuses of the stellar zoo. While a few astronomers pondered and committed their thoughts to paper, most ignored the peculiar objects. The principal characteristics of the symbiotic stars – the simultaneous display in an apparently single object of two temperature regimes differing by a factor of 30 or more – then was unique. Today, having a broader base in the electromagnetic spectrum, we take such matters for granted: witness radio galaxies, the X-ray binaries, or the Vega dust cloud.

It was just as apparent half a century back as now that the 3000 K component is a stellar photosphere. But we remain ignorant of the luminosity class of that star in most cases. The origin of the 100,000 K radiation, however, exercises but so often eludes us.

In the 1980's interacting binaries have become fashionable to describe the idiosyncracies of a range of stellar systems that previously seemed utterly bizarre. The symbiotic stars fall readily into the binary bin, and I doubt whether more than 1% of attendees at this conference will argue wholeheartedly for a single-star interpretation of the majority of symbiotic stars. I hope that some do, because it is essential not to blinker ourselves too much. But there is an undeniable elegance in the binary interpretations, and I will offend few if I posit as an introduction to this conference that *symbiotic stars are interacting binary systems*.

3. SUBCLASSES OF SYMBIOTIC STARS

I should qualify that statement. More than most taxons, that of the symbiotic stars has become the dustbin of stellar eccentrics. It is a peculiarity of our species that we need to categorise things. Consequently, we put into the symbiotic bin objects about which we know depressingly little, simply because they show a subset of the characteristics of the well-studied specimens. Many probably belong, but some equally probably do not. And some of the misfits may indeed be single objects. I therefore propose the following: *of the objects currently classified as symbiotic stars, only those which are interacting binaries belong in the class.*

Of course, lack of adequate data on the majority forces us to maintain a lengthy list of probable members of the class, pending a better description of each. The intention of this somewhat brutal definition is not to reduce the class to a handful of examples, but to encourage an alternative classification for any that are shown to be single objects. We may, as a result of this conference, identify a few examples that quite clearly are not binaries, or are binaries in which interaction appears irrelevant to the appearance; if we do, then let us classify them differently. My aim is to remove from the symbiotic classification some of the litter of stars that has tarnished it in the past. My personal belief is that we will lose very few from the present catalogue by this tighter definition, and that we will gain unity in our research.

Even within the binaries, however, we must recognise that there are several types of interaction that can stimulate the apparition we call a symbiotic star. At the start of the decade these various possibilities were hotly debated. We tended at that time to seek a single mechanism to account for *all* members of the class. In the intervening years our views have matured. It is important when approaching this subject to recognise that we are working on several distinct types of interaction, and to realise that not all papers necessarily refer to all the subclasses. Nor is it certain that we have yet identified all possible subclasses. We are nearing the stage at which we should agree on definitions for some of the subclasses, perhaps naming them after their best-studied type stars.

The physical conditions that appear able to produce objects classifiable as symbiotic stars are as follows:

- A main sequence star accretes by Roche-lobe overflow from a late-type giant. The source of ultraviolet luminosity is purely gravitational, and originates in an accretion disk. Variability is caused by changes in the mass transfer rate, as in cataclysmic binaries. Example (and ideal type star): CI Cyg.

- A white dwarf accretes directly from the wind of a late-type giant or mira variable. Much of the hydrogen is burnt as it accretes, providing the ultraviolet flux. The accretion rate undergoes relatively small changes, so the variability is not extreme. Example: RW Hya.

- A white dwarf, that has accreted from the wind of a late-type giant or mira variable and so accumulated unburnt hydrogen, undergoes a shell flash. An ultra-slow nova results. Example: AG Peg. The flash appears capable of persisting for many decades, raising the possibility that some underwent their slow nova eruptions before observations began. A probable example of the latter is H1-36.

- A neutron star accretes from the wind of a late-type giant. The only known example is V2116 Oph, the optical counterpart of GX1+4.

Even amongst these groups the question of the luminosity class remains open. Is the mass donor making its first ascent of the giant branch, or is it on the asymptotic branch? Are supergiants ever involved? And what of the small number of systems (such as BD $-21°3873$) which seem to involve a G star?

It will be apparent that there is a danger at this stage of opening up the symbiotic class too widely. The second of the configurations listed above includes Mira (o Ceti) which has a distant white dwarf companion accreting from its wind. There are probably valuable details to be learnt from a study of Mira, but it would confuse our field to call Mira a symbiotic star. Consequently, I favour retaining an earlier definition that symbiotic stars exhibit emission of $\lambda4686$ He II or perhaps of some equivalently ionized species. In effect, this is an untidy attempt to define a minimum temperature and luminosity for the hot companion.

The different types of symbiotic stars I listed have grossly different evolutionary states. It is possible, even, that some binaries pass through two symbiotic phases, as each member star in turn becomes the mass donor. Are the numbers of the various types compatible with evolutionary expectations? We can only answer this important question if we can classify the 130 or so known symbiotic stars into their appropriate interaction configurations. Unfortunately it has only been possible to classify a very few examples, and then by time-consuming individual study. Are there quicker ways to classify them?

It has been sugested, for example, that the velocity widths of the unidentified emission bands at 6830, 7088 Å (1000 km s^{-1} is typical) are too large to be produced in the wind from or the disk around a main sequence star, and thus imply the presence of a white dwarf? Is this a valid way to recognise the white dwarf accretors? Again, the presence of forbidden lines in the spectrum indicates that considerable amounts of gas at relatively low densities are illuminated by the hot star. The absence of forbidden lines implies that all the illuminated gas is of high density, a circumstance that is unlikely to arise if a wind operates from either star. Can we therefore argue that systems without forbidden lines have Roche-lobe overflow? If we make both these plausible assumptions then why do we see the 6830, 7088 Å bands in many symbiotic stars which are devoid of forbidden lines, for Roche lobe overflow from a cool giant would surely exceed the accretion capabilities of a white dwarf.

4. MORE QUESTIONS

I have just raised a few of the questions for which I hope answers will soon be forthcoming. Other questions come to mind, and I list a few now. Some require further observational study, others need theoretical treatment. In some cases the present body of observational data may provide an answer if we can think it through with sufficient clarity.

- To what extent is the cool component influenced by its energetic friend? Can our classification of it be seriously confused?

- Do magnetic fields, especially in white dwarfs, contribute to the symbiotic phenomenon as they do in cataclysmic variables?

- What can we learn from the 6830/7088 bands, seen in no other types of object? A challenge remains here for the atomic physicists. I stuck out my neck 8 years ago with the suggestion that they are permitted transitions in highly-ionized iron, and to my surprise nobody has challenged that view.

- How important is the wind from the accreting star? What are its effects on the accretion process? How many of the emission lines can be accounted for by a standing shock where the winds from both stars balance? Can the complex emission-line profiles in AG Peg, RX Pup and others be explained by wind interactions?

- In how many cases does an accretion disk around a white dwarf contribute appreciably to the visible and ultraviolet flux? Do we really understand accretion from a stellar wind, in particular with regard to disk formation? Since the dimensions of any such disk have a steep dependence on the wind speed of the cool star, can we improve on estimates of that parameter, for example using SiO masers?

- What clues have we neglected from other interacting binary systems?

- Can we find better methods than those already attempted to deduce the luminosity classes of the cool components, and hence the distances of more symbiotic stars?

- The two symbiotic stars that lie closest to us, R Aqr and CH Cyg, include jet-like radio structures. Dare we extrapolate to infer similar behaviour in other symbiotic stars?

5. AND SOME THOUGHTS ON WORK TO BE DONE

Judging by the lengthy collection of papers that follows, and by their varied titles, symbiotic stars have become fashionable again. More research on them is being performed than ever before, and by more people. I suspect that symbiotic stars have become recognised as one of the topical challenges to follow the major strides that have been made in recent years in the field of cataclysmic variables. Indeed, the two recurrent novae T CrB and RS Oph are claimed as both symbiotics and cataclysmics. Symbiotic stars are, undoubtedly, much more difficult subjects than cataclysmics, if only because of their grossly longer orbital periods. The recent outburst of RS Oph attracted much interest among researchers who had not previously studied symbiotic stars, and many beautiful data emerged. It is still unclear, however, whether the outburst was a shell flash on a white dwarf or an accretion event onto a main-sequence star.

Although the number of symbiotic stars with known periods is increasing, we still desperately need more. That work requires long, patient monitoring of light curves and radial velocity variations. We should not ignore other possibilities, including the use of microwave masers associated with the cool stars, and searches for periodic changes of optical polarization. There is a wealth of work for those with access to large amounts of time on small telescopes.

I see a great need for better modelling of the emission nebulae, with regard to both line strengths and velocity structure. The full treatment of a shock between the stars needs to be incorporated into these models, and with several types of symbiotic system to model there is work for a number of energetic souls. We still await treatment of the zeroth order approximation wherein an isotropic mass-loss nebula is illuminated by an external source of ionizing radiation.

The question of distances and luminosities of both components remains thorny. There is scope for some deep surveys in regions of known distance. Baade's window is one such location, and the Magellanic Clouds another. To date only carbon symbiotic stars have been located in the Clouds, but I can report the recent discovery by Morgan, Good and myself of an oxygen-rich symbiotic star in the LMC which, from luminosity arguments, must involve an asymptotic branch K5 giant.

6. A CATALOGUE OF BIZARRE QUALITIES

I have concentrated somewhat on optical matters, and it is time to redress the balance. I do so by listing some of the unexpected, or unexplained features that have shown up in other wavebands, and make no apology that a few of them have already been noted above. Scott Kenyon, in the next paper, will elaborate much more on the multifrequency nature of these stars, and subsequent papers will highlight different aspects.

- V2116 Oph is a hard X-ray pulsar and the only hard X-ray symbiotic.

- Slow X-ray flares occur during the outbursts of some symbiotic stars; otherwise only a few symbiotics are stable X-ray sources.

- The ultraviolet continua often show more components than can readily be modelled.

- The ultraviolet emission lines of AG Peg and RX Pup are broad and complex, and differ from ion to ion.

- R Aqr is enveloped in a vast bipolar nebula.

- The thermal emission at radio frequencies has a spectral index (~ 0.9) that is not readily explained.

- Systems containing Mira variables show greatly reduced maser emission in OH, H_2O and SiO than normal Miras.

- Nonthermal radio flares have been seen in at least three stars.

- Radio jets are seen in CH Cyg and R Aqr; the R Aqr jet has an optical counterpart.

7. INTRODUCTION

I end with an introduction, not to this paper but to this volume. Symbiotic stars are challenging objects, rewarding to pit oneself against. More than almost any other stellar types their radiation spans the electromagnetic spectrum, from low-frequency radio to hard X-ray. There is something in them for almost every style of astronomy and astronomer. The attendance at this conference alone testifies to the burgeoning interest they are generating.

This panchromatic characteristic of the symbiotic stars does, however, mean that we must guard against blinkering ourselves. The history of the subject is sprinkled with examples where research on one facet of the stars has been described and discussed with inadequate reference to available data at other wavebands; on more than one such occasion the existing literature violated the conclusions reached.

With the publication of this volume we shall have taken another valuable step forward in understanding the symbiotic stars. Covered within its pages are virtually all facets of the subject. It should be compulsory reading for all who work on these fascinating systems.

REFERENCES

Friedjung, M. & Viotti, R. (eds.), 1982. *"The Nature of Symbiotic Stars"*, *Int. Astr. Union Coll.* **70** (Reidel, Dordrecht).
Kenyon, S. J., 1986. *"Symbiotic stars"* (Cambridge University Press).
Stencel, R. E. (ed), 1981. *"Proc. N. Amer. Workshop on Symbiotic Stars"* (Joint Institute for Laboratory Astrophysics, National Bureau of Standards and University of Colorado).

MULTIFREQUENCY OBSERVATIONS OF SYMBIOTIC STARS

Scott J. Kenyon
Smithsonian Astrophysical Observatory
Harvard-Smithsonian Center for Astrophysics
60 Garden Street
Cambridge, MA 02138 USA

ABSTRACT. This paper reviews the discovery of symbiotic stars, and introduces multifrequency observations made during the past two decades.

1. Historical Introduction

In 1912, Mrs. W.P. Fleming published a remarkable paper entitled "Stars Having Peculiar Spectra" (Fleming 1912). Aside from providing detailed lists of new novae, gaseous nebulae, and O-type stars, Fleming noted several long period variables (including R Aqr and RW Hya) with bright emission lines and unusually small ranges in brightness. Miss A. Cannon later noted the existence of several M stars with bright H I and He II lines, such as Z And and CI Cyg, in work that went generally unrecognized (Shapley 1922). Merrill and Humason (1932) rediscovered CI Cyg as an anomalous star having bright He II λ4686 *in combination* with the TiO bands characteristic of M4 stars, and noted that RW Hya and AX Per possessed similarly striking features. Merrill and Humason remarked that T CrB and R Aqr also displayed peculiar spectra, and suggested that spectrograms of planetary nebulae and other emission-line stars might be examined for TiO absorption bands. Over a dozen *stars with combination spectra* had been discovered by 1941, when Merrill coined the phrase *symbiotic stars* to describe objects which show evidence for cool M-type photospheric absorption lines and very high temperature emission lines.

Merrill and Humason's short paper produced a great deal of interest in peculiar emission stars, and much effort was spent in obtaining new spectra and monitoring the light variations of these unusual systems. Of special interest were the discoveries by Lindsay (1932) and Greenstein (1937) that AX Per and CI Cyg had undergone 2-3 mag nova-like eruptions prior to Merrill and Humason's observations. Searches of the photometric archives revealed similar eruptive activity in Z And, BF Cyg, and RS Oph, demonstrating that outbursts were normal events for many symbiotic systems (see Figure 1). In addition, a few objects were shown to have quiescent photometric fluctuations which repeated on time scales of 600-900 days (see Payne-Gaposchkin 1957; Kenyon 1986 and references therein). Spectroscopic observations by Merrill at Mt. Wilson Observatory and by Swings and Struve at McDonald Observatory showed that radial velocity variations were correlated with these intensity changes in RW Hya and AG Peg, and lent support to the idea that these objects might be binary systems. Periodic radial velocity variations were not observed in some symbiotics, because their violent eruptive activity made it difficult to follow absorption and emission features from one orbital cycle to the next.

11

J. Mikolajewska et al. (eds.), The Symbiotic Phenomenon, 11–22.

12

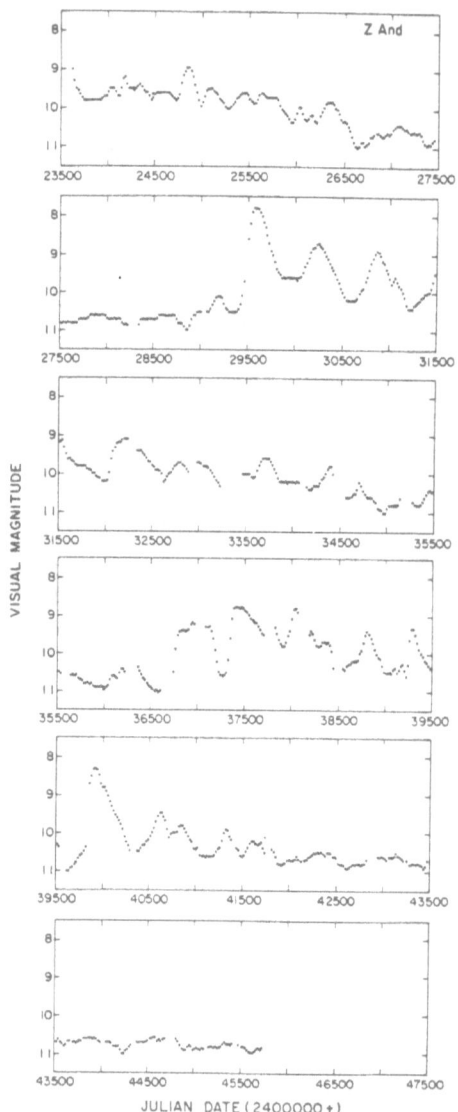

Figure 1 - Optical light curve of Z And as observed by members of the AAVSO (from Kenyon 1986).

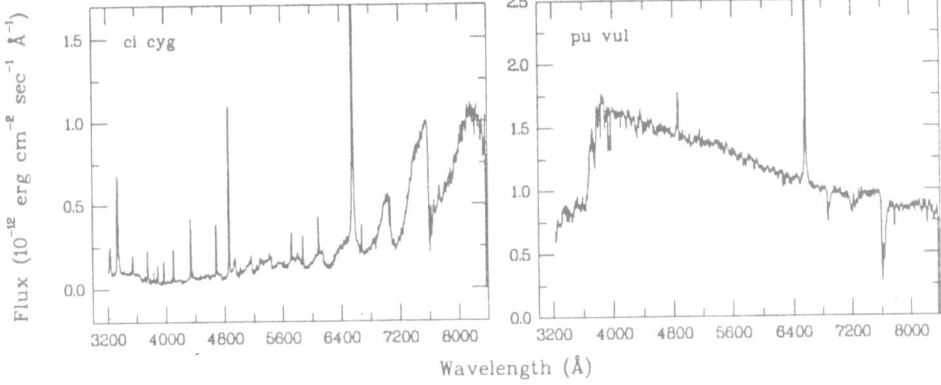

Figure 2 - Optical spectra of CI Cyg and PU Vul obtained at Kitt Peak National Observatory. Strong emission lines (H I, He I, He II, [Ne V], and [Fe VII]) and prominent TiO absorption bands are visible on the quiescent spectrum of CI Cyg. The eruptive spectrum of PU Vul shows evidence for H I emission and the absorption lines characteristic of F stars.

Explanations for the symbiotic phenomenon were debated soon after Merrill and Humason's paper. Both Berman (1932) and Hogg (1932) suggested that these objects might be binaries comprised of a rather normal M star and a hot star similar to the central star of a planetary nebula. A detailed model for the eruptions of symbiotic stars was not available at this time, but Berman noted that instabilities in the hot object (similar perhaps to those in novae) might give rise to the observed behavior. Kuiper (1941) later introduced Roche geometry into stellar astronomy, and proposed a model for peculiar emission objects in which a star filling its critical Lagrangian surface loses material to its binary companion. Kuiper correctly reasoned that the kinetic energy of infalling matter might give rise to bright emission lines similar to those observed in AX Per and T CrB, and noted that the orbital periods deduced for symbiotic stars (P_{orb} ~ years) required the lobe-filling star to be an M giant. The discovery of radial velocity variations lent some support to these natural binary interpretations for symbiotic stars, although the lack of similar velocity changes in other systems led to the development of rather novel single star mechanisms (see Kenyon 1986).

By the late 1960's, astronomers had achieved an immense amount of descriptive material concerning the behavior of symbiotic stars. It was now well-established that emission lines from a wide range of species, such as H I, He I, He II, O III, Ne V, and Fe VII, were typical of many symbiotic stars, and that these lines (and the TiO bands of the giant) fluctuated in intensity on time scales ranging from several days to several years. In spite of this wealth of information, the physical mechanism which causes a star (or stars) to become symbiotic was not identified unambiguously. Advances in technology have led to observations in other wavelength regions and important breakthroughs in understanding the complex symbiotic phenomena observed in the optical. A brief introduction to these data is the subject of this review.

II. Optical Data and the Defining Features of Symbiotic Stars

An optical spectrum of a fairly normal quiescent symbiotic star, CI Cyg, is shown in the left panel of Figure 2. The characteristic features of the class are readily apparent, and may be briefly summarized:

1. absorption features (TiO, VO, and Na I) observed in red giant stars and an associated red continuum (identified as the *cool giant component*);

2. strong emission lines from fairly highly ionized species (H I, He I, and [O III]) found in planetary nebulae and a blue continuum (identified as the *hot component*).

Periodic photometric variations are typical of many systems with amplitudes ~ 0.5-1.0 mag and periods of 200-1000 days (see Figure 1). These brightness changes usually are interpreted in terms of orbital motion in a binary system, either as a result of an eclipse of the hot component by the giant, or because the hot star illuminates the facing hemisphere of its giant companion (see Kenyon 1986, Chapter 3).

Aside from the periodic light variations, symbiotic stars display irregular eruptions with amplitudes ~ 2-7 mag. The optical spectrum of an outbursting symbiotic star usually resembles an A-F supergiant, as shown by the spectrum of PU Vul in the right panel of Figure 2. It is obvious that spectral features commonly observed in quiescent symbiotic stars are not present in outburst: Balmer lines are the only emission features. The higher ionization emission lines and the late-type absorption features reappear during the decline from an outburst, and most systems (including the recurrent nova symbiotics T CrB, RS Oph, and V1017 Sgr) return to minimum in a few months or a few years. *Symbiotic novae* require several decades to decline from visual maximum, as discussed by Viotti in this volume.

The motivation for identifying symbiotic stars as binary systems is apparent from inspection of Figure 2. It is fairly natural to associate the absorption features and red continuum with an evolved red giant star and to place the blue continuum and associated emission features with a hot companion star. Direct verification that symbiotic stars are binaries was difficult with the techniques available prior to 1970, because the small radial velocity amplitudes expected in a binary containing a red giant (~ 5-10 km s^{-1}) cannot be measured easily with photographic plates. Modern observations with photon counting detectors and cross-correlation analysis have demonstrated that all bright symbiotics are binary stars (Garcia, this volume), but estimating physical parameters of the two components require data to supplement optical photometry and spectroscopy. Given the rising red continuum of CI Cyg, it makes sense to study the red giant star in the infrared and the information contained in these data is the next topic of discussion.

III. Infrared Data

A. Photometry

Pettit and Nicholson (1933) first demonstrated that infrared (IR) photometry might be a useful probe of the giant components in symbiotic stars with observations of R Aqr. Other symbiotics were too faint to be observed with their bolometer, but modern techniques have shown that most symbiotics appear to be either normal red giants or normal Mira variables when studied at wavelengths of 1-5 μm. The majority of systems have near-IR colors consistent with stellar photospheric temperatures of 2500 K to 3500 K, and have been called

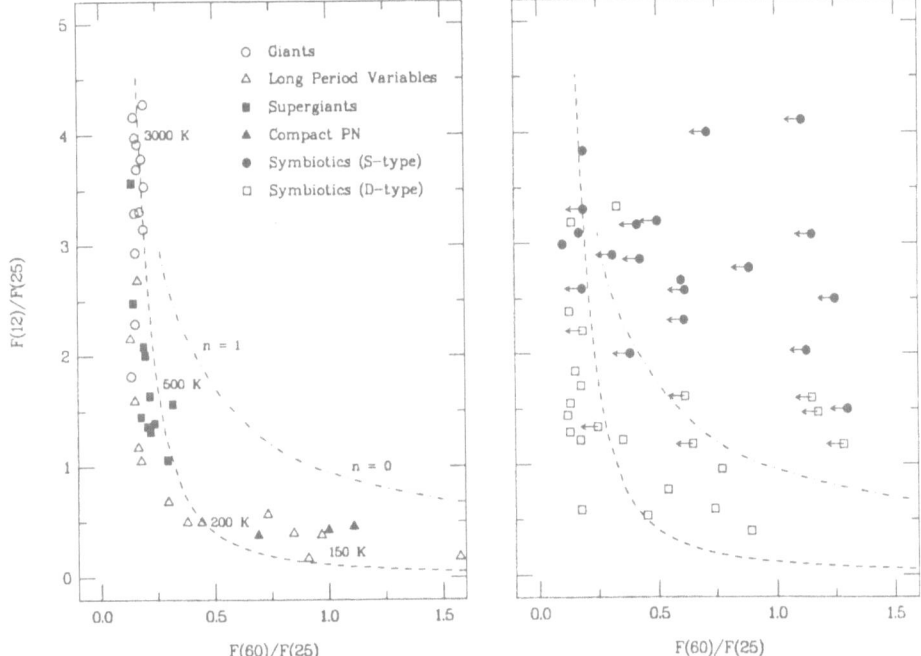

Figure 3 - Flux ratio diagram based on *IRAS* observations of symbiotics and other M-type stars. The dashed and dot-dashed curves indicate the loci of blackbody and power law sources, respectively (temperatures and power law indices, $F_\nu \propto \nu^n$, are indicate beside each curve). A key for interpreting the symbols is given in the legend. Upper limits to the 60 μm flux are indicated by arrows.

S-type (stellar) systems (Webster and Allen 1975). A few other symbiotics have near IR colors more typical of blackbodies with temperatures ~ 1000 K and are known as *D-type* systems (Webster and Allen 1975). These D-type objects have the strong H_2O absorption bands and periodic variability at 2 μm ($\Delta K \sim 1$ mag, $P \sim 300$-600 days) that typifies Mira variables, although their energy distributions are significantly redder than an average Mira (Allen 1982; Whitelock 1987). Allen (1983) and Kenyon, Fernandez-Castro, and Stencel (1986) showed that the IR continua of several D-type symbiotics could be made to agree with that expected for a Mira *if the cool components in these binaries are heavily reddened* (extinctions at K, $A_K \sim 0.5$-2 mag).

Aside from their near-infrared emission, symbiotic stars are prodigious far-infrared sources. Over 50% of known symbiotic stars were detected by the *Infrared Astronomical Satellite (IRAS)*, and a flux ratio diagram based on observed fluxes at 12, 25, and 60 μm is presented in Figure 3. It is apparent that the S-type systems are significantly bluer than typical D-type systems in the *IRAS* survey, although there is some overlap near $F_\nu(12 \, \mu m)/F_\nu(25 \, \mu m) \sim 2$. As might be expected from near-IR observations, the S-type systems have far-IR colors similar to those of normal giant stars, while the D-type objects more closely resemble Mira variables.

The *IRAS* data are very useful for estimating mass loss rates from the cool components of symbiotic stars providing accurate fluxes are available at 12-60 μm (Whitelock 1987, and this volume; Kenyon, Fernandez-Castro, and Stencel 1988). The very large 12 μm excesses

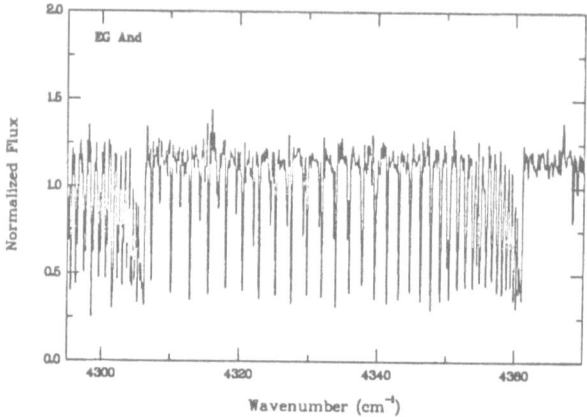

Figure 4 - High resolution IR spectrum of EG And. All of the absorption features in this spectrum are real, and are caused by CO absorption in the red giant atmosphere.

of the typical D-type object imply mass loss rates, $\dot{M} \sim 10^{-5}$ $M_\odot$ yr^{-1}, which is somewhat larger than that of most Mira variables (Whitelock, Pottasch, and Feast 1987). Since most S-type systems were not detected by *IRAS* at 60 μm, mass loss rates can be estimated only for UV Aur (10^{-6} $M_\odot$ yr^{-1}), CH Cyg (6 x 10^{-8} $M_\odot$ yr^{-1}), and AG Peg (2 x 10^{-7} $M_\odot$ yr^{-1}). These mass loss rates are close to those expected for normal M giants (AG Peg and CH Cyg) and carbon stars (UV Aur).

B. Spectroscopy

High resolution IR spectra of symbiotic stars have not been obtained as routinely as in the optical or ultraviolet (see below), but significant progress has been made with low resolution ($\lambda/\Delta\lambda \sim 50$-200) spectrometry. The broad CO and H_2O absorption bands present on near-IR spectra of symbiotic stars are typical of M giants and Mira variables, and can be resolved into individual absorption lines with high resolution spectra. An example of the data that can be obtained with the KPNO 4-m Fourier Transform Spectrometer appears in Figure 4. This spectrum of EG And has been normalized to unity in a standard way (see Hartmann and Kenyon 1987), but is not flux-calibrated. The strong absorption features are CO lines intrinsic to the M star in EG And, and have a strength comparable to CO lines observed in other M2.5 giants.

Spectroscopic data also have been secured in the mid-IR with large, ground-based telescopes (Roche, Allen, and Aitken 1983) and with *IRAS*. Examples of *IRAS* spectra for three D-type systems are shown in Figure 5, and these data are similar in most respects to mid-IR spectra obtained for very late-type giants and Mira variables. Silicate grain emission at 10 μm is very prominent on the *IRAS* spectra, and is indicative of a large mass loss rate. It appears from the broad band *IRAS* measurements that symbiotic stars generally display more dust emission than normal red giants of similar spectral types, and their silicate features seem stronger as well.

This brief introduction to the IR observations confirms that the cool components of symbiotic binaries are fairly normal red giant stars, as suggested by optical spectroscopy.

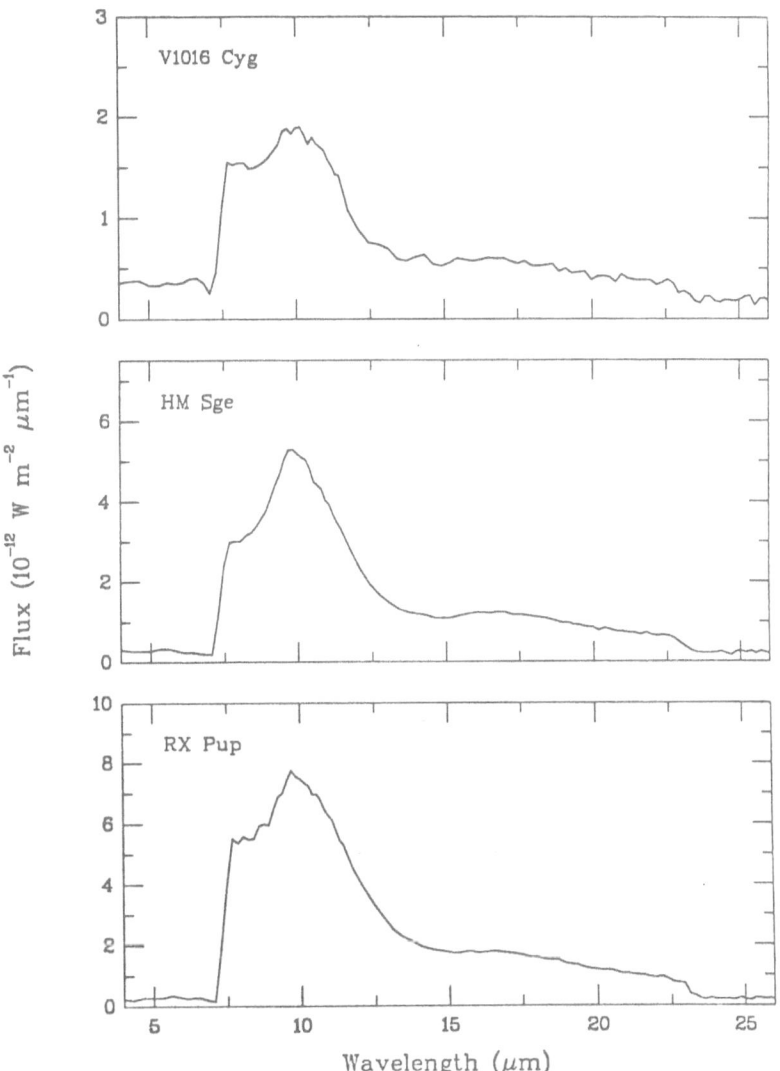

Figure 5 - Low resolution *IRAS* spectra of several symbiotics.

The division of these objects into S-types and D-types is a major physical result, and will be discussed in more detail later in this volume. Aside from the H I Paschen and Brackett emission lines observed in several systems, the IR data reveal little information concerning the physical nature of the hot component. The weak blue continuum present on optical spectra provides one clue that ultraviolet spectroscopy might yield important insights into the energetics of symbiotic stars, and observations obtained in the satellite ultraviolet are the subject of the next section.

IV. Ultraviolet Spectroscopy

Initial ultraviolet (UV) observations of symbiotic stars were made by the OAO-2 and TD-1 satellites in distinct photometric bands somewhat analogous to UBVRI optical photometry. The symbiotic nova AG Peg was detected by both UV instruments, and Gallagher *et al.* (1979) demonstrated that the hot component in this system emitted a continuum similar to that of a Wolf-Rayet star with an effective temperature of 30,000 K. Only one other symbiotic, SY Mus, was detected by TD-1 (and only in two band passes, 1565 Å and 1965 Å), and OAO-2 failed to detect any other bright symbiotic.

The launch of the *International Ultraviolet Explorer (IUE)* revolutionized the study of symbiotic stars and demonstrated the validity of the basic binary models proposed in the 1930's and 1940's (see Nussbaumer and Stencel 1987). A spectrum for AG Peg is presented in the left panel of Figure 6, and the very intense blue continuum confirms that the hot component in AG Peg is a stellar source. There is good evidence for the presence of the 2200 Å interstellar absorption feature on this spectrum, and a *simultaneous* fit for the color temperature of the hot component, T_h, and the interstellar reddening, E_{B-V}, results in $T_h \approx$ 30,000 K to 60,000 K and $E_{B-V} \approx$ 0.11-0.19 using well-exposed *IUE* spectra (Kenyon and Webbink 1984).

The majority of symbiotics display UV spectra similar to that shown for AG Peg, and their continua can be understood simply in terms of the Rayleigh-Jeans tail of a hot blackbody and free-bound emission from an ionized nebula. However, several symbiotic stars, as in the spectrum of CI Cyg shown in the right panel of Figure 6, have the fairly flat UV continua more typical of A or B-type stars. It is apparent that an A or B star does not produce sufficient high energy photons to power the emission spectrum, so this simple interpretation of the UV observations cannot be correct. An alternative picture is to suppose such continua arise from an accretion disk surrounding a compact star (see Pringle and Wade 1985). Kenyon and Webbink (1984) showed that the UV data were consistent with this idea only if the accreting star is a low mass main sequence star, and could not identify any symbiotic star with an accreting white dwarf. This model is similar to that proposed by Kuiper (1941) to account for the optical behavior of the same systems.

Emission lines are fairly prominent on UV spectra of symbiotic stars, confirming the presence of the large ionized nebula inferred from optical data. The strongest lines in the spectra usually are from highly ionized species such as He II, C III, C IV, Si IV, and N V, although strong [O I] and Fe II features have been observed in several instances. Individual emission features have been resolved in the brighter symbiotic objects, and several systems show evidence for gas velocities of ~ 1000 km s^{-1} (e.g., AG Peg). These velocities are significantly larger than the ~ 20-100 km s^{-1} expected from material lost by the red giant, and presumably represent gas near the hot component. It is fairly obvious that gas has been *ejected* at these large velocities in AG Peg, but present observations have not been sufficient to determine if the hot components of other symbiotics are ejecting material or accreting portions of the wind lost by the red giant.

Aside from yielding dynamic information about the ionized gas in symbiotics, high resolution *IUE* spectra may be used as a probe of the physical conditions in the nebulae (see Nussbaumer and Stencel 1987; Nussbaumer, this volume). Reliable estimates for the electron density, n_e, have been extracted successfully from line ratios of intercombination doublets such as C III] $\lambda\lambda1907,1909$, N IV] $\lambda1483,1487$, and Si III] $\lambda\lambda1883,1892$, while the O III] and [O III] optical/UV lines provide a useful measure of the electron temperature, T_e. Typical results for the electron temperature, T_e ~ 10,000 K to 20,000 K, suggest that symbiotic nebulae are photoionized rather than shock heated, and the density diagnostics indicate that the gas is much denser, n_e ~ 10^6 to 10^{10} cm^{-3}, than material in typical planetary

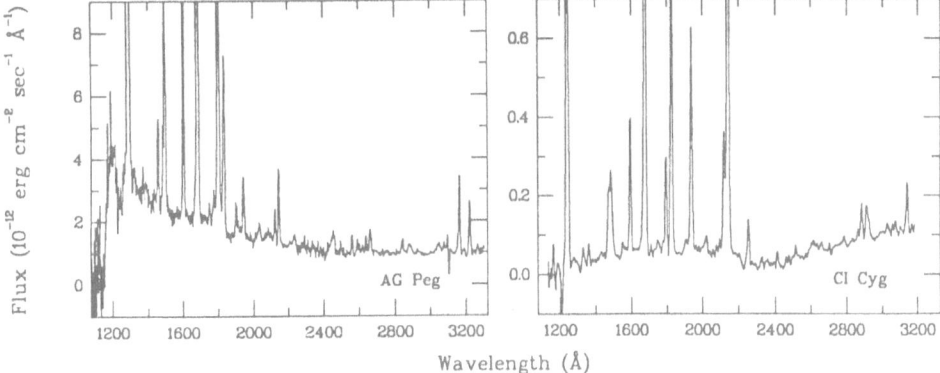

Figure 6 - Ultraviolet spectra for AG Peg (left panel) and CI Cyg (right panel).

nebulae, $n_e \leq 10^4$ cm^{-3}.

The UV/optical/IR observations have thus confirmed the natural binary models proposed by Berman, Hogg, and Kuiper when symbiotics were first discovered. By analogy with other interacting binaries, such as cataclysmic variables, symbiotics might be expected to emit soft X-rays as a result of accretion processes. And if the ionized nebula surrounding the binary is sufficiently large, we might anticipate observable amounts of radio emission. Brief summaries of initial X-ray and radio observations of symbiotics are described in the next two sections.

V. X-rays

The high blackbody temperatures inferred for several symbiotic stars ($T_{hot} \gtrsim 150,000$ K) suggest that these systems might be soft X-ray sources, but initial searches for X-ray bright symbiotics have been discouraging. Aside from GX1+4 (which is a neutron star accreting material from the wind of a red giant companion; Doty, Hoffman, and Lewin 1981), symbiotics observed with the *Einstein* satellite were fairly weak sources. Most notable among these first detections was the "yellow symbiotic" AG Draconis. Einstein observations obtained before the 1981-1985 series of eruptions recorded ~ 0.27 IPC count s^{-1}, as described by Anderson, Cassinelli, and Sanders (1981). These data have been reprocessed, and are consistent with a blackbody source (kT = 0.016 ± 0.003 keV, R = 1500 km) in addition to the bremsstrahlung source (kT ~ 0.1 keV) suggested by Anderson, Cassinelli, and Sanders. The X-ray temperature of ~ 200,000 K is in fair agreement with the source temperature of ~ 100,000 K to 150,000 K inferred from *IUE* observations, although the radius deduced from *IUE* data, R ~ 10^4 km, is much larger than the X-ray value. Similar disagreements in source parameters result in an analysis of the other *Einstein* detections, V1016 Cyg and HM Sge, which led to plausible interpretations for the X-rays involving shocks in stellar winds (Kwok and Leahy 1984; Willson *et al.* 1984).

Several symbiotics were observed by the *EXOSAT* satellite, which was much more sensitive to soft X-rays than was *Einstein*. Among the normal symbiotics observed by *Einstein*, *EXOSAT* made unambiguous detections of AG Dra and HM Sge, but failed with V1016 Cyg. Observations of R Aqr indicated that high energy photons are associated with the jet in this unusual Mira symbiotic, and are consistent with a bremsstrahlung spectrum having kT ~ 0.04 keV if the hydrogen column density is fairly low (Viotti *et al.* 1987). *EXOSAT* also detected CH Cyg following its decline from a recent eruption (Leahy and Taylor 1987).

It is unfortunate that the luminosity and temperature derived from fits to the X-ray data are so very sensitive to the values adopted for the interstellar hydrogen column density, N_H, and the local electron density, n_e. For example, the derived source luminosity for the relatively unreddened symbiotic AG Dra changes by a factor of ~ 150 when the hydrogen column density is increased from $N_H = 0$ cm^{-3} to $N_H = 1.5 \times 10^{20}$ cm^{-3} (corresponding to values for the B-V color excess of $E_{B-V} = 0$ and 0.025, respectively). The reason for this problem is that symbiotic X-ray spectra are very soft (photon energies < 0.1 keV), and most of the X-rays are efficiently absorbed by interstellar hydrogen and helium atoms along the line of sight.

Finally, it is important to remember that symbiotic stars have significant **local** hydrogen and helium column densities. For the case of AG Dra, the observed He II λ4686 flux indicates ~ 2×10^{44} He$^+$-ionizing photons are absorbed every second by the AG Dra nebula. The luminosity in absorbed He$^+$ photons, ~ 2×10^{34} erg s^{-1}, **exceeds the luminosity of the source deduced from the X-ray flux by a factor of two!** The degradation of the X-rays by the surrounding nebula thus causes an underestimate in the actual source luminosity, which can be recovered only by an accurate estimate of the soft X-rays absorbed locally. Since V1016 Cyg and HM Sge also have very strong He II λ4686 emission lines, the X-rays emitted by these objects are also poorly determined. Thus, a good understanding of symbiotic X-ray spectra awaits better models for their gaseous nebulae.

VI. Radio Emission

The bright emission lines observed on symbiotic ultraviolet, optical, and infrared spectra suggest the presence of large ionized nebulae surrounding symbiotic binaries, and estimated sizes for the H II regions are $R_{H\,II}$ ~ 1-100 AU (Kenyon 1986). If the ionized gas is optically thick and spherically symmetric, the predicted radio emission at a frequency of ν is

$$S_\nu \sim 0.2 \left[\frac{\nu}{5 \text{ GHz}} \right]^2 \left[\frac{R_\nu}{d} \right]^2 \text{ μJy} , \tag{1}$$

where R_ν is the radius of the nebula in AU and d is the distance in kpc. Since the current detection limit at the VLA is ~ 0.1 mJy at 6 cm, only those symbiotics with large radio nebulae, R_ν ~ 100 AU for d ~ 1 kpc, are expected to be radio sources.

Initial detections of symbiotic stars were reported by Wright and Allen (1978; RR Tel), Purton, Feldman, and Marsh (1973; V1016 Cyg), and Altenhoff and Wendker (1973; V1329 Cyg). The most luminous radio source among symbiotics was later found to be H1-36 ($S_{6\,cm}$ ~ 45 mJy), and several other systems have been observed to have radio fluxes ≥ 10 mJy at 6 cm. It is apparent that these symbiotics have very extensive ionized regions, with $R_{6\,cm}$ ~ 100-1000 AU. As pointed out by Seaquist (this volume), the association of D-type symbiotics with strong radio sources has important consequences for the physical structure of these objects.

Radio maser emission (primarily SiO and OH) is characteristic of many late-type giant stars, but searches for these transitions in symbiotic stars have been discouraging (see Kenyon 1986). The Mira symbiotic R Aqr is a fairly strong SiO source, but the hot components of other symbiotics appear to be fairly efficient in maintaining conditions that are unfavorable for maser emission. There is still some hope that masers might be observed in other systems, because the upper limits reported in the literature remain fairly high. More sensitive searches may be rewarded with new detections.

VII. Summary

Before this colloquium moves on to a physical interpretation of the symbiotic phenomenon, it is important to note that the basic observational data require symbiotic stars to be long period binary systems. Three components are typically observed:

1. The cool giant component has an effective temperature of 2500-4000 K. In most cases, strong TiO absorption bands and a bright red continuum signify the nature of the giant; these features are not observed on optical spectra of several systems, but infrared CO and H_2O bands serve to identify the giant in these instances. IR spectral classification divides the cool components into normal M giants (S-types) and heavily reddened Mira variables (D-types). Both classes have 10-20 μm excesses signifying large mass loss rates ($\sim 10^{-8}$ to 10^{-7} $M_\odot$ yr^{-1} for S-types; $\sim 10^{-6}$ to10^{-5} $M_\odot$ yr^{-1} for D-types). However, unlike many evolved stars symbiotics typically do not show evidence for maser emission.

2. The hot companion displays a bright blue continuum at UV wavelengths and sometimes is an X-ray source. Most hot components have UV continua that can be fit by the Rayleigh-Jeans tail of a hot ($T_h \sim 100{,}000$ K), compact ($\lesssim 0.1$ $R_\odot$) blackbody in combination with nebular continuum radiation. Several objects appear to contain main-sequence stars accreting at rates $\sim 10^{-5}$ $M_\odot$ yr^{-1}.

3. A gaseous nebula envelops the binary. The nebula is denser $n_e > 10^6$ cm^{-3} than similar material found in planetary nebulae ($n_e < 10^4$ cm^{-3}), and gives rise to substantial radio emission. Local absorption of the X-rays by this nebula probably prevents many symbiotics from being detected by X-ray satellites.

The observations further demonstrate a division into two basic *physical* classes of interacting binary systems.

1. Detached symbiotics containing a red giant (or a Mira variable) in combination with a hot white dwarf or subdwarf.

2. Semi-detached symbiotics containing a lobe-filling red giant and a solar-type main sequence star.

The hot component accretes material lost by the red giant in both classes: wind accretion dominates the detached systems, while semi-detached objects interact through tidal overflow and the formation of a viscous accretion disk.

The author thanks Mike Garcia for reanalyzing *Einstein* data for V1016 Cyg, AG Dra, and HM Sge. Portions of this work were supported by NASA through grants NAG5-87 (*IUE*), 957278 (*IRAS Data Analysis Program*), and NAG8-605 (*HEAO-2 Guest Investigator Program*).

22

References

Allen, D.A. 1982. in *IAU Colloquium No. 70, The Nature of Symbiotic Stars*, ed. M. Friedjung and R. Viotti (Dordrecht: Reidel), p. 27.

Allen, D.A. 1983. *Mon. Not. Roy. Astr. Soc.*, **204**, 113.

Altenhoff, W.J. and Wendker, H.J. 1973. *Nature*, **241**, 37.

Anderson, C.M., Cassinelli, J.P., and Sanders, W.T. 1981. *Astrophys. J. (Letters)*, **247**, L127.

Berman, L. 1932. *Pub. Astr. Soc. Pac.*, **44**, 318.

Doty, J.A., Hoffman, J.A., and Lewin, W.H.G. 1981. *Astrophys. J.*, **243**, 257.

Fleming, W.P. 1912. *Ann. Harv. Coll. Obs.*, **56**, 165.

Gallagher, J.S., Holm, A.V., Anderson, C.M., and Webbink, R.F. 1979. *Astrophys. J.*, **229**, 994.

Greenstein, N.K. 1937. *Bull. Harv. Coll. Obs.*, No. 906.

Hartmann, L., and Kenyon, S.J. 1987. *Astrophys. J.*, **312**, 243.

Hogg, F.S. 1932. *Pub. Astr. Soc. Pac.*, **44**, 328.

Kenyon, S.J. 1986. *The Symbiotic Stars* (Cambridge Univ. Press: Cambridge).

Kenyon, S.J., Fernandez-Castro, T., and Stencel, R.E. 1986. *Astr. J.*, **92**, 1118.

Kenyon, S.J., Fernandez-Castro, T., and Stencel, R.E. 1988. in preparation.

Kenyon, S.J., and Webbink, R.F. 1984. *Astrophys. J.*, **279**, 252.

Kuiper, G.P. 1941. *Astrophys. J.*, **93**, 133.

Kwok, S., and Leahy, D.A. 1984. *Astrophys. J.*, **283**, 675.

Leahy, D.A., and Taylor, A.R. 1987, *Astr. Astrophys.*, **176**, 262.

Lindsay, E.M. 1932. *Bull. Harv. Coll. Obs.*, No. 888.

Merrill, P.W., and Humason, M.L. 1932. *Pub. Astr. Soc. Pac.*, **44**, 56.

Nussbaumer, H., and Stencel, R.E. 1987. in *Exploring the Universe with the IUE Satellite*, ed. Y. Kondo (Dordrecht: Reidel), p. 203.

Payne-Gaposchkin, C. 1957. *The Galactic Novae* (Amsterdam: North Holland).

Pettit, E., and Nicholson, S.B. 1933. *Astrophys. J.*, **78**, 320. Webster, B.L. and Allen, D.A. 1975. *Mon. Not. Roy. Astr. Soc.*, **171**, 171.

Purton, C.R., Kwok, S., and Feldman, P.A. 1983. *Astr. J.*, **88**, 1825.

Roche, P.F., Allen, D.A., and Aitken, D.K. 1983. *Mon. Not. Roy. Astr. Soc.*, **104**, 1009.

Shapley, H. 1922. *Bull. Harv. Coll. Obs.*, No. 778.

Viotti, R., Piro, L., Friedjung, M., and Cassatella, A. 1987. *Astrophys. J. (Letters)*, **319**, L7.

Whitelock, P.A. 1987. *Pub. Astr. Soc. Pac.*, in press.

Whitelock, P.A., Pottasch, S.R., and Feast, M.W. 1987. in *Late Stages of Stellar Evolution*, ed. S. Kwok and S.R. Pottasch (Dordrecht: Reidel), p. 269.

Willson, L.A., Wallerstein, G., Brugel, E., and Stencel, R.E. 1984. *Astr. Astrophys.*, **133**, 154.

Wright, A.E., and Allen, D.A. 1978. *Mon. Not. Roy. Astr. Soc.*, **184**, 893.

V641 CAS - A SYMBIOTIC CANDIDATE ?

T. Tomov
National Astronomical Observatory Rozhen,
4700 Smoljan, PB 136, Bulgaria

M. Mikolajewski
Institute of Astronomy, N. Copernicus University,
Chopina 12/18, 87-100 Torun, Poland

The light variability of V641 Cas was discovered by Guinan et al. (1982). Spectral investigations were made earlier by Barbier(1971,1975), wich classified it as a VV Cep-type star and suggested that the V641 Cas is a binary system consisting of M3Ia supergiant and a B type star. Shaw and Guinan(1985) estimated the spectral class of the companion as B2.5. Recently Guinan et al. (1986) reported about semiregular changes with maximal amplitude of about 1.0^mduring 1979-1986. Slow variations in the brightness with a mean period 310d and variable amplitude of 0.3^m-0.9^m ph are established by Berthold(1983).

Our photometric and spectroscopic observations of V641 Cas are made during Nov.-Dec. 1986. The UBV photometry shows small increase in V from 8.50^m(Nov.12) to 8.46^m(Dec.11), while the star brightened by about 0.2^min U. This may be interpreted in terms of presence of a hot, variable component, observed in UV by Shaw and Guinan (1985). The position of V641 Cas on the (B-V)/(U-B) diagram (Fig.1) indicates a relatively low contribution of the hot component to the optical light. It is conspicuous, that assuming a luminosity class I for the M component the resulting reddening is remarkably low for an object lying in the galactic plane (l=118.3, b=+1.5). Thus, we can expect that it is either a reddened bright giant or a giant with additional circumstellar reddening. This implies that the hot component is about 100-1000 times fainter than a B2-3 main sequence star. The patrol observations show possible rapid light changes in U band with an amplitude of 0.1^m-$0.15^m$and on a time scale of 5-15 min. These are similar to the flickering observed in CH Cyg (Tomov et al.,1986) and in o Cet (Warner, 1972).

Our spectral observations of moderate (18 A/mm) dispersion show, that the absorption spectrum is very similar to one of a M2-3II star. The emission spectrum (especially the forbidden lines) is rather similar to this of CH Cyg than to the VV Cep (Fig.2). The Balmer emission lines, as well, as the broad absorptions (reported early by Barbier,1971,1975) are absent here. The average radial velocities of the M absorption lines and of the emission lines of [FeII] are close to those published by Barbier and the differences does not exceed 3-4 km/s.

Additionally we have found 6 spectra of V641 Cas in the Torun collection of objective prism plates, in 1964-1972. On these spectra the

23

J. Mikolajewska et al. (eds.), The Symbiotic Phenomenon, 23–24.

24

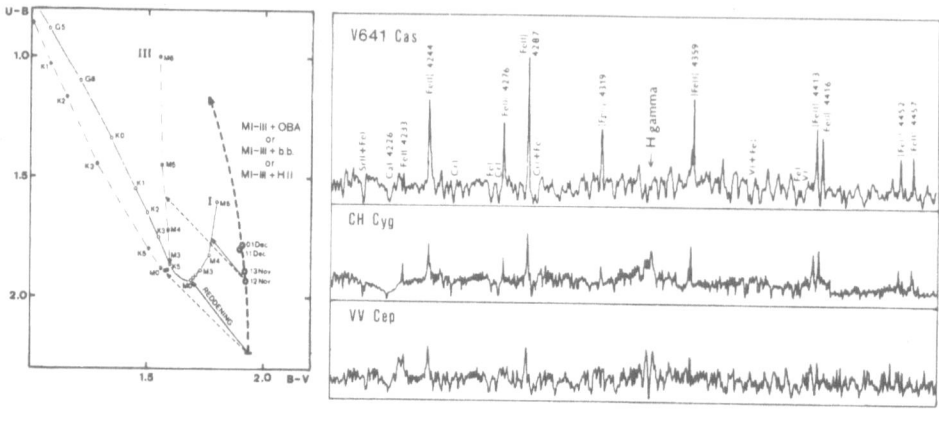

Fig.1   Fig.2

Figure 1. UBV observations of V641 Cas on the color-color diagram (large open circles). The sequences of giants (dots) and supergiants (small open circles) and interstellar reddening vectors are marked. Thick dashed line reflects positions of an object with composite spectrum.
Figure 2. A sample of the spectra of V641 Cas, CH Cyg and VV Cep.

Balmer lines (H_γ and higher) show remarkable changes. The behaviour of H_γ is especially interesting. Relatively strong emission is seen on all the spectra, except this from 17 Feb. 1966, where no emission is seen.
    In spite of the fact that the observational data about V641 Cas are still quite scanty, it seems that the possible binary period of the system is about 1900d. The last deep minimum in broad-band H_α photometry (Guinan et al.,1986) occured about 1900d before our observations. On our spectra the emissions and absorptions of HI are absent, and the objective prism spectra show, that these are not seen also in 1966 - about 4x1900d before our observations.
    In conclusion we suggest two possibilities about V641 Cas:
1. It is a long-period binary (possible eclipsing) consisting of M2-3 bright giant and a main sequence star - similar to VV Cep-type stars.
2. The V641 Cas is an object related to CH Cyg and o Cet, with accreting white dwarf as a secondary. If the companion is a subluminous object one may expect symbiotic phenomenon in this star.

REFERENCES

Barbier,M.,1971,Astron. Astrophys.,14,396
Barbier,M.,1975,Astron. Astrophys. Suppl.,20,305
Berthold,T.,1983,I.B.V.S.,No.2443
Guinan,E.,F.,McCook,G.,P. and Weisenberger,A.,G.,1982,I.B.V.S.,No.2227
Guinan,E.,F. Wacker,S.,W.,McCook,G.,P.,Harris,W.,T.,Robinson,C.,
        Dombrowski,E.,G. and Donahue,R.,A.,1986,I.B.V.S.,No.2925
Shaw,J.,S. and Guinan,E.,F.,1986,B.A.A.S.,17,875
Tomov,T.,Mikolajewski,M. and Mikolajewska,J.,1986,I.B.V.S.,No.2921
Warner,B.,1972,M.N.R.A.S.,159,95

ULTRAVIOLET FE II ABSORPTION LINES IN HD 59643

Joel A. Eaton
Indiana University
Bloominton, IN U.S.A.

ABSTRACT. The peculiar red giant HD 59643 (C6,2) has a composite spectrum like a symbiotic star's except for the absence of highly ionized species. UV observations show a highly variable hot continuum with superimposed absorptions of Fe II and the common emission lines of C IV, Si III], C III], C II], and Mg II. These Fe II absorptions, which are practically ubiquitous in interacting binaries, are an effective diagnostic of turbulence and distribution of scattering gas, even at low resolution.

In general symbiotic stars are long-period binaries consisting of a mass-losing red giant and a small hot companion. Ultraviolet radiation from the companion ionizes the cool giant's wind which then emits like a planetry nebula. Charactistic lines of symbiotics require ionization by a source of $\sim 10^5$ K. There are, however, other interacting binaries with similar physical properties but without the emission of highly ionized species. One of these symbiotic-like objects may be the carbon star HD 59643 (Querci et al. 1986). Optical and UV spectra show a blue continuum, Balmer emission lines, unusually strong chromospheric emission lines (Mg II, Fe II, O I), very strong C IV $\lambda1550$, and the intersystem lines C III] and Si III].

In IUE observations at eight epochs, this star's UV continuum has undergone apparently cyclic variations in strength. Figure 1 shows the strongest UV spectrum of HD 59643. Longward of 1550 Å, almost all of the absorption dips are Fe II. We see, extending from 1550 to about 1725 Å, a group of depressions produced by Fe II multiplets UV38-45, UV68, and UV8. Features resembling those seen here are also detected to various degrees in ultraviolet spectra of interacting binaries and symbiotic stars. Some examples are HD 35155 (Ake and Johnson 1987), 5 Ceti (Eaton and Barden 1987), SX Cas (Plavec et al. 1982), RY Gem (Plavec and Dobias 1987), 22 Vul (Ake et al. 1985), and AR Pav (Hutchings et al. 1983).

The bottom panel of Figure 1 gives a theoretical spectrum for a 10,000 K model atmosphere in which the opacity is primarily scattering by individual lines (a few strong ones of C II, Si II, and O I and the very many of Fe II from Kurucz 1981), ionization and excitation are in LTE, and line broadening is by turbulence (40 km/s) and radiative damping. This approach agrees well with the observations, although simple chromospheric attenuation of a hot star's light would serve just as well if the turbulence and column density were chosen properly.

25

J. Mikolajewska et al. (eds.), The Symbiotic Phenomenon, 25–26.
© *1988 by Kluwer Academic Publishers.*

26

Figure 1. HD 59643 in the UV. At top is the strongest continuum yet seen with IUE (1985/37). Most features are absorption by a rather turbulent gas, especially the characteristic features from 1550 to 1750 A, due almost entirely to Fe II. C IV λ1550 is in emission, as are Si III] and C III] near 1900 A. The dip at 1860 A is probably Al III UV1. Below is a theoretical spectrum discussed in the text.

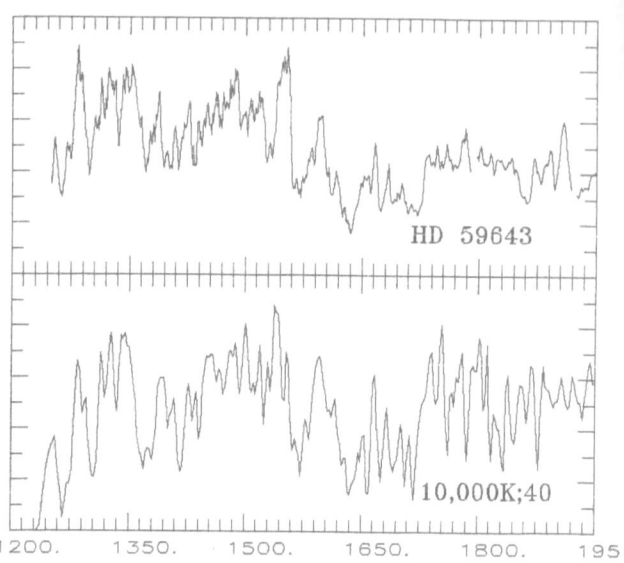

WAVELENGTH (A)

The importance of calculations like these is that they show the crucial role of velocity spread (turbulence?) in determining the strength of UV absorption features. They also suggest a way of studying the distribution of gas about the hot source: If gas is localized in a ~ plane parallel atmosphere or in an isolated cloud along the line of sight, photons will simply be scattered away from us. But if the gas surrounds the star in a detached shell, about as many photons will be scattered into the line of sight as out of it, and the line will be weak (cf. Che et al. 1983). We may be seeing this effect in the strong Fe II line UV41 λ1658 which is much more pronounced in the model than in the observations. Generally, however, the spectrum of HD 59643 seems to be formed by practically pure attenuation. In contrast, 5 Cet shows filling in of the stronger features of Fe II, such as the dip at 1535 A, from scattering by a chromospheric shell, and AR Pav is surrounded by so much scattering material that it does not seem to eclipse at all in the UV.

REFERENCES

Ake, T. B., and Johnson, H. R. 1987, Ap. J., in preparation.
Che, A., Hempe, K., and Reimers, D. 1983, Astron. Ap., 126, 225.
Eaton, J. A., and Barden, S. J. 1987, in preparation.
Hutchings, J. B., Cowley, A. P., Ake, T. B., and Imhoff, C. L. 1983, Ap. J., 275, 271.
Kondo, Y., McCluskey, G. E., and Parsons, S. B. 1985, Ap. J., 295, 580.
Kurucz, R. L. 1981, S. A. O. Special Report, No. 390.
Plavec, M. J., and Dobias, J. J. 1987, A. J., 93, 440.
Plavec, M. J., Weiland, J. L., and Koch, K. H. 1982, Ap. J., 256, 206.
Querci, M.&F., Johnson, H.R.,and Baumert, J.H. 1986, Adv.Space Res.,6,215.

ORBITAL RADIAL VELOCITY CURVES OF SYMBIOTIC STARS

Michael R. Garcia and Scott J. Kenyon
Harvard/Smithsonian Center for Astrophysics,
60 Garden St., Cambridge, MA., USA

ABSTRACT. For the past 6 years we have been measuring the radial velocity of the M-giant component of Symbiotics Stars with high dispersion echelle spectrographs. Because these velocities are based on the **absorption line** spectrum of the M-star, very reliable orbital radial velocity curves can be constructed. The orbital solutions derived from these velocity curves allow the approximate size of the binary to be determined, and also provide insight into the origin of the photometric variations seen in these stars. For most of the stars, the photometric variations are due to eclipse or reflection effects (two notable exceptions are Z And and RW Hya). Stars which may be near to filling their Roche lobes are TX CVn, Z And, EG And, UV Aur, and CI Cyg. Stars which do not appear to fill their Roche lobes are AG Dra, RW Hya, and AG Peg.

1. INTRODUCTION

As our understanding of interacting binary stars is based largely on measurements of stellar masses, periods, and mass transfer rates (see, for example, the study of cataclysmic variables by Patterson 1984), measurements of these most basic parameters for symbiotic stars would seem to be a prerequisite to obtaining a clear understanding of these systems. Modern echelle spectrographs and reticon readout devices allow radial velocity (V_R) determinations to $\pm 1/2$ km s^{-1}, so observations over several years should allow the orbital V_R curves to be measured. With these curves in hand, the size of the Roche lobe (R_L) can be computed for a given total system mass and mass ratio(q). Optical and IR observations allow the spectral type and luminosity class of the giant to be determined (Kenyon and Gallagher 1983), and therefore one can determine whether a symbiotic is likely to be filling its Roche lobe.

2. TECHNIQUE

The echelle spectra which we have used to measure V_R consist of a single, ~ 40Å wide order centered at ~ 5200Å with ~ 0.2Å resolution. After emission lines are excised from the spectrum, V_R is determined with a computerized cross-correlation technique (Tonry and Davis 1982). We have concentrated on S-type symbiotics with known photometric periods and which have not had a recent outburst, thereby ensuring that the M giant photosphere is visible at 5200Å. Once the V_R are determined, orbital solutions are found using the Fourier decomposition technique described by Monet (1979).

J. Mikolajewska et al. (eds.), The Symbiotic Phenomenon, 27–32.
© *1988 by Kluwer Academic Publishers.*

3. RESULTS.

TX CVn. The large quantity of data for this star allows an 11σ level detection of the eccentricity ($e = 0.2$). The detection of a significant eccentricity in the orbit of TX CVn is surprising, as it is not expected on evolutionary grounds. For the remaining stars there is not enough data to justify fitting of an eccentric orbit, and we have set $e = 0.0$. The orbital parameters derived for TX CVn are shown in Table 1, and the data and best fit eccentric orbit are plotted in Figure 1. Optical and IR observations of the giant in TX CVn indicate it is a M0 III with a radius(R_G) of ~ 35 $R_\odot$ (Kenyon and Gallagher 1983). If the mass of the giant is ~ 1 $M_\odot$, and the mass of the hot star is $0.25 - 0.5$ $M_\odot$, then the fraction of the Roche lobe filled, $R_G/R_L \sim 0.5$. The slightly larger giant radius required for the giant to fill its Roche lobe is most likely allowed by current observations.

EG And. The orbital solution in Table 1, which is based on data taken over 3.5 orbital cycles, confirms that of Oliversen *et al.* 1985. The phase of the radial velocity curve, when compared with the photometric phase for Hα suggested by Oliversen *et al.*, is within ± 0.1 of that expected for an eclipse. The period determined from this radial velocity data is 5σ larger than that derived by Smith (1980) from the Hα and photometric variations. While Smith does not quote and error in his derived 470 day period, examination of the data indicates it may be as large as 10 days, thus rendering the two periods nearly consistent. The IR and optical observations of Kenyon and Gallagher (1983) indicate that the giant component may have a radius of ~ 100 $R_\odot$. Assuming the companion is a typical white dwarf with mass of 0.5 $M_\odot$, we find that $R_L/R_G < 1.5$ for $q < 3$, which indicates that the giant is likely to be filling its Roche lobe.

Z And. The orbital parameters deduced from these measurements (which cover two orbital cycles) are shown in Table 1, and the radial velocity curve is shown in Figure 1. The relative phasing of the radial velocity curve and photometric variations is a bit mysterious — spectroscopic conjunction occurs at $\phi = 0.27 \pm 0.07$, where $\phi = 0.0$ is defined as minimum light (Kenyon and Webbink 1984). This phasing is not that expected for (simple) reflection or eclipse phenomena, suggesting that rather complex geometric effects may be involved in producing the photometric variations. Optical and IR spectra indicate that the giant is an M3 II with a probable radius of ~ 150 $R_\odot$ (Kenyon and Gallagher 1983). Given the orbital parameters in Table 1, the giant fills its Roche lobe if the mass of the accreting component(M_H) is 1 $M_\odot$ and $q = 1$, or if $M_H = 0.5$ $M_\odot$ and $q = 2$. The giant may therefore be filling its Roche lobe.

UV Aur. These data cover 3 orbital cycles, but are sparse near phase 0.5 due to the proximity of the orbital period to 1 year. The orbital parameters derived from the radial velocity curve are shown in Table 1, and the data themselves are shown in Figure 1. Spectroscopic conjunction occurs at phase 0.4 of the ephemeris of Zakarov (1951), which defines phase zero to be at maximum light. Given the probable errors in the derived period, the relative phasings are consistent with that expected for eclipse or reflection phenomena. If the giant has a radius of ~ 100 $R_\odot$ (Kenyon 1986), then it fills its Roche lobe for $M_H \sim 1$ $M_\odot$ and $q \sim 1$. Assuming $M_H = 1$ $M_\odot$ and $q = 2$, then $R_L/R_G = 1.5$, so the giant in this system is likely to be filling its Roche lobe.

CI Cyg. The orbital parameters derived from these data (Table 1), which cover 2 full cycles, are substantially different from those reported by Iijima (1982). Spectroscopic conjunction occurs at phase 0.98 on the ephemeris of Aller (1954), confirming the widespread belief (see Kenyon 1986 and references therein) that the photometric variations are caused by eclipses. The long duration of the eclipse in CI Cyg indicates that the giant, which has a radius of ~ 175 R$_\odot$, fills its Roche lobe (Kenyon and Gallagher 1983). For $q = 1$ to 2, the radial velocity data are consistent with a lobe filling giant with mass ~ 1 M$_\odot$.

AG Dra. The orbital parameters derived from over 3 cycles of observations (see Figure 2) are shown in Table 1. Using the ephemeris and period found by Meinunger (1979), we find that spectroscopic conjunction occurs at $\phi = 0.48 \pm 0.08$, as expected for eclipse or reflection effects (maximum light is defined to occur at $\phi = 0.0$). The spectral type of the red star in AG Dra is somewhat uncertain as various authors have found spectral types ranging from G5 (Mirzoyan and Bartaya 1960) to K0 I (Huang 1983). The average seems to be about K3 III (Kenyon 1986), but the luminosity class is rather uncertain. Assuming a spectral type of K3 III, the radius of the giant would be ~ 30 R$_\odot$. For reasonable values of q and system mass, the giant is well within its Roche lobe ($R_L/R_G \sim 4$).

RW Hya. The orbital parameters derived from this more extensive data set (Table 1) tend to confirm the orbital parameters found earlier (Garcia 1986). Because the orbital period is so close to one year our phase coverage is rather poor (see Figure 2) even though we have observed ~ 5 orbital cycles. The phase of spectroscopic conjunction is $\phi = 0.2 \pm 0.1$ on the ephemeris of Kenyon and Webbink (1984), which defines phase zero as maximum light. This relative phasing is rather difficult to understand, as it is nearly opposite that expected for reflection or eclipse phenomena. The giant in RW Hya is most likely a M2 III with a radius of ~ 60 R$_\odot$(Kenyon and Fernandez-Castro 1987). In this case the giant fills its Roche lobe only for $q \sim 1$ and $M_G \sim 0.2$ M$_\odot$, which seems improbably low. Thus it appears that the giant substantially underfills its Roche lobe ($R_L/R_G \sim 2.0$).

AG Peg. The orbital parameters found from these radial velocity measurements (Table 1) are very similar to those found by Hutchings, Cowley and Redman (1975). The data (which cover 1.3 orbital cycles) and best fit solution are shown in Figure 2. Using the ephemeris for minimum light found by Meinunger (1981) we find that spectroscopic conjunction occurs at phase 0.9 ± 0.1, consistent with that expected from eclipse or reflection effects. If the giant is a normal M3 III, as implied by the optical and IR observations (Kenyon and Gallagher 1983), then it must have a radius < 100 R$_\odot$. The radial velocity measurements then require that the system have unreasonably small mass ($M_G \sim 0.2$ M$_\odot$, $q \sim 1$) if the giant is to fill its Roche lobe. We conclude that a giant with $R_G < 100$ R$_\odot$will not fill its Roche lobe in this system.

AX Per. The orbital parameters derived from these data (which cover 2.4 cycles, see Figure 2) are shown in Table 1. The orbital period derived from these radial velocity measurements alone is $\sim 8\sigma$ below the photometric period (Kenyon 1982). However, the photometric period is based on ~ 600 measurements covering 7 cycles, therefore must be considered more reliable. This view is further supported by the observation that the scatter in the radial velocities measured for AX Per about the fit curve is larger than that seen in other symbiotics (See Figures 1 and 2). Using the ephemeris for minimum light defined by Kenyon (1982), we find that spectroscopic conjunction occurs at phase 0.96 ± 0.13, i.e., consistent with that expected for eclipse or reflection effects. As the radius of the giant is somewhat unclear at present (see Kenyon and Gallagher 1983), we cannot determine whether it fills its Roche lobe or not.

Table 1: Summary of Orbits from V_R Measurements

Star	Best P (days)	Other P	K (km/sec)	e	f(m)	γ	N_{points}
EG And	492 ± 4.3	470 ± 10	7.1 ± 0.2	0.0	.017	-94	19
Z And	750 ± 8	756.85 ± 1.0	8.1 ± 0.5	0.0	.042	+2	19
UV Aur	388 ± 5	$395.2\pm?$	5.1 ± 0.3	0.0	.005	+6	19
TX CVn	199.0 ± 0.5	none	6.1 ± 0.1	0.2	.003	+2	37
CI Cyg	812 ± 10	855.25	6.5 ± 0.3	0.0	.024	+18	25
AG Dra	530 ± 5	554 ± 5	5.3 ± 0.3	0.0	.009	-146	17
RW Hya	366 ± 2	372.45 ± 0.3	8.7 ± 0.9	0.0	.025	+14	10
AG Peg	796 ± 15	827.0 ± 4	6.4 ± 0.2	0.0	.022	-14	17
AX Per	601 ± 10	681.6 ± 7.2	6.7 ± 0.4	0.0	.021	-114	19

The γ velocities are in km s^{-1}, and have errors of approximately ± 1 km s^{-1}. The errors in the mass functions f(m) = $K^3P/2\pi G$ can be derived directly from the quoted errors on P and K. The number of radial velocity measurements used in the orbital solution is N_{points}.

4. CONCLUSIONS.

Continuing measurements of the V_R in Symbiotics will allow more accurate determination of the orbital periods and perhaps the measurement of orbital eccentricities in more stars. Comparisons of the V_R curve of the giant with those derived for emission lines should allow the structure of the emission line region to be mapped out. While the radial velocity curves allow the approximate size of the binary to be determined, the critical question of whether or not the giant fills its Roche lobe depends strongly upon the often uncertain nature of the giant. The IR and optical studies of Kenyon and Gallagher (1983) go a long way toward solving this problem, but clearly there is more work to be done.

REFERENCES.

Aller, L.H., *Publ. DAO* Victoria, B.C., **9**, 321.

Garcia, M.R., 1986, *Astr. Ap.* **91**, 1400.

Huang, C.C., 1983, in *IAU Colloquium No. 70, The Nature of Symbiotic Stars*, ed. M. Friedjung and R. Viotti, (Dordrecht:Reidel), p. 151.

Hutchings, J.B., Cowley, A.P., and Redman, R.O., 1975 *Ap.J.* **201**, 404.

Iijima, T., 1982 *Astr. Ap.* **116**, 210.

Kenyon, S.J., and Fernandez-Castro, T., 1987 *Ap.J.* **316**, 427.

Kenyon, S.J., 1986, *The Symbiotic Stars*, (Cambridge: Cambridge University Press).

Kenyon, S.J., and Webbink, R.F., 1984, *Ap.J.* **279**, 252.

Kenyon, S.J., and Gallagher, J.S., 1983 *A.J.*, **88**, 666.

Kenyon, S.J., 1982 *P.A.S.P.* **94**, 165.

Mirzoyan, L.V., and Bartaya, R.A., 1960, *Izv. Abastumani Astrofiz. Obs.*, No. 25, 121.

Meinunger, L., 1981, *Inf. Bull. Var. Stars*, No. 2016.

Meinunger, L., 1979 *Inf. Bull. Var. Stars*, No. 1611.

Monet, D.G., 1979 *Ap.J.* **234**, 275.

Oliversen, N.A., Anderson, C.M., Stencel, R.E., and Slovak, M.H., 1985 *Ap.J.* **295**, 620.

Patterson, J., 1984 *Ap.J. Suppl.* **54**, 443.

Payne-Gaposchkin, C., 1957, *The Galactic Novae* (Amsterdam: North Holland).

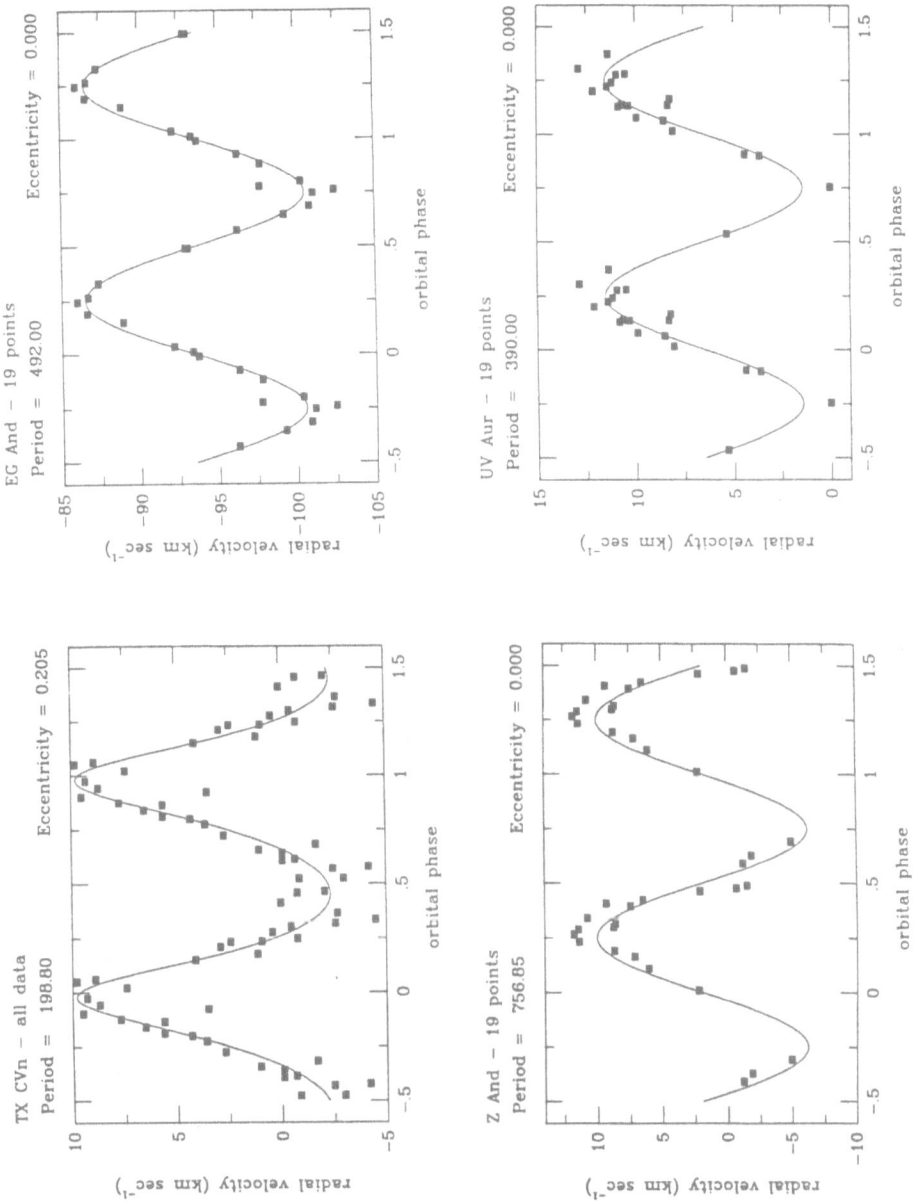

Figure 1: Radial Velocity Curves of Symbiotic Stars
The measured radial velocities and orbital solutions based on photometric periods are shown for
TX CVn, EG And, Z And, and UV Aur.

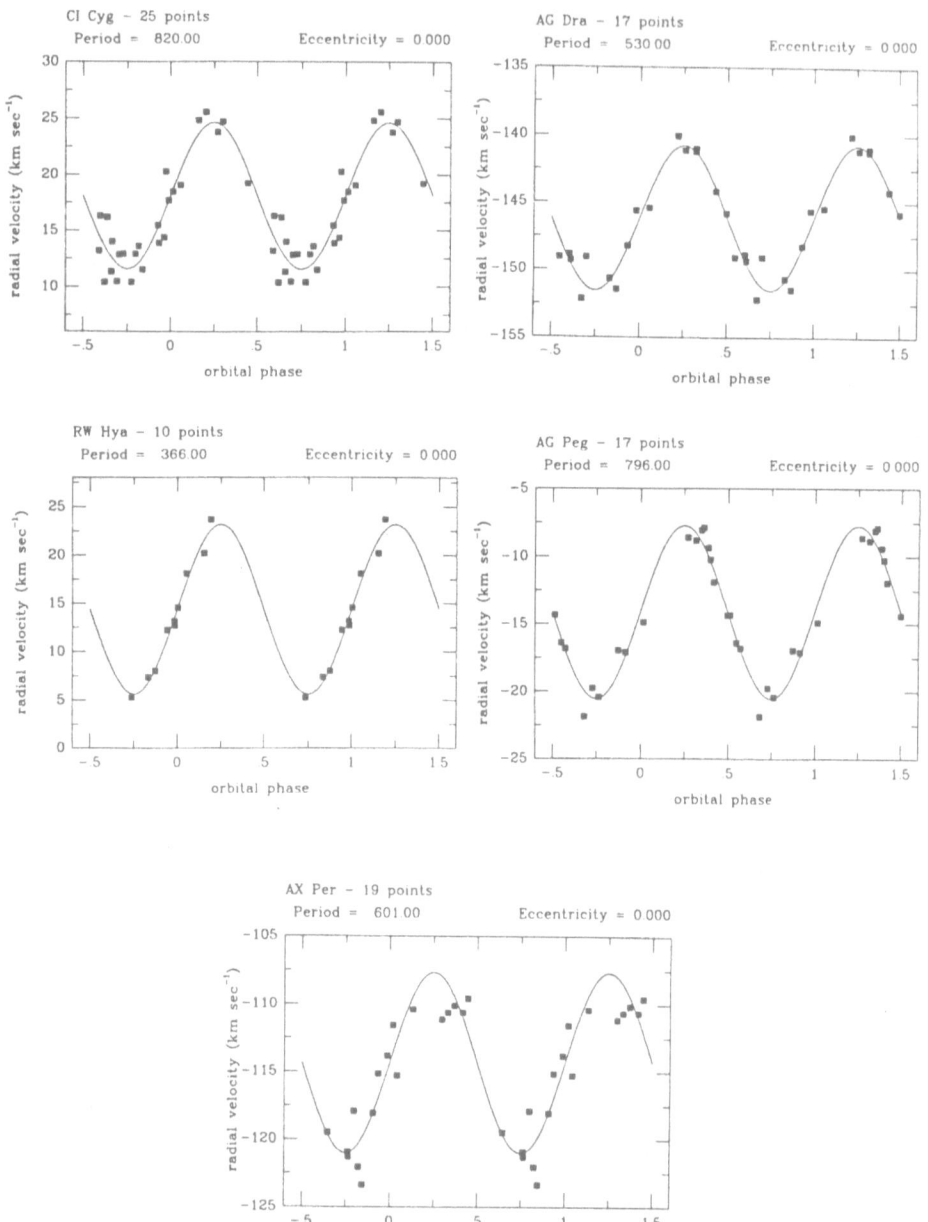

Figure 2: Radial Velocity Curves of Symbiotic Stars

The measured radial velocities and orbital solutions based on photometric perids are shown for CI Cyg, AG Dra, RW Hya, AG Peg, and AX Per.

ORBITAL PARAMETERS OF THREE SYMBIOTIC STARS

Elia M Leibowitz and Liliana Formiggini

Wise Observatory
The Sackler Faculty of Exact Sciences
Tel Aviv University
Tel Aviv 69978, Israel.

ABSTRACT.   The optical light curves of the three symbiotic stars AG Dra, AX Per and
AG Peg, in the U, B and V photometric bands, in the quiescent states of these stars, are
interpreted on the basis of a "reflection" model for the observed light variations.   A least
squares fit of a numerical model that is constructed to the observed magnitudes of the stars
yields the value of few of the orbital elements of these binary systems.   Some predictions
are made on the basis of the model and its parameter values about the detailed structure of
the light curves of the three stars, and of the expected structure of radial velocity curves
of the cool components in these systems.

1. INTRODUCTION

The reflection effect was mentioned in this colloquium more than once as an
interpretation or a possible interpretation for the observed light curves (LC's) of a few
symbiotic stars (SS's). Also in the very first talk of this meeting, Dr. David Allen stressed
the need for orbital elements of SS's. Our work is an attempt to test whether the reflection
effect can indeed account for observed LC's of SS's.   At the same time we examine
whether with a quantitative analysis of the effect one can perhaps also derive the value of
a few orbital elements of symbiotic star systems from their observed optical LC's.

We have found photometric data for 3 SS's, AG Dra, AX Per and AG Peg, that show
in their quiescent state a smooth light variation with one minimum and one maximum
point per cycle. The three system are also characterized by LC's that are dominated by a
periodic component for a long time. Our analysis is applied to these periodic variations,
which   according to the reflection interpretation are due to the orbital revolution of the
system.

II. DATA AND MODEL

The discrete points in Figure 1 represent the observed LC of AG Dra in the U, B and
V photometric bands, folded onto the known photometric period of this star of 554 days.
The points in Figure 2 are the B and V observed LC's of AX Per, folded onto its period
of 681 days. The discrete points in Figure 3 represent the U, B and V observed LC's of
AG Per, folded onto the period of 815.7 days. This period was chosen by us on the basis

J. Mikolajewska et al. (eds.), The Symbiotic Phenomenon, 33–36.
© *1988 by Kluwer Academic Publishers.*

of a power spectrum that we have computed for all the data points that we had on our bands. Note that in all 3 figures the variation cycle is displayed twice. Data for the figures were taken from Meinunger 1979, Taranova and Yudin 1982, Kenyon 1982, Belyakina 1970, Fernie 1972, Oliversen and Andersen 1982 and Meinunger 1983.

We assume that the observed flux in each photometric band is composed out of 3 components: 1) The flux of the cool star in the system. 2) The flux of the hot star. 3) The "reflected" flux from the hemisphere of the cool star that is facing the hot star.

Component (1) is phase independent. Component (2) is also constant, except when eclipse occurs. If the apparent disk of the cool star eclipses the hot star, which in our model is treated as a point source, the contribution from this star vanishes. If there is no eclipse during the entire orbital cycle, the second component of the observed flux adds up to the first one and both terms are considered together as a single contribution of a DC component to the LC.

Most of the phase dependence is contained in the 3rd. contribution. In deriving an explicit expression for this term we assume that a plane parallel beam of the hot star radiation is illuminating the near hemisphere of the cool star. A fraction β of the infalling energy is reradiated from this hemisphere in the photometric band under consideration, with an angular distribution that can be represented by a limb darkening law, with a limb darkening parameter μ. The LC is given by the model in a parameter form, as a pair of functions $\mu(\theta)$ - the magnitude of the system, and $x(\theta)$ - the observed phase corresponding to a phase angle θ. The two functions depend on the model parameters that include the orbital inclination angle, the eccentricity of the orbit, the radius of the cool star in the system, the direction angle to the periastron point in the orbital plane, the limb darkening coefficient and a few others.

Further details of the mathematical representation of the model are given in Formiggini and Leibowitz 1987.

III. RESULTS

We applied a least squares procedure to find the set of parameter values that give the best fit of the model LC to the observed measurements. We have found 2 minimum points for AG Dra, 2 for AX Per and 1 for AG Peg. The solid curves in Figures 1, 2, and 3 are the plots of the (first) best solution for each star. Figure 4 displays again the observed LC's of AG Dra, with the second model solution for this star. Note that the two solutions for AG Dra imply that this system is undergoing an eclipse phase. The second solution for AX Per which predicts an eclipse of the hot component in this system too, may probably be rejected because it requires that the luminosity of the hot component is larger than that of the cool star, in the entire optical spectral region. This requirement is not supported by the observed optical spectrum of this star.

IV. DISCUSSION

Figures 1,2,3 and 4 show that the reflection effect may indeed account for the observed periodic variations in the LC of the three stars AG Dra, AX Per and AG Peg. The model also enables one to derive the values of a few orbital elements of these systems, some of them within a rather limited range of uncertainty. In particular we feel that the values of the eccentricities are well established because they persist in all local minima that we have encountered in our optimization procedure. It seems that if the periodic variations are indeed caused by the orbital motion in the system, the geometry of the orbit determines the structure of the LC in a unique or in a close to unique way.

The discrete points in Figure 5 are the measured radial velocities of the cool star in AG Dra, as measured by Dr. Mike Garcia and presented by him in this colloquium. The smooth curve in the figure is the radial velocity curve that is computed with the orbital elements from our model. It is clear that these elements are consistent with the radial velocity observational results.

The two solutions for AG Dra that are plotted in Fig. 1 and 4 differ from each other mainly by the value of the parameter r. The first one has a small r, namely a giant as the cool star in this binary system. The large value of r in the second solution implies that the cool star in AG Dra is a supergiant. There is no way to distinguish between these solutions on the basis of the photometric data alone, because both have the same statistical significance. The supergiant solution may be slightly questionable because a star with the dimensions implied by this solution is expected to be tidally distorted. The ellipsoidal geometry of such a star is generally noticeable in the LC's of close binary systems, e.g. massive x-ray binaries. No indication for an ellipsoidal effect can be found in the LC of AG Dra.

Our claim of an eclipse in the LC of AG Dra can be checked by high quality photometric measurements near phase 0.5 of this system.

ACKNOWLEDGEMENT

We thank Dr. M. Garcia for providing us with his radial velocity measurements. This work is partially supported by the Israeli Academy of Sciences.

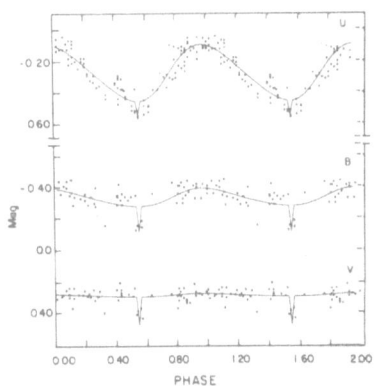

Figure 1. Light curves of the star AG Dra in the three photometric bands U,B and V, folded onto its binary period of 554 days. Discrete points are observations. Smooth curve is the best fitted "giant" solution of our reflection model.

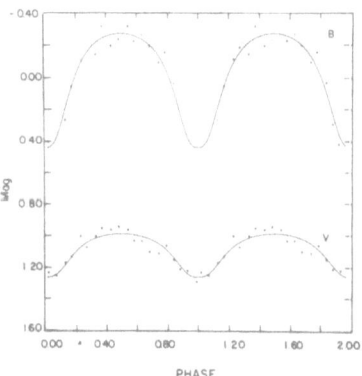

Figure 2. Same as in Figure 1 for the star AX Per.

36

REFERENCES

Belyakina, T. S., 1970, *Astrophysics* **6**, 22.
Fernie, J. D. 1972, *P.A.S.P.*, **84**, 528.
Formiggini L. and Leibowitz, E. M., 1987, *Astron. & Astrophys.*, in preparation.
Kenyon, S. J., 1982, *P.A.S.P.*, **94**, 165.
Meinunger, L., 1979, *IAU IBVS*, No. 1611.
Meinunger, L., 1983, *IAU IBVS*, No. 2016.
Oliversen, N. A. and Anderson, C. M., 1982, *'The Nature of Symbiotic Stars,'* Proceedings of *IAU* Coll. No. 70, M. Friedjung and R. Viotti, eds., P. 180.
Taranova, O. G. and Yudin, B. F., 1982, *Soviet Astronomy - AJ*, **26**, 57.

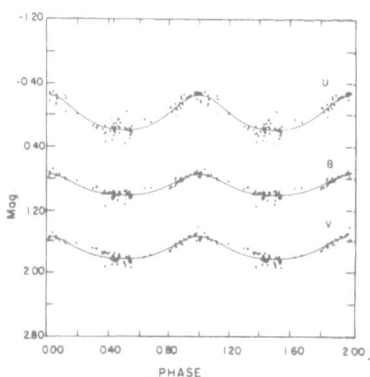

Figure 3. Same as in Figure 1
for the star AG Peg.

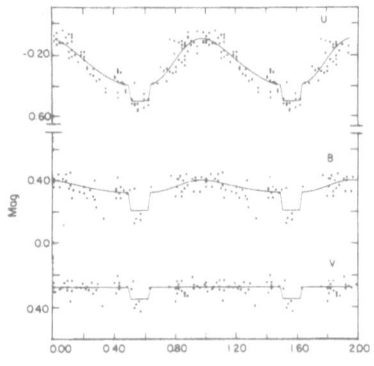

Figure 4. Same as in Figure 1
Smooth curve, the supergiant
solution.

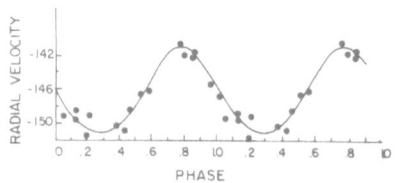

Figure 5. Radial velocity curve of
the cool component in the binary system
AG Dra. Points are observations. Smooth
line is the theoretical curve computed
with the orbital elements from our model.

PROPERTIES OF COOL STELLAR COMPONENTS IN S-TYPE SYMBIOTIC STARS

O. G. Taranova and B. F. Yudin
Sternberg State Astronomical Institute,
Moscow, U.S.S.R.

In 1975 Webster and Allen (1975) divided all symbiotic stars into two groups-those in which the 1-4 μ m continuum show only the presence of a cool star (type S),and
those in which dust emission dominates (type D). With the
exception of some of yellow symbiotic stars, the dust presence in others correlates with the spectral type of
their cool components. That is why one can say that S-type symbiotics contain red giants with spectral type earlier than M6-M7.

At the IAU Colloq. N 70 Allen (1982) noted that it is
difficult to escape the conclusion that symbiotic stars
contain normal cool giants. Nowadays it is certaiu to be
correct because the modern observations of S-type symbiotic stars have not yet discovered any specific distincti-
ons between their cool stellar components and normal red
giants. At the same time it should be noted that some of
these, for example, Z And, CI Cyg may be interacting binaries in which the cool component apparently fields its
Roche lobe and unstable accretion of gas from the red giant onto its hot companion leads to the out bursts of the
latter (Kenyon and Webbink 1984; Yudin 1987).

IR brightness of the cool components in S-type symbiotics fluctuate intrinsically but with small amplitude
($\Delta J \sim 0^m.5$). No definite connection has yet been found between light variation of the hot emission sources exceeding
for example, 2^m in the U filter for BF Cyg, AG Dra
and Z And in 1980. ,and the variations in the brightness
of cool stellar components. However, it should be pointed
out that the unusually prominent minimum in visual brightness of the cool component of Z And was observed half a
year before the outburst of its hot companion in March
1984 (Yudin 1985). Such phenomenou was also observed before the out burst of the recurrent nova T CrB in 1946.
It was supposed that this event was connected with the development of dynamical instabilities in the cool star at-

37

J. Mikolajewska et al. (eds.), The Symbiotic Phenomenon, 37–41.

mosphere resulting in dynamical ejection of a blob of gas
by the latter and to accretion-powered outburst of its hot
companion (Webbink 1976). However, before other outbursts
of Z And its cool component did not demonstrate the essential decrease in its visual brightness.

It is possible that variations in IR brightness in
the eclipsing binary CI Cyg have a regular component with
$P = P_{orb}/2$, due to ellipsoidal deformation of its tidally
distorted red giant (Taranova 1987). This giant has a very
strong 2.3 μm CO absorption band (Kenyon and Gallagher
1983) and thus it may have bright giant dimensions and
fills its tidal lobe. On the other hand Kenyon and Fernandez-Castro (1987) possibly detected the periodic changes
in the spectral type of CI Cyg's red giant determined by
TiO and VO absorption bands in the red with $P = P_{orb}$,
due to heating of its hemisphere facing towards its hot
companion. Verification of existence of these effects requires further systematic observations, since in other similar symbiotic stars they have not been yet detected,with
the only one exception of AG Peg in which the effect of
heating of one hemisphere of the cool star by its hot companion is definitely observed. So, our spectrophotometric
observations of AG Peg show that, when its cool giant lies
in front of or behind the hot component, the depth of TiO
absorption bands in the red is comparable to that found in
normal red giants having spectral types slightly later
than M3 or earlier than M2 respectively (see also Kenyon
and Fernandez-Castro 1987). As mentioned above, the same
phenomenon is possibly observed in CI Cyg although in this
binary system the disk-like gas envelope surrounds its hot
component sothat the latter cannot directly illuminate the
cool companion. Therefore, the reasons of heating of the
red giant's hemisphere facing towards the hot companion
may be partly different in AG Peg and CI Cyg, if of course
this phenomenon really exists in the latter.

Comparing red giants' spectral types derived for five
symbiotic stars both from two-micron and optical spectroscopy Allen (1980) noted that the agreement between them is
not as good as one might hope, the optical types being later. Our photometric and spectrophotometric observations
of symbiotic stars also show that, as a rule, the estimates of the red giants' spectral types in S-type symbiotics,
determined using TiO absorption bands in the red and the
(R-J) colour index, turn out to be later than thein (J-K)
colour estimates (Taranova and Yudin 1987). Similar result
(Fig. 1) can be derived proceeding from the works of Kenyon and Gallagher (1983), and Kenyon and Fernandez-Castro
(1987).

Determination of the luminosity of the cool components
in S-type symbiotics using their spectral parameters is of
great interest but it is still open to question. It is

clear that optical spectra cannot provide an accurate estimate for the luminosity of the red giants in symbiotic
stars. The chemical composition of their cool components
may be also different. Therefore, it is not surprising
that the yellow giant in the high latitude star AG Dra is
classified by Huang (1982) as KOIb. Preliminary attempts
to derive luminosity classes for the same symbiotics'cool
components by measuring the strength of the 2.3 μ m CO
band have been carried out by Kenyon and Gallagher (1983).
They showed that Z And, CI Cyg and T CrB possibly contain luminosity class II bright giants. On the other hand,
the presence of the strong CaII triplet in absorption in
the spectra of AG Peg and EG And possibly indicates that
their cool components are of luminosity class III (Andrillat 1982).

However, as mentioned above, the works in this direction are only at their begining and they must be certainly continued. Therefore, at present we should try to find
some indirect approach in solution of the problem of determination of the luminosity of symbiotic's cool component.

Its luminosity can be calculated for the symbiotic
systems with known binary period in which the cool component fills its tidal lobe. Thus, the problem to be solved
consists in determination of such a parameter in symbiotic star's emission which should be necessary condition
for filling by the cool component its tidal lobe.

On our opinion, the recurrent nova-like outbursts observable on the light curves of a number of symbiotic
stars, such as Z And, CI Cyg, AG Dra, AX Per, BF Cyg, may
be such parameter (Yudin 1987), It is nova-like Z And type outbursts that are distinct and determing feature of
the subgroup of the so-called classical symbiotic stars
among the objects classified by their as symbiotic.

Briefly speaking, the time-dependent and energetic
characteristics of these outbursts demonstrate that their
energy source appears to be the gravitational energy of
the matter accreted by the hot component. Therefore, their appearance is connected with a sharp increase of the
mass transfer in a binary system ($\dot{M} \geqslant 10^{-6}$ $M_{\odot} yr^{-1}$). However, since the hot component of the classical symbiotic
stars is the source of fast wind itself with $\dot{M}_h \cdot v_h \sim$
$\dot{M}_e \cdot v_e$ ($L_{h,bol} \sim L_{e,bol}$ between outburst for above mentioned symbiotic stars), then it cannot accrete matter from
the stellar wind of its companion. Therefore, the cool
components of classical symbiotic stars with necessity
must fill tidal lobes, and this property appears to be an
inalienable signpost distinguishing them from other symbiotic stars.

Taking into account this hypothesis one can understand why in such symbiotic systems as Z And, AG Peg, EG

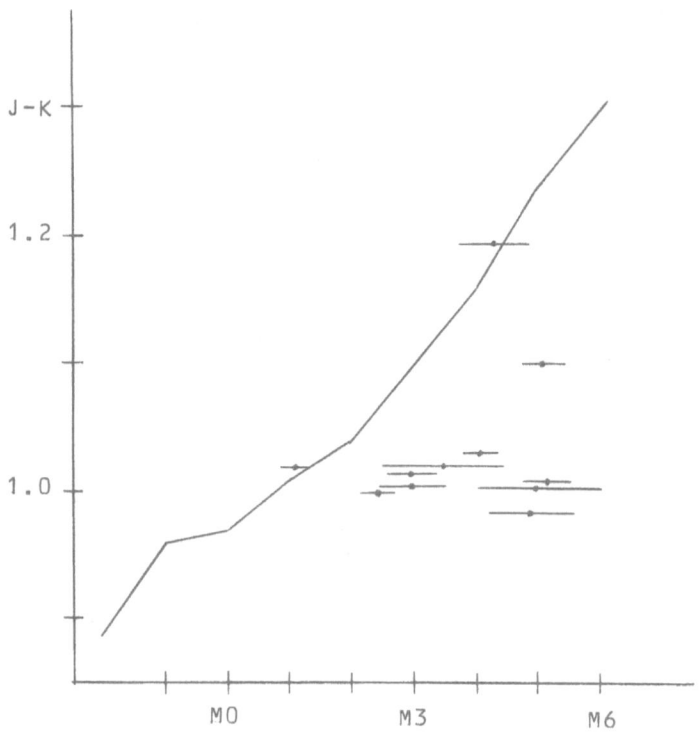

Figure 1. Plots of colour index (J-K) against spectral type for S-type symbiotic stars. (J-K) colour indices are from Kenyon and Gallagher (1983). Spectral types derived using TiO and VO bands in the red are from Kenyon and Fernandez-Castro (1987). The continuous curve is the intrinsic field giants.

And, which are similar in many respects, the recurrent nova-like outbursts of the hot component are observed only in Z And. All this becomes clear if we assume that the cool component in Z And has the luminosity class II, that is classified as asymptotic branch giant M3 II, whereas in AG Peg and EG And as M3 III.

IRAS pointed observations of five S-type symbiotic have shown that EG And, T CrB, RS Oph, AR Pav, AX Per have had IR energy distributions which are very similar to those of normal M giant (Kenyon et al. 1986). The same result have been received for Z And (Schaefer 1986). However, it should be noted that the ground-base observations of CI Cyg, Z And, BF Cyg, V443 Her have revealed at some moment the presence of a small IR excess at 10 μm due to emission from the dust (Taranova and Yudin 1983).

Finally, now that we have yet no systematic photometric and spectroscopic data on the cool stellar components in the majority of S-type symbiotic stars, we should turn our attention to them.

REFERENCES

Allen,D.A.: 1980, Monthly Notices Roy. Astron. Soc., 192, 521.
Allen,D.A.: 1982, IAU Colloq. N 70, p. 27.
Andrillat,Y.: 1982, IAU Colloq, N 70, p. 47.
Kenyon,S.J. and Gallagher,J.S.: 1983, Astron. J. 88, 666.
Kenyon,S.J. and Webbink,R.F.: 1984, Astrophys. J. 279, 252.
Kenyon,S.J., Fernandez-Castro,T. and Stencel,R.F.: 1986, Astron. J. 92, 1118.
Kenyon,S.J., Fernandez-Castro,T.: 1987, Astrophys. J. in press.
Schaefer,B.E.: 1986, Publ. Astron. Soc. Pacific 98, 556.
Taranova,O.G. and Yudin,B.F.: 1983, Soviet Astron. Letters 9, 332.
Taranova,O.G.: 1987, Soviet Astron. Letters, in press.
Taranova,O.G. and Yudin,B.F.: 1987, Astron. Zh., in press.
Webbink,R.F.: 1976, Nature 262, 271.
Webster,B.L. and Allen,D.A.: 1975, Monthly Notices Roy. Astron. Soc. 171, 171.
Yudin,B.F.: 1985, Astrophys. and Space Sci. 110, 277.
Yudin,B.F.: 1987, Astrophys. and Space Sci., in press.

Absolute Energy Distributions for Selected Quiescent Symbiotic Stars

M. H. Slovak and A. D. Code
Washburn Observatory
University of Wisconsin, Madison, Wisconsin, USA

ABSTRACT. Absolute continuum energy distributions for 1200 A to 100μ have been derived for eight quiescent symbiotics. These data are a combination of contemporaneous satellite and ground-based observations, including low resolution IUE and optical spectrophotometry, broadband JKLM infrared photometry and IRAS Point Source Catalogue fluxes. The resulting multifrequency energy distributions represent a merged, uniformly calibrated set of observations and provide a unique opportunity for comparison with quiescent theoretical models.

1. Multifrequency Observations of Symbiotics

The now nearly universal interpretation of the symbiotic stars as interacting binary systems reinforces the importance of obtaining multifrequency observations in order to successfully deconvolve the overall energy distributions. At this juncture, various surveys of the symbiotics have obtained contemporaneous sets of data spanning the X-ray through the radio wavelength domains. We have merged the available data for eight S-type symbiotics in quiescence to provide uniformly calibrated continuum energy distributions. These data permit a detailed comparison with theoretical predictions for the nature of the hot primary , the cool secondary and the circumbinary gas. When combined with recent orbital solutions from radial velocity studies, unique interpretations for the masses of the components, the nature of the mass exchange process and the evolutionary status will be possible.

2. Quiescent Continuum Energy Distributions

2.1. Low resolution IUE spectrophotometry

A significant number of the S-type symbiotics brighter than V = 14th mag have now been observed at low resolution with the IUE satellite. We have produced an atlas of merged SWP and LWR low resolution data for selected S-type systems in quiescence. Representative data for BF Cygni are seen in Figure 1, where the observed fluxes are displayed. These data reveal the large diversity of continua and emission-line spectra which characterize the symbiotics in the ultraviolet. Estimates of E(B-V) have been derived using the interstellar 2175 A depression and additional determinations of (n_e, T_e) can be derived from suitable line ratios, as discussed by Nussbaumer (1982) and Nussbaumer and Storey (1981).

2.2. Cassegrain optical spectrophotometry

Contemporaneous with IUE observations, efforts have been made to obtain comparable ground-based optical spectrophotometry. Calibrated optical data have been published by Slovak (1982), Blair, Stencel, Feibelman, and

43

J. Mikolajewska et al. (eds.), The Symbiotic Phenomenon, 43–45.

Michalitsianos (1983) and Ipatov and Yudin (1986). We have combined high signal-to-noise Digicon spectrophotometry with low resolution IUE data for selected quiescent S-type symbiotics, adjusting the data to a common flux scale. Such data are shown for Z Andromedae in Figure 1. These data provide a critical constraint on model predictions in that both the primary and secondary contribute nearly equally in this regime.

2.3. Broadband JKLM infrared photometry

Broadband JKLM near-infrared photometry for many symbiotics has been provided by Allen and Glass (1974), Feast, Robertson and Catchpole (1977) and recently by Whitelock (1987) and her coworkers. For the S-type systems in quiescence, little variability is seen ($\Delta m_J \sim 0.2$ mag) and these data may be reasonably combined with other quiescent observations. The addition of low resolution IR spectra (e.g. Kenyon and Gallagher 1983) and the continued monitoring (Whitelock 1987) have continued to improve the ground-based IR results. Typical KLM results for AG Dra are seen in Figure 1.

2.4. IRAS Point Source Catalogue fluxes

The IRAS satellite detected a number of the symbiotic stars in the far infrared (Kenyon, Fernandez-Castro, and Stencel 1986; Slovak, Cassinelli, Anderson, and Lambert 1987; Whitelock 1987). Various of the S-type symbiotics have far IR distributions consistent with normal M giants (Kenyon *et al.* 1986). However, detailed modeling of AG Dra (Slovak *et al.* 1987) in quiescence reveals that the free-free emission from the circumbinary nebula may begin to contribute and even dominate the energy distribution longwards of 25μ. Unfortunately, many of the fluxes for the S-type symbiotics detected in the Point Source Catalogue (PSC) are upper limits, especially at 60μ and 100μ. The IRAS PSC color-corrected fluxes for AG Dra are seen in Figure 1, combined with other quiescent observations.

3. The Necessity of Simultaneous versus Contemporaneous Multifrequency Data

The essential requirement for *simultaneous* (at the *same* time) as opposed to *contemporaneous* (during the *same period*) data must be emphasized, even for studies of quiescent systems. While the uncertainty in integrating disparate data sets is minimized by carefully selecting quiescent systems, intrinsic variability of sensitive UV-optical line ratios and even continua argue that secure conclusions can only come from coordinated observing campaigns. Such efforts are even more essential to follow and understand the detailed behavior of the symbiotics during their activity phases.

4. References

Allen, D. A. and Glass, I. S. 1974, *M.N.R.A.S.*, **167**, 337.
Blair, W. P., Stencel, R. E., Feibelman, W. A., and Michalitsianos, A. G. 1983, *Ap. J. Suppl.*, **53**, 573.
Feast, M. W., Robertson, B. S.C., and Catchpole, R. M. 1977, *M.N.R.A.S.*, **179**, 499.
Ipatov, A. P. and Yudin, B. F. 1986, *Astr. Ap. Suppl.*, **65**, 51.
Kenyon, S. J. and Gallagher, J. S. 1983, *A. J.*, **88**, 666.
Kenyon, S. J., Fernandez-Castro, T., and Stencel, R. E. 1986, *A. J.*, **92**, 1118.
Slovak, M. H. 1982, PhD Thesis (University of Texas: Austin).
Slovak, M. H., Cassinelli, J. P., Anderson, C. M., and Lambert, D. L. 1987, *Ap. Sp. Sci.*, **131**, 765.
Whitelock, P.A. 1987, *Pub. A.S.P.*, **99**, 573.

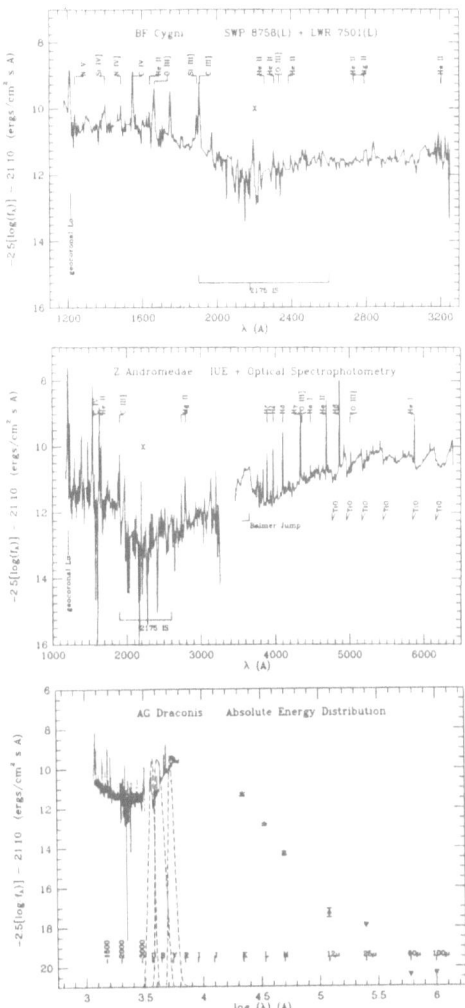

Figure 1. Representative multifrequency observations of three quiescent S-type symbiotics BF Cygni, Z Andromedae and AG Draconis. (Upper panel) Combined low resolution SWP and LWR **IUE** spectrophotometry for BF Cyg is displayed on a magnitude versus linear wavelength scale. Strong permitted and forbidden emission lines are identified and a pronounced 2175 interstellar reddening feature is evident, corresponding to E(B-V) ≈ 0.40. (Middle panel) Low resolution **IUE** and optical Cassegrain Digicon spectrophotometry are shown for Z And, displayed as for BF Cyg. A gap of 300 A exists between the satellite and ground-based data near 3300 A. The observed fluxes are moderately reddened [E(B-V) ≈ 0.45] and a strong Balmer Jump appears in emission. The optical data are dominated by hydrogen and helium emission lines, as well as a number of TiO molecular bandheads. (Lower panel) Merged IUE, optical, ground-based infrared and **IRAS** fluxes are seen for AG Dra displayed on a magnitude versus a logarithmic wavelength scale. The **IRAS** 25μ, 60μ, and 100μ fluxes are only upper limits. The dashed curves represent the Johnson **UBV** filter bandpasses and clearly reveal the limited portion of the energy distribution about which classical photometry provides information.

INFRARED OBSERVATIONS OF SYMBIOTIC MIRAS

Patricia Whitelock
S.A. Astronomical Observatory
P.O. Box 9
Observatory 7935   Cape
South Africa

ABSTRACT.   Symbiotic Miras are identified by their infrared characteristics.   It is shown how an understanding of the evolutionary position of normal Mira variables together with the empirically established period luminosity relation can be used to derive various physical parameters for similar objects in symbiotic systems.   The pulsation periods of symbiotic Miras measured so far fall between 280 and 580 days, their ages must be in the 5-10 Gyr range while main sequence masses of the order 1 to 1.5 $M_\odot$ are indicated.   The obscuration events seen in the IR light curves of several symbiotic Miras are highlighted as potentially important and possible causes are discussed.

1. INTRODUCTION

A small subgroup of the symbiotic stars have the infrared variability and colours which identify their cool giant components as Mira variables.   These are largely objects which have more usually been designated "D-type" symbiotics in the past (e.g. Allen 1984). Observational data for these objects and their interpretation have recently been reviewed in detail (Whitelock 1987, hereafter Paper I). Owing to restrictions in space it has not been possible to give full references in the present paper, they can however be found in paper I.
This paper is an updated summary of the IR observations of symbiotic Miras and their interpretation.   It concentrates particularly on the application of our knowledge of normal (i.e. single) Miras to their symbiotic counterparts.   o Cet (Mira) is included in the symbiotic Mira category as it appears to exhibit weak symbiotic activity and has a number of similarities with the other objects under discussion.   Their relationship with o Cet is in fact probably closer than with the other, S-type, symbiotics.
The objects under discussion are listed in Table I together with other data which will be described below.

J. Mikolajewska et al. (eds.), The Symbiotic Phenomenon, 47–56.
© 1988 by Kluwer Academic Publishers.

TABLE I   SYMBIOTIC MIRAS

Name	RA	(2000)	Dec	ℓ	b	P (day)	K (mag)	A_K (mag)	d (kpc)
o Cet	02 19 21	-02	58.5	167.8	-58.0	332	-2.5	0	0.12
RX Pup	08 14 12	-41	42.4	258.5	- 3.9	580	2.4	0.7	1.3
KM Vel	09 41 14	-49	22.7	274.2	+ 2.6	370	5.4	1.1	3.0
He2-38	09 54 43	-57	18.9	280.8	- 2.2	433	4.4	0.4	2.8
BI Cru	12 23 27	-62	38.2	299.7	+ 0.1	280	4.8	0.9	2.0
SS38	12 51 26	-65	00.0	302.9	- 2.1		6.1	1.2	(4.2)
V704 Cen	13 54 56	-58	27.3	311.2	+ 3.4		8.7	-	-
He2-104	14 11 52	-51	26.4	315.5	+ 9.5	400	6.7	1.7	4.3
V835 Cen	14 14 09	-63	25.7	312.0	- 2.0	450	4.7	0.9	2.8
He2-127	15 24 49	-51	49.9	325.5	+ 4.2		7.8	0.3	(14)
He2-139	15 54 45	-55	29.6	366.9	- 1.4		5.8	0.7	(5.0)
He2-147	16 14 01	-56	59.5	327.9	- 4.3	370-380	4.7	0.1	3.4
He2-171	16 34 04	-35	05.4	346.0	+ 8.6		6.5	1.3	(4.7)
AS210	16 51 21	-26	00.4	355.5	+11.6		6.5	1.2	(4.9)
V2110 Oph	17 43 32	-22	45.6	5.0	+ 3.6		7.6	-	-
H1-36	17 49 48	-37	01.5	353.5	- 4.9	450-500	7.4	1.5	7.6
W16-312	17 50 17	-30	57.6	358.8	- 1.9		7.6	1.3	(7.8)
AS245	17 50 58	-22	19.4	6.3	+ 2.4		7.5	0.2	(12)
SS122	18 04 41	-27	09.2	3.7	- 2.7		6.4	-	-
H2-38	18 06 01	-28	17.1	2.8	- 3.5		6.8	0.4	(8.1)
Hen 1591	18 07 32	-25	53.7	5.1	- 2.6		8.9	-	-
He2-390	18 20 59	-26	48.4	5.7	- 5.8		7.4	2.0	(5.2)
HM Sge	19 41 57	+16	44.7	53.6	- 3.2	540	4.2	1.2	2.1
V1016 Cyg	19 57 05	+39	49.8	75.2	+ 5.7	450	5.1	0.9	3.3
RR Tel	20 04 16	-55	43.3	342.2	-32.2	387	4.1	0.2	2.6
R Aqr	23 43 49	-15	17.0	66.5	-70.3	387	-1.1	0.1	0.25

2. MIRA VARIABLES

Mira variables are long-period ($P \geqslant 100$ d), large-amplitude ($\Delta V \geqslant 2.5$ mag), pulsating red giants with spectral types M, S or C.   Single Miras can be distinguished from other red variables, usually without ambiguity, due to their relatively regular periods, large visual light amplitudes and at certain pulsation phases low excitation emission lines.   Unfortunately in symbiotic Miras these distinguishing characteristics are often masked by the symbiotic phenomenon; frequently to the extent that there is no observational indication of the presence of a Mira in the visual spectral region.   It is therefore of considerable importance that Miras have near-IR colours and spectral features which are distinct from those of other stars of the same spectral type.*   This, together with the fact that their energy distribution peaks in the near-IR makes IR photometry a powerful way of identifying them in the presence of either a hot star or high extinction.

Before going on to discuss symbiotic systems it is useful to examine what we know about the evolutionary position and galactic distribution of Miras.   This subject was recently reviewed by Feast & Whitelock (1987).   Studies of Miras in globular clusters and the LMC have been crucial to our understanding of their evolutionary condition. Miras occur in the more metal-rich clusters where they are the coolest and brightest objects present.   Their luminosities exceed that of core helium-flash, establishing that they are on the Asymptotic Giant Branch (AGB).   Indeed, the available evidence points to Miras as the most luminous and terminal phase of AGB evolution.   Their next evolutionary step must be the loss of the remaining hydrogen envelope to form a planetary nebula, which will be energised by the remnant stellar core.

Knowledge of the pulsation period of a Mira is important because it has been shown to be indicative of the stellar population group to which the star belongs.   Evidence for this is provided by: (1) the good correlation between the value of $[Fe/H]$ for a cluster and the period(s) of the Mira(s) it contains; (2) the fact that the kinematic properties of Miras in the solar neighbourhood are a function of their period. Stars with P ~ 200 d belong to an intermediate population II or thick disk population as do the metal-rich globular clusters.   Longer period stars belong to kinematically younger populations, with an age of the order of 5 Gyr at a period of 350 d.

The existence of a period luminosity (PL) relation among Mira variables is now well established.   Current evidence is consistent with the same PL relation applying to Miras in the rather different environments of the globular clusters, the LMC and the galactic bulge. It has also been suggested that the long-period (up to 2000 d) large-amplitude OH/IR sources fall on an extrapolation of the PL relation to higher luminosities.   This would be consistent with their being younger (<1 Gyr), more massive ($1.5-2.0$ $M_\odot$) and more metal-rich analogues of the Mira variables.   According to stellar evolution theory the maximum luminosity which a star reaches on the AGB is a function of the mass of the helium core which is in turn a function of the initial mass and metallicity.   Although in general we have no direct information on the luminosity of a symbiotic Mira, if its pulsation period has been determined then its luminosity can be derived from the PL relation.   A comparison of this luminosity with predictions of stellar evolution theory should provide some insight on the mass and metallicity of the red giant progenitor.   This is discussed further in section 5.

Footnote: *Although there are certain, rare, supergiant variables which have similar colours to Miras, considerations of distance and height above the galactic plane make it exceedingly unlikely that any of the D-type symbiotics contain this type of object.

It seems likely that the symbiotic phenomena exhibited by Miras are a direct result of, among other factors, the high mass loss rates of these stars. The mechanism for mass loss in Miras and its relationship to the stellar parameters of mass, luminosity, composition etc. are very poorly understood. It is interesting that recent work has shown that the mass of circumstellar dust (m) around a Mira is a function both of its pulsation period (P) and pulsation amplitude (ΔM_{bol}):

$$\log (m/M_\odot) = 2.17 \log P + 1.32 \, \Delta M_{bol} - 12.36.$$

The dependence on amplitude clearly establishes pulsation as a major cause of mass loss. Presumably pulsation-driven shock waves move the outer atmosphere to a sufficiently large distance that conditions are suitable for dust grain formation. Mass loss will then be driven by radiation pressure (a function of luminosity and hence period) on the grains.

3. NEAR-INFRARED

Early JHKL photometry of Mira symbiotics showed an apparent excess over the flux expected from a normal stellar atmosphere. This was attributed to thermal emission from dust (hence the D-type classification). More recently it has become clear that the colours of certain Miras and of most symbiotics are at least as much a function of circumstellar reddening as of circumstellar emission. This realisation has significantly changed our view of these objects.

The presence of Mira variables in a few of the objects under discussion has been established from their large amplitude ($\Delta J \sim 1$ mag), long period (P > 250 d) IR variability. The Mira identification is supported in some cases by the detection of strong H_2O absorption

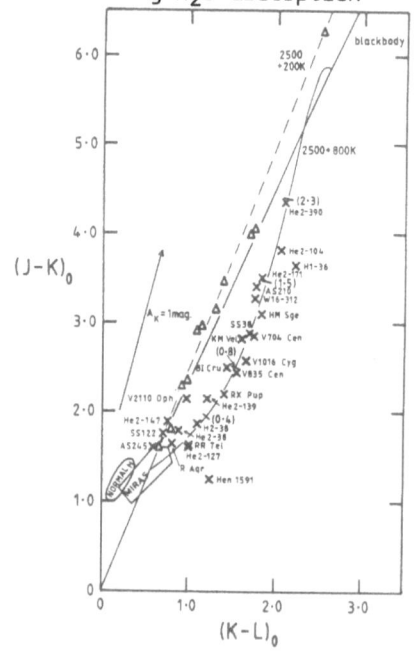

Fig. 1.   Near IR two-colour diagram.   Crosses: normal mean colours of symbiotic Miras; triangles: extreme Miras. The numbers in parenthesis are the extinction at K for a 2500K Mira with an 800K dust shell.   Most observations are SAAO unpublished data, others are referenced in paper I, Fig. 3.

features in the 1-3 μm spectra.   It is then tempting to assume that all
symbiotics with similar IR colours contain Miras.   However this may not
prove to be the case and further observations are important.

The characteristic colours of the D-type symbiotics are illustrated
in the near-IR two-colour diagram (Fig. 1).   Mean points (excluding the
obscuration phases discussed below), are shown for objects which have
been measured repeatedly.   An approximate correction has been applied
for interstellar reddening.   Shown for comparison are some of the
extreme Miras discovered from the IRAS survey.   They are single stars
with thick dust shells and pulsation periods between 350 and 600 d.

Most of the symbiotic Miras fall close to the locus representing
the combination of a 2500K Mira with an 800K dust shell.   This locus is
calculated on the assumption that a silicate dust shell surrounds the
Mira whose light it absorbs and re-emits at the given temperature.   The
further up the locus the thicker is the dust shell (approximate values
of the extinction at K (A_K) are noted along the line).   This is
unlikely to be a very realistic model of the dust shell around a
symbiotic Mira given the presence of a second star and the complexity of
the interaction between the two stars.   However, it serves to
illustrate the difference between the symbiotic Miras and the extreme
Miras whose dust shells are predominantly at much lower temperatures and
therefore produce very little emission in the near-IR even at
L(3.45μm).   It also indicates that reddening as well as dust emission
has a strong influence on the observed colours of symbiotic Miras.   The
"extra" heating for the symbiotic dust, over that of the dust around the
extreme Miras, is almost certainly provided by the hot star and/or an
accretion disk.   Hen 1591 has distinctly unusual colours which require
further investigation.

One of the most interesting and potentially important characteristics of the symbiotic Miras is their tendency to undergo faint phases
in addition to the modulation associated with Mira pulsation.   These
phases typically last between a year and several years and involve a
decrease in brightness at J of 1 to 2 mag from the normal mean (see
Fig. 4, Paper I).   The change at longer wavelengths is less.   A point
representing the colours of the symbiotic during the faint phase falls
higher up the star plus dust shell locus in Fig. 1 (see also Fig. 3,
Paper I).   Thus it appears that the most reasonable explanation of this
phase is obscuration due to dust.   It should be emphasised that no
comparable effect is seen in single M-type Miras.   <u>The obscuration must
be a consequence of the symbiotic nature of the Miras.</u>

4.   IRAS   OBSERVATIONS

The IRAS observations of symbiotics have been discussed by Whitelock
(1985), Kenyon <u>et al</u>. (1986) and in Paper I.   As one might predict from
their near-IR characteristics the symbiotic Miras are as a group
stronger far-IR emitters than are the S-type objects.   IRAS fluxes are
reproduced in Table II for those Miras which appear in version 2 of the
point source catalogue.   Upper limits are indicated by an L.   Most of
the 100 μm and some of the 60 μm observations of the objects near the
galactic plane were overwhelmed by background emission.

TABLE II   IRAS DATA FOR SYMBIOTIC MIRAS

Name	IRAS No	Flux (Jy)			
		12 μm	25 μm	60 μm	100 μm
o Cet	02168-0312	4881.	2261.	301.	884.
RX Pup	08124-4133	182.	132.	150.	7.49
KM Vel	09394-4909	11.8	7.81	1.36	9.5L
He2-38	09530-5704	9.26	4.28	3.8L	38L
BI Cru	12206-6221	17.3	15.3	12L	119L
SS38	12483-6443	7.71	3.29	1.9L	19L
V704 Cen	13515-5812	1.43	1.16	0.5L	19L
He2-104	14085-5112	8.56	9.09	6.83	9.8L
V835 Cen	14103-6311	32.5	26.1	5.95	29.8:
He2-127	15210-5139	0.49	0.29	0.7L	9.2L
He2-139	15508-5520	5.87	2.99	10.3L	140L
He2-147	16099-5651	4.04	2.84	0.5L	31L
He2-171	16307-3459	7.47	4.58	0.75	5.0L
AS210	16482-2555	3.93	1.21	0.4L	2.1L
H1-36	17463-3700	18.1L	28.2	5.75	36L
W16-312	17470-3056	5.86	3.34	22L	291L
AS245	17479-2218	1.67	0.62	2.3L	34L
SS122	18015-2709	1.44	1.00	13L	212L
H2-38	18028-2817	3.39	2.06	4.0L	41L
He2-390	18178-2649	4.68	4.37	2.10	6.8:
HM Sge	19396+1637	119	82.6	9.45	15L
V1016 Cyg	19553+3941	42.9	34.2	4.03	5.8L
RR Tel	20003-5552	19.9	16.4	2.62	1.0L
R Aqr	23412-1533	1577.	544.	66.6	16.6

Fig. 2 is an IRAS two-colour diagram on which are marked the symbiotic and extreme Miras as well as the region occupied by normal Miras with P > 300 d.   Fig. 3 is a composite IRAS/near-IR two-colour diagram for the same groups of objects.   These figures indicate that the majority of the symbiotic Miras have similar far-IR flux distributions to those of normal or extreme Miras.   For normal Miras the 12 μm emission arises from the combination of photophere and dust while at longer wavelengths it is predominantly from dust.   Dust completely dominates the far-IR emission at all wavelengths for most of the symbiotic Miras.   The objects, He2-390 and He2-104, which fall on the right hand side of Fig. 2, seem to show distinctly different dust shell characteristics from Miras or the other symbiotics.   It is possible to explain their IRAS colours as arising from a combination of blackbodies of various temperatures.   They do not appear to contain a significant quantity of material with the strongly wavelength dependent emissivity characteristic of normal Mira dust shells (Paper I).

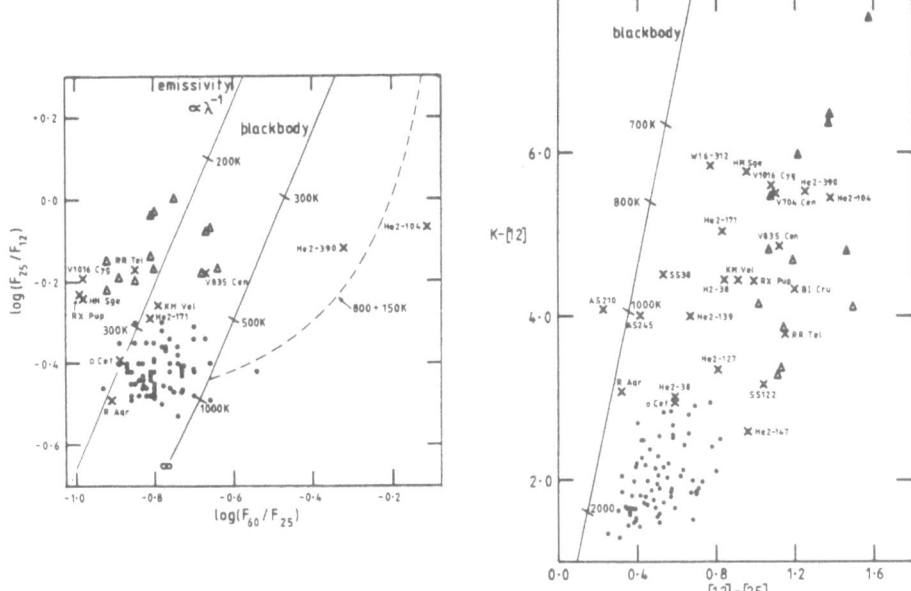

Fig. 2.   IRAS 2-colour diagram.   Closed circles: normal Miras with
P > 300 d, other symbols as Fig. 1.   Shown for comparison are the
blackbody locus and the locus of dust with emissivity ∝ λ⁻¹.   The
dashed line represents the combination of blackbodies with temperatures
of 800 and 150 K.

Fig. 3.   Combined IRAS-near-IR 2-colour diagram.   Symbols as for Fig.
1 and 2.   Note that some of the scatter in the diagram will be due to
the lack of simultaneity of the IRAS and K observations.

Further insight into the nature of the dust emission from these
stars may be gained from the 10μm spectroscopy performed by Roche et
al. (1983).   They found that the spectra of the majority of symbiotic
Miras arises from two components: First the normal silicate particles
and secondly an unknown material which exhibits a featureless 10 μm
spectrum.   However, four objects, He2-390, He2-104, BI Cru and AS210,
showed a featureless spectrum only, with no sign of silicate emission or
absorption.   Note that although BI Cru was not detected at 60 μm by
IRAS its [12]-[25] colour is consistent with it having the same shell
characteristics as He2-390 and He2-104.   AS210 has quite different
colours which might indicate that its Mira is carbon-rich (paper I).

It is interesting that He2-390 and He2-104 have extremely red
near-IR colours (Fig. 1) which are more typical of the other symbiotics
during an obscuration phase.   We have very little data on He2-390, but
He2-104 has been monitored for several years and it shows no gross
changes in its magnitude or colours beyond those associated with
pulsation.   Perhaps the peculiar IRAS colours and featureless 10 μm
spectrum are associated with the obscuration phase and He2-104 may be
undergoing a prolonged event of this kind.

5. DISCUSSION

Pulsation periods are given in Table 1 for the 13 objects for which they have been determined. They range from 280 to 580 d with a mean value of 420 ± 23 d. Following the procedure described by Feast & Whitelock (1987) we can use this period to establish an estimate of the age, initial mass (M_i), final white dwarf mass (M_F), present mass (M_P) and planetary nebula mass (M_{PN}) as given in Table III.

TABLE   III

P (d)	M_{Bol} (mag)	Age (Gyr)	M_i	M_F	M_P ($M_\odot$)	M_{PN}
280	−4.36	10	1.07	0.59	0.75	0.16
580	−5.07	5	1.29	0.63	1.01	0.38

Obviously these estimates are model dependent and in particular are a function of the very uncertain assumed mass-loss rates. The initial mass estimates are somewhat lower than are sometimes derived for symbiotics. Lund and Leedyarv (1986) recently estimated $M_i \sim 3M_\odot$ for a group of symbiotics including o Cet, R Aqr, RR Tel and RX Pup. Such high masses are not consistent with the parameters of normal Miras discussed in section 2. Also, given that several of the symbiotic Miras lie in the galactic bulge, masses much higher than those listed here would seem inappropriate.

Miras or OH/IR sources with periods in excess of 580 d are very rare among single stars, so it is not surprising to find them unrepresented among the symbiotics. The absence of short-period symbiotics however requires some explanation. It may be that binary systems are rare among older stars; alternatively, if the binaries do exist then they may not be obviously symbiotic. This latter condition might arise because short-period Miras lose mass at a slower rate than their long-period counterparts. We might therefore expect a lower, possibly unobservable, level of symbiotic activity in such systems except for relatively close binary pairs.

The PL relation discussed in section 2 should provide one of the best methods for determining distances to symbiotic Miras. However severe problems arise due to uncertainties in the circumstellar extinction, including the possibility that significant quantities of neutral extinction are present (paper I). Table I contains distance estimates for the objects with known periods derived from the PL relation using the K magnitude after correcting for extinction (A_K). The corrections are derived from the simple dust shell model described earlier and are very uncertain for the redder objects. Distance estimates are given in brackets for objects without measured periods but with sufficient IR measurements to confirm their Mira nature. For the

purpose of this estimate they are all assumed to have P = 400 d.   The
error on a distance determined from the PL relation excluding the
extinction uncertainties is less than 10%.   The distances derived for
objects such as o Cet and R Aqr for which there is little uncertainty in
the extinction are therefore probably the best currently available (they
should be superseded by parallax measurements in due course).   It must
however be emphasised that the distance estimate for objects such as
He2-390 and He2-104 are highly uncertain.

I would like to stress particularly the importance of the light
level changes, described in section 3, for the understanding of this
type of symbiotic system.   Willson et al. (1981) have suggested that
the obscuration seen in R Aqr around 1977 and similar effects seen in
the optical light curve at earlier epochs are eclipses by an accretion
disk or cloud associated with the companion star.   They thereby infer a
binary period of 44 yr.   If this interpretation is correct and
comparable events in other symbiotics are similarly caused then we have
been provided with a relatively simple method of measuring orbital
periods.   It is hardly necessary in a meeting such as this to stress
the importance of establishing the binary periods of these objects.
However if the effect is due to material confined in the orbital plane
it is perhaps surprising that it has been seen in such a large fraction
of those symbiotic Miras which have been monitored over long time
periods.

An alternative explanation for these events may be that they
represent spontaneous large-scale mass loss from the Mira.   Such mass
loss might be caused by an effective reduction in surface gravity of the
Mira due to the proximity of the second star.   If the orbits of these
systems are elliptical then periastron passage could trigger the
increased mass loss; thus timing the events is still potentially a
measure of the orbital period.   Strong support for the spontaneous mass
loss mechanism is provided by the recent work of Seaquist & Taylor
(1987).   They interpret increased radio emission from RX Pup during
1985, when the IR was faint, to indicate enhanced mass loss over that in
the mid-1970's, when the IR was bright.   Within a few months of the
time the IR started to fade in RX Pup the re-appearance of high
excitation emission lines was noted by Klutz & Swings (1981).   The
appearance of such lines is very probably symptomatic of increased mass
accretion associated with the hot star.   In that case we can make some
estimate of the separation of the two stars from the fact that the
material made the transition in a time not significantly exceeding 6
months.   Thus if the material was travelling at the 60 km s^{-1} estimated
by Seaquist and Taylor, the orbital period cannot be in excess of 20
yr.   It is obviously well worth monitoring in detail the IR-photometric
and optical spectroscopic development of the other symbiotic Miras.

Despite the above discussion one of the biggest gaps in our
knowledge of symbiotic Miras is the lack of any definite information on
the orbital parameters of these systems.   Various analyses, including
radial velocities, of S-type, i.e. non-Mira, symbiotics have yielded
orbital periods in the range 0.5 to 4 yrs.   In marked contrast, the
suggested periods for symbiotic Miras are much longer - over 20 yr.

Given reasonable assumptions about the masses of the component stars (say, $M_1 + M_2 = 2 M_\odot$ or slightly less), the Mira will fill its Roche lobe if the orbital period is less than about 20 yr. The low mass-transfer rates thought to be necessary to produce the observed phenomena in these systems suggest Roche-lobe overflow to be unlikely. However the stellar wind of the Mira, which to some extent behaves as an extension of its atmosphere, may well be channelled onto the other component via the inner Lagrangian point.

Observations indicate that the mildly symbiotic star o Cet transfers mass to a white dwarf via an accretion disk. The difference between it and those stars normally called symbiotic probably lies only in the larger separation (> 45 AU) of the o Cet binary components resulting in a lower level of interaction. The period of this system must be over 200 yr.

An important consequence of the suggested difference between the orbital periods of Mira and non-Mira (S-type) symbiotics is to highlight the fundamental difference between these two types of objects. It seems very unlikely that those containing Miras would have been recognised as symbiotic prior to onset of the high mass-loss rates associated with Mira variability. The mass transfer at the suggested binary separation would be too small. Equally it is not clear what a typical S-type symbiotic with an orbital period < 5 yrs will evolve into. As the giant component nears the Mira stage, i.e. approaches the top of the AGB, it will overfill its Roche lobe and the subsequent evolution will probably be extremely rapid. It therefore seems very unlikely that there is any evolutionary relationship between Mira and non-Mira symbiotics.

ACKNOWLEDGEMENTS

I am grateful to M W Feast and J W Menzies for helpful discussions and to all my colleagues in the SAAO IR-group for permission to use data in advance of publication.

REFERENCES

Allen, D.A., 1984.   Astrophys. S.S., **99**, 101.
Feast, M.W. & Whitelock, P.A., 1987.   The Late Stages of Stellar
   Evolution, ed. S. Kwok & S.R. Pottasch, Reidel, Dordrecht p. 33.
Kenyon, S.J., Fernandez-Castro, T. & Stencel, R.E., 1986.   Astron. J.,
   92, 1118.
Klutz, M. & Swings, J.P., 1981.   Astr. Astrophys., **96**, 406.
Lund, L. & Leedyarv, L., 1986.   Astrophys., **24**, 154.
Roche, P.F., Allen, D.A. & Aitken, D.K., 1983.   Mon. Not. R. astr.
   Soc., **204**, 1009.
Seaquist, E.R. & Taylor, A.R., 1987.   Astrophys. J., **312**, 813.
Willson, L.A., Garnavich, P. & Mattei, J.A., 1981.   IBVS no. 1961.
Whitelock, P.A., 1985.   1st IRAS Conf., Light on Dark Matter, ed. F.P.
   Israel, Reidel, Dordrecht p. 323.
Whitelock., P.A., 1987.   Publ. astr. Soc. Pacif., **613.**

The Environments of Cool Stars

Robert E. Stencel
Center for Astrophysics and Space Astronomy
University of Colorado
Boulder, CO 80309-0391 USA

ABSTRACT: This review describes recent conclusions about the physical environment of red giant and supergiant stars. This includes coronae, chromospheres, dust formation and stellar winds. This knowledge can provide the boundary conditions for considering what role such objects play as members of binary star systems, where tidal forces and companion behavior alter observed characteristics.

1 Observed Properties of Single Stars

The study of binary stars can help us to understand the behavior of single stars by allowing for determination of stellar masses, internal structure (apsidal motion) and atmospheric properties (eclipses). In this review, I want to outline present knowledge concerning single post-main sequence late-type stars, so that we might better understand how the natural environment of cool stars can affect the interaction in binary stars. New multi-spectral observations and theoretical work is yielding more precise information about the boundary conditions involved with mass transfer and accretion.

Observationally, there are numerous aspects to the environments of cool stars. Ultimately, much of the phenomena is driven by interior conditions. S-type symbiotic systems often contain early M giants which probably evolved vertically upward in the Hertzprung Russell Diagram (HRD), from solar-like main sequence stars. We have gained some insight into the atmospheric transformations associated with this evolution by presenting the statistical results of multi-spectral observations of cool stars on the HRD, and thereby illustrating several "dividing lines". These are indicative of atmospheric transitions between groups of stars which differ in their atmospheric diagnostics (Figure 1). Although represented in the HRD with narrow lines, the transitions are gradual and some admixture near the locus is found. More important than whether these lines are sharp or diffuse, is the significance of what they imply. A strong case can be made that we are observing the atmospheric responses to interior changes.

Access to the soft x-ray and ultraviolet regions of spectra has made possible the identification of the key dividing lines. Ayres et al. (1981), Haisch and Simon (1982)

J. Mikolajewska et al. (eds.), The Symbiotic Phenomenon, 57–64.

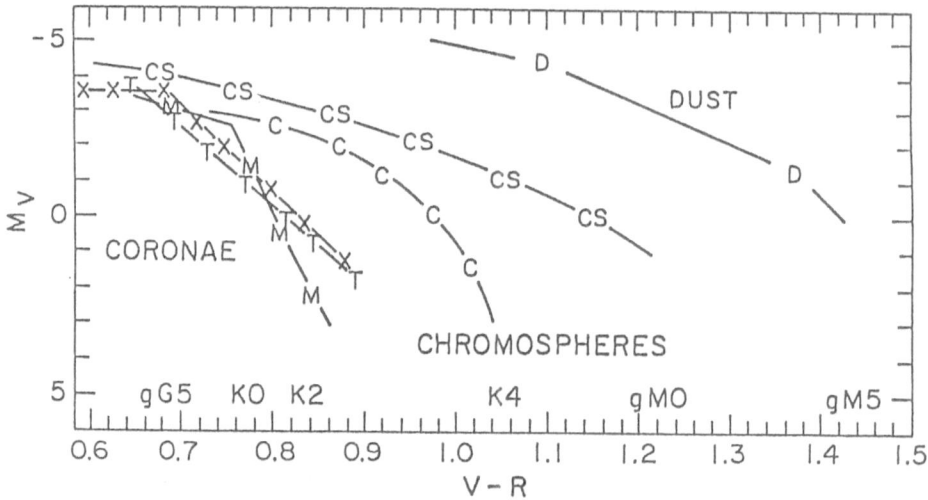

Figure 1: Divisions between major atmospheric conditions in the H–R Diagram. See text for identification of symbols.

and Gondoin et al. (1987) pointed out the rapid decrease in the soft x-ray flux with increasing B–V color along the giant sequence (transition at K1 III, labelled X in Fig. 1). These authors argue that the soft x-ray emission is a good diagnostic of solar-like coronal output (at temperatures of 10^{6-7}K), and, therefore the lack of soft X-rays implies the lack of such coronal material among cooler giants. The coronal volume may be replaced by an "extended chromosphere" (Carpenter et al. 1985) dominated by fluorescent radiative processes (Judge 1986) among the cooler giants. O'Brien and Lambert (1986) have argued that this is indeed the case, presenting a changeover in the observed character of the He I 10830A spectral line between between strong-steady absorption and much weaker, variable absorption at the same spectral type location along the giant sequence.

Early observations with the *IUE* satellite suggested to Linsky and Haisch (1979) that there was a clear separation between stars showing or not showing ultraviolet emission lines corresponding to Transition Region (TR) gas (at 10^5K) in their outer atmospheres. More intensive observations (Simon et al. 1983; Haisch 1987) appear to support the initial impression. This dividing line, labelled T in Fig. 1, occurs near the coronal dividing line. This coincidence makes physical sense if the outer atmospheric structure of the Sun, where the TR underlies coronal gas, is used as the archetype. There is an important group of stars, the K-type bright giants ("hybrids"), which violate this division by showing transition region emission lines and strong mass loss (see below), but these are now associated with older, helium core-burning objects, evolving blueward (Brown 1986).

What is remarkable about these two essentially coincident dividing lines is that they occur at the base of the nearly vertical ascent in luminosity, up the red giant branch, for

1–2 solar mass stellar models, at the point of predicted first dredge up (Iben 1981) and an observed rapid change in C/N surface abundances (Brown 1987). The additional lines in Figure 1 can be viewed in terms of additional envelope and atmospheric responses to interior changes.

The three remaining dividing lines illustrated in Figure 1 can be viewed in terms of the effects of all the previously outlined physics. Two of the lines (marked C and M, respectively) relate to the observed asymmetry in the emission cores of the Ca II and Mg II resonance lines. These chromospheric features show a "doubly reversed" emission core as a consequence of their formation under optically thick conditions (cf. Mihalas 1970). Static atmospheres produce symmetric doubly reversed profiles. Although there are ambiguities in interpreting the profile asymmetry, it is reasonable to ascribe asymmteries to systematic velocity fields in the chromosphere. Stencel (1978), and Stencel and Mullan (1980) found, statistically, that the emission core asymmetries changed systematically from the solar-like blue peak dominated shape to a red peak dominated shape, for stars cooler than along the coronal dividing line. The offset between these dividing lines may reflect their relative optical depths (Mg II being greater). The Mg II survey was done early in IUE's lifetime and should be re-examined with more complete data. Earlier, Reimers (1977) had identified an additional locus above which stars exhibited blueshifted lines due to cool, neutral gas – ascribed by Deutsch (1956) to circumstellar shells (labelled CS in Fig. 1). Taken together, these dividing lines suggest a transition from the circulation patterns observed in the solar chromosphere, to predominantly outflow velocities involving increasing amounts of cool material. Indeed, mass loss derivations suggest a gradual increase in the same direction, consistent with the simple velocity field change concept outlined here.

The third and final division is for stars which show evidence for dust production in their infrared spectra. Woolf and Ney (1969) pointed out the match between the 9.7 micron spectral emission feature in a number of cool giants and the infrared spectrum of minerals like olivine (e.g. $[Mg,Fe]_2 SiO_4$). The working hypothesis is that such stars produce silicate dust grains in association with enhanced mass loss. Numerous observations support this view (cf. Volk and Kwok 1987). Merrill and Stein (1986), citing unpublished work by Gilman and Woolf, described this dust formation locus (labelled D in Figure 1). IRAS data largely agrees with its placement. The dust dividing line largely coincides with the onset of pulsation along the red giant branch: stars cooler than M4-M5 show marked, periodic light variation (cf. Brugel, this volume). Many of these are, of course, Mira variables. Jura (1986) has demonstrated a strong correlation between pulsation, dust production and high rates of mass loss among such stars.

Also largely coincident with stars above this dividing line is the occurrence of stellar masers (Olnon et al. 1980; Bowers 1985). Masers are observed to occupy a large volume, and indicate the circumstellar chemistry associated with dust production and heavy mass loss. According to Allen (this volume), only R Aqr and H1-36 among Symbiotics show any maser emission (SiO). Apparently, tidal interaction can disrupt H_2O and OH maser occurance among Symbiotic binary stars.

The environments of cool stars are strongly affected by their ongoing mass loss processes. Recently, it has been recognized in the IRAS data that several late-type objects possess spatially extended far infrared emission (Hacking et al. 1985). R CrB shows an 18 arc minute shell (Gillett et al. 1986). Stencel et al. (1987) have discussed the evidence for similar shells surrounding a number of visually bright red supergiants. Such cool shells (at roughly 50°K) are simply interpreted in terms of the remnants of long term mass outflow, implying several tenths of a solar mass lost over timescales of 10^5 years.

The application of Charge Coupled Devices (CCDs) to imaging planetary nebulae has produced an astonishing array of structures associated with former red giant stars. Balick (1987) has summarized some of the discoveries, including a surprisingly high frequency of bipolar symmetries and multiple shells. The global symmetry argues for more importance of rotation and/or magnetic fields than usually assumed at these advanced stages of evolution (Gray 1986), particularly when helium shell flashes are occuring.

Finally, overlaid on all of this, is the fact of variability on almost every observed timescale, even among the non-pulsating stars. The extent of data varies, but it is clear that the mass loss process is episodic in some stars. Chromospheric emission profiles and associated circumstellar features vary on a few month timescale (Brosius et al. 1985). Dust emission is observed to vary in intensity from one season to the next (Bloemhoef et al. 1985). SiO maser cores closest to the star come and go on a few hundred day timescale (McIntosh 1987). Interpolating between high resolution observations of the turbulent solar surface and the bipolar maser maps (Chapman and Cohen 1987) ought to caution anyone who imagines that static, homogeneous models of stellar surfaces are adequate approximations.

Although I have not dealt much with abundance and isotopic effects, these too are important clues to evolution-driven atmospheric changes. Nor have I mentioned polarimetry, which will prove increasingly important as the technique matures (cf. Doherty 1986; Magalhaes, this volume). The challenge before us is to place all of the above information concerning the environments of single cool stars into the context of interacting binaries, like the Symbiotics.

2 Model Outer Atmospheres

In order to interpret the dividing lines, we need a realistic magnetohydrodynamic description of the outer atmospheres of cool stars. There are several good references to developments in model outer atmospheres, including the recent series of *Cool Star Workshops* published by Springer-Verlag, and the IUE Conferences published by NASA and ESA. These can serve as initial guides to the literature. Before any realistic model atmosphere can be constructed, a knowledge of fundamental parameters is required in order to set correct boundary conditions.

In terms of model stellar atmospheres, the most well determined case is probably the Sun (Vernazza et al. 1981). For evolved stars, the most prolific sources of new models have been the Indiana (Johnson) and Texas-Stockholm groups (Lambert-Gustafsson). There are still disagreements of several hundred degrees Kelvin in the temperature scale for M giants (cf. Johnson 1987) and the determination of stellar gravities for red giants desperately needs work. Models for chromospheres and transition regions have been developed by Ayres with semi-empirical multi-level radiative transfer calculations, and by Jordan using emission measure techniques. All of these models emphasize the commonality of physics among solar-like stars. In particular, for the lower gravity stars, an increased mass column density above the base of the chromosphere is implied, consistent with the observation that stellar winds involve more matter flux in such stars. Despite the dividing lines, there is not enough known about the winds of coronal giants to indicate whether there is a smooth or rapid increase in mass loss rates between the thin, hot, fast coronal winds and the thick, slow non-coronal ones. In addition to the observations, some of the improvements being made to atmospheric models involve inclusion of sphericity, comoving frame methods and more complete treatments of atomic and molecular opacities. In addition, new diagnostics continue to be developed for temperature and density (cf. Feldman 1981; Lennon, et al. 1985).

An excellent review of the possible mass loss mechanisms for cool stars has been given by Holzer and MacGregor (1985). There are four major processes: thermal pressure gradient; Alfven waves; pulsation, and, radiation-driven dust. The thermal pressure gradient is the Parker mechanism, operative in solar-like coronae. Alfven waves are important in the solar wind and may be crucial to understanding most stellar winds. "Molecular catastrophes", as proposed by Muchmore et al. (1987) can help alter the outer atmospheric opacity for stars above the Dust onset line, and give rise to masers and dust formation. Careful evaluation of the optical depths and geometry in extended atmospheres indicates that the radiation force in extended atmospheres does not reach its maximum for several stellar radii above the surface (Abbott and Friend 1986). This suggests that chromospheric gas may persist below substantial dust shells. Among Miras, pulsation can levitate matter to the necessary altitude for dust formation to begin. Because there is measureable mass loss for non-coronal stars below the Dust onset line in Fig. 1, and for stars warmer than Mira variables, then pulsation, dust and probably thermal pressure gradients can be ruled out as causes of mass loss among such stars. Alfven waves hold great promise, but magnetic fields are required. Recently, Cohen (1987) has observed Zeeman splitting in the OH maser lines of the red supergiant VX Sgr, suggesting surface magnetic fields of several Gauss.

3   Behavior of Red Giants/Supergiants in Binary Systems

Space does not permit a complete elaboration of the implications of placing a cool star and its naturally occuring circumstellar environment within the tidal influence of a hot companion star. This is in fact part of the purpose of this meeting.

To provoke some discussion and thought, I have placed in Table 1 a cursory set of possibilities. Any comparison of a Symbiotic's UV and X-ray output to a single late type giant immediately reveals that the Symbiotic output is greatly enhanced. This is why EXOSAT, HEAO-2 and IUE could observe a number of visually faint Symbiotics compared with rather few visually bright single red giants. The ouput is probably not standard chromospheric or coronal emission. The possible causes for this enhancement include the illuminated nebular material over a large apparant surface area, possibly some shock excited gas as is seen in the ζ Aur systems (Ahmad et al. 1983), or even because of magnetic enhancement of the output as a result of tidal synchronization analogous to the short period RS CVn stars (Rutten and Schrijver 1987; Savonije and Papaloizou 1985). Densities derived using emission line diagnostics (Nussbaumer and Stencel 1987) tend to

Table 1: Comparison of Cool Star Environments

Observable	Binary-vs-Single	Physical Cause
UV,X emission	much enhanced	not chromospheric prob. Stromgren, shock structure? rotation-activity?
mass loss, dust production	tidally constrained	$f(P)$, plus disk IR
variability	same?	internally driven

be somewhat larger than determined for single cool giant chromospheres (Lennon et al. 1985).

Second, mass loss is enhanced by dust production in the circumstellar environment. Building on a suggestion by Allen (1983), Kenyon et al. (1986) have argued on the basis of IRAS data that D-type Symbiotics are distinguished from their S-type counterparts by heavy reddening of the Mira variable member, with the hot component lying outside this dust shell. I would like to take this one step further and argue that the distinction between Dusty and Stellar Symbiotics is largely due to orbital separations. The longer period systems provide the volume necessary for the cool component to run through more of its intrinsic evolution prior to tidal interaction. This provides the time for Mira characteristics to develop, including substantial dust production and increase mass loss. That this occurs is suggested by the influence of orbital separation on derived mass loss rate among the ζ Aur binaries. Dupree and Reimers (1987) summarize such rates for a set of these objects, but the trend in mass loss rate is consistent with $P^{2/3}$ and hence orbital separation. The occurrence of SiO masers is generally on scales of 10^{15} cm among single stars, and their unimpeded development in Symbiotics would imply very long periods.

Finally, we have mentioned that mass loss is episodic in some of the single stars. Assuming this characteristic is unmodified in a binary situation, the temporal variation helps understand some of the fine structure in Symbiotic light curves. Short term changes could reflect the stochastic production of dust clouds ejected and intercepting the accretion field of the hot companion, while the slow nova characteristics could reflect the result of helium shell flashes producing planetary nebula shells late in the life of Miras. Perhaps, simply speaking, the dust free s-types evolve to their Roche limits on the first ascent of the red giant branch, the d-prime types while on the Helium burning main sequence and the d-types while on the AGB.

I wish to thank the organizers of this Colloquium for their support, encouragement and hospitality. I also acknowledge NASA grants NAG5-816 and JPL 957632 for partial support in this work. It is a pleasure to thank Joseph Pesce for additional help in the preparation of the manuscript. I hope the conference will bring about an harmonic convergence of ideas about Symbiotics.

REFERENCES

Abbott, D. and Friend, D. 1986 Ap.J. 311, 701.

Ahmad, I., Chapman, R. and Kondo, Y. 1983 A&A 126, L5.

Allen, D. 1983 M.N.R.A.S. 204, 113.

Ayres, T., et al. 1981 Ap. J., 250, 293.

Balick, B. 1987 Astron. J. in press (October).

Bloemhoef, E., Danchi, W. and Townes, C. 1985 Ap.J. 299, L37.

Bowers, P. 1985 in Mass Loss from Red Giants, ed. M. Morris and B. Zuckerman (Reidel; Dordrecht), p.189.

Brown, A. 1986 Adv. Space Res. 6, 195.

Brown, J. 1987 Ap.J. 317, 701.

Brosius, J., Mullan, D. and Stencel, R. 1985 Ap.J. 288, 310.

Carpenter, K., Brown, A. and Stencel, R. 1985 Ap.J. 289, 676.

Chapman, J. and Cohen, R.J. 1986 M.N.R.A.S. 220, 513.

Cohen, R.J. 1987 M.N.R.A.S. 225, 491.

Deutsch, A. 1956 Ap.J. 123, 210.

Doherty, L. 1986 Ap.J. 307, 261.

Dupree, A. and Reimers, D. 1987 in Exploring the Universe with the IUE Satellite, ed. Y. Kondo et al. (Dordrecht; Reidel), p. 321.

Feldman, U. 1981 Physica Scripta 24, 681.

Gondoin, P., Mangenay, A. and Praderie, F. 1987 A&A 174, 187.

Gillett, F., Backman, D., Beichman, C. and Neugebauer, G. 1986 Ap.J. 310, 842.

Gray, D. 1986 Adv. Space Res. 6, 161.

Gray, D. and Toner, C. 1986 Ap.J. 310, 277.

Hacking, P., al. 1985 Publ. Astr. Soc. Pacific 97, 616.

Haisch, B. 1987 in The Fifth Cambridge Workshop on Cool Stars, Stellar Systems and the Sun, ed. J. Linsky and R. Stencel, (Springer-Verlag; Heidelberg), in press.

Haisch, B. and Simon, T. 1982 Ap.J. 263, 252.

Iben, I. 1981 in The Physics of Red Giant Stars, ed. I. Iben and A. Renzini (Reidel; Dordrecht), p.25.

Johnson, H. 1987 in The Fifth Cambridge Workshop on Cool Stars, Stellar Systems and the Sun, ed. J. Linsky and R. Stencel, (Springer-Verlag; Heidelberg), in press.

Judge, P. 1986 M.N.R.A.S. 221, 119.

Jura, M. 1986 Irish Astron. Journal 17, 322.

Kenyon, S. 1986 The Symbiotic Stars (University Press; Cambridge).

Kenyon, S., Fernandez-Castro, T. and Stencel, R. 1986 Astron. J. 92, 1118.

Lennon, D., Dufton, P., Hibbert, A. and Kingston, A. 1985 Ap.J. 294, 200.

Linsky, J. and Haisch, B. 1979 Ap.J. 229, L27.

Holzer, T. and MacGregor, K. 1985 in The Physics of Red Giant Stars, ed. I. Iben and A.

Renzini (Reidel; Dordrecht), p.25.

McIntosh, G. 1987 Dissertation, University of Massachusetts.

Merill, K. and Stein, W. 1976 *Publ. Astr. Soc. Pacific* **88**, 285.

Mihalas, D. 1970 *Stellar Atmospheres*, Freeman, San Fransisco.

Muchmore, D., Nuth, J. and Stencel, R. 1987 *Ap.J.* **315**, L141.

Nussbaumer, H. and Stencel, R. 1987 in *Exploring the Universe with the IUE Satellite*, ed. Y. Kondo et al. (Dordrecht; Reidel), p. 203.

O'Brien, G. and Lambert, D. 1986 *Ap.J.Suppl.* **62**, 899.

Olnon, F., Winnberg, A., Matthews, H. and Schultz, G. 1980 *A&A Suppl.* **42**, 119.

Reimers, D. 1977 *A&A* **57**, 395.

Rutten, R. and Schrijver, C. 1987 *A&A* **177**, 155.

Savonije, G. and Papaloizou, J. 1985 in *Interacting Binaries*, ed. P. Eggleton and J. Pringle (Dordrecht: Reidel), p. 83.

Simon, T., Linsky, J. and Stencel, R. 1983 *Ap.J.* **257**, 225.

Stencel, R. 1978 *Ap.J.* **223**, L37.

Stencel, R. and Mullan, D. 1980 *Ap.J.* **238**, 221 & **240**, 718.

Stencel, R. 1980 in *Cool Stars, Stellar Systems and the Sun*, ed. A. Dupree; Smithson. Astrophys. Obs. Special Report 389, p. 183.

Stencel, R., Pesce, J. and Bauer, W. 1987 *Astron. J.* submitted.

Vernazza, J., Avrett, E. and Loeser, R. 1981 *Ap.J. Suppl.* **45**, 635.

Volk, K. and Kwok, S. 1987 *Ap.J.* **315**, 654.

Woolf, N. and Ney, E. 1969 *Ap.J.* **155**, L183.

Postscript: Two additional important articles which may interest the reader are ones by R. Gilliland 1985 Astrophys. Journ. 299, 286 entitled "The Relation of Chromospheric Activity to Convection, Rotation and Evolution Off the Main Sequence", and by S. Drake, J. Linsky and M. Elitzur 1987 Astron. Journ. (in press) entitled "A Radio Continuum Survey of the Coolest M and C Giants".

DUST EMISSION FROM SYMBIOTIC STARS: INTERPRETATION OF IRAS OBSERVATIONS

B.G. Anandarao[1], A.R. Taylor[2] and S.R. Pottasch[3]
1. Physical Research Laboratory, Navrangpura, Ahmedabad-380 009, India
2. Nuffield Radio Astronomy Laboratories, University of Manchester, Jodrell Bank, Macclesfield, Cheshire, SK 11 9DL, U.K.
3. Kapteyn Laboratorium, University of Groningen, Postbus 800, 9700 A.V. Groningen, The Netherlands.

SUMMARY. Symbiotic stars are strongly believed to be binary systems - a hot component presumably a white dwarf ionizing and exciting gas from a cool primary star undergoing mass-loss. Silicate dust temperatures for six symbiotic stars namely CH Cyg, UV Aur, HM Sge, RX Pup, V1016 Cyg and RR Tel are determined for the first time using IRAS-LRS data. Some of these stars showed an outburst activity which was caused probably due to hydrogen shell flash in the secondary white dwarf as a result of mass transfer from the primary M giant; while others had been relatively quiescent. All the stars showed indications of the presence of silicate dust grains except perhaps UV Aur. The dust temperatures in the symbiotic stars which had in the past an outburst activity are in general much smaller (by 400 K) than those in the symbiotics which had been quiescent and not shown any nova-like activity. Also there is an indication from our analysis that in the case of the former the dust masses are larger and the shells are spatially more extended than in the case of the latter. These results are given in detail in Table 1. Our results also indicate that the grain sizes could be larger, the earlier the outburst was, as revealed by the relative strengths of the 10 and 18 μ features in the case of the symbiotic novae. Our results support the hypothesis that during an outburst there would be a common envelope of dust blown farther and farther away by the nova wind resulting in the dust parameters described in the paper; while in the case of the quiescent symbiotics, the Mira dust envelopes get constantly heated up by the hot companion star.

(Full paper to be submitted to Astronomy and Astrophysics)

J. Mikolajewska et al. (eds.), The Symbiotic Phenomenon, 65–66.

Table I : Silicate Dust Parameters in Symbiotic Stars

Star	Type	Dust Shell 1		Dust Shell 2		Distance (p.c)	Dust mass M_d ($10 \mu m$) $M_\odot$	Outburst year
		T_1(°K)	Radius r_1 (AU)	T_2(°K)	Radius r_2 (AU)			
CH Cyg	S	750	4.4	67	440	400	1.6E-6	–
UV Aur	S	735	3.6	32	3,400	1000	1.4E-7	–
HM Sge	D	345	46.0	29	12,800	2000	2.0E-6	1975
RX Pup	D	344	42.0	65	10,600	1000	6.3E-7	1972/80
V1016 Cyg	D	334	17.0	30	19,500	3000	1.2E-6	1964
RR Tel	D	270	60.0	65	1,025	2500	3.5E-6	1944
R Agr*	D	800	4.6	87	184	300	3.0E-8	–

* Taken from Anandarao and Pottasch (1986) and given here for comparison.

Atmospheric Shocks in Mira Variables - MgII Emission

Edward W. Brugel
Center for Astrophysics and Space Astronomy
University of Colorado
Boulder, CO 80309 (USA)

T. E. Beach, L.A. Willson and G. H. Bowen
Physics Department
Iowa State University
Ames, Iowa 50011 (USA)

ABSTRACT. We have undertaken an extensive program to investigate atmospheric shock dynamics in Mira variables through monitoring the MgII emission strengths versus phase. Theoretical calculations of Bowen (1987), which represent the structural variations in the extended atmosphere of a pulsating Mira, are presented in addition to the MgII observations. The model calculations generate density, temperature and velocity structures as determined by stellar parameters and by interactions of the driving amplitude, dust opacity and thermal relaxation processes. The main diagnostic provided by the models is the total emitted shock luminosity versus phase, which can be compared to the observed MgII emission. Models and observations are presented for three Miras: S Car, R Car and T Cep.

1 OBSERVATIONS AND DYNAMIC ATMOSPHERE MODELS

The LWP camera on the *International Ultraviolet Explorer (IUE)* has been used to monitor several Mira variables. The essential ultraviolet data acquired through this long-term monitoring program are the flux in the h and k lines of MgII. In addition, the *IUE* FES-magnitudes are used to ascertain an approximate visual magnitude and corresponding phase of the visible light curve.

Dynamic atmosphere models for comparison with the MgII luminosity data were generated using a code by Bowen (1987). The code emphasizes the modeling of the dynamic nature of the atmospheres, and utilizes simplified radiative transfer - a grey spherical atmosphere with constant opacity. Although detailed light curves are not generated by the models, the actual light curves can be interpreted in terms of the shock structure shown. The peak in the MgII luminosity of a Mira typically occurs between phase 0.2 and 0.4 of the visual light curve, and can be interpreted as occuring when the main shock reaches an optical depth of $\tau(2800\text{Å}) = 2/3$.

J. Mikolajewska et al. (eds.), The Symbiotic Phenomenon, 67–68.

2 COMPARISON OF MODELS TO MGII DATA

Model atmospheres were prepared for comparison with S Car, R Car, and T Cep using the parameters listed in Table 1. All models assume fundamental mode pulsation. Three driving amplitudes were used for each model: Model 1a, 1b, and 1c were driven at 1.5, 2.5, and 3.5 km/s, respectively. Models 2 and 3 used 2, 3, and 4 km/s for a, b, and c.

Table 1
Model Parameters

	Mass ($\mathcal{M}_\odot$)	T_{eff} (K)	Radius ($\mathcal{R}_\odot$)	Period (days)
Model 1 / S Car	0.8	3143	150	149
Model 2 / R Car	1.2	3067	255	308
Model 3 / T Cep	1.4	3030	305	387

The models give us the total post shock luminosity L_{shock} as a function of phase, but only a fraction of this luminosity is from MgII. To compare the model with the data we would need to know the ratio L_{MgII}/L_{shock} as a function of phase. Unfortunately, this ratio is a function of temperature, density, and thermodynamic history of the post-shock gas, and is not easily calculated for the non-LTE conditions in the post-shock region. When fitting the data to the model curves, priority was given to matching the data points for which the shock was at low optical depths, since this region of the model suffers least from the simplified radiative transfer calculations used.

The best correspondence between models and data was found using the parameters listed in Table 2. A range of acceptable fits can be found near these "best fits" by using a proportional increase or decrease of both the opacity and luminosity ratios.

A good correspondence could not be found between the S Car data and Model 1 at any of the driving amplitudes. This lack of agreement between model and data may be the result of using an unsuitable model (perhaps S Car pulsates in an overtone mode); or the simplifications used are most inaccurate for the shorter period pulsators.

Table 2
Model Parameters For "Best" Data Fits

	Driving amplitude (km/s)	$\kappa(2800\text{Å})/\kappa(Rosseland)$	L_{MgII}/L_{shock}	$\Delta\phi$ (light phase - driving phase)
T Cep/Model 3c	4	1000	0.009	0.075
T Cep/Model 3b	3	2000	0.060	0.075
R Car/Model 2c	4	1000	0.070	0.100
S Car/Model 1	-	-	-	-

REFERENCES

Bowen, G.H. 1987, Ap.J. (submitted).

THE RADIO PROPERTIES OF SYMBIOTIC STARS

E. R. Seaquist
Department of Astronomy
University of Toronto
Toronto, Ontario
Canada M5S 1A7

ABSTRACT. Radio thermal bremsstrahlung emission has now been detected at centimeter wavelengths from about thirty symbiotic stars. These data combined with optical and IR data show that the radio emission in most systems may be understood in terms of the ionization of part of the wind of a red giant by a hot companion. The radio properties of symbiotic stars are reviewed and the agreement with and the limitations of this picture are examined.

1. INTRODUCTION

Insofar as present studies show, all radio emitting symbiotic stars (except possibly the 1985 outburst from RS Oph) emit by free-free or thermal bremsstrahlung radiation. Therefore the size of the circumstellar environment probed in the radio is 10^{15} cm or greater, limited by the maximum brightness temperature of thermal emission from a plasma at $T=10^4$K. Thus radio observations provide information about the mass outflows on spatial scales one to two orders of magnitude larger than the binary separation for most symbiotics, typically a few AU. It is important to emphasize that the radio data are therefore complementary to the optical data on the circumstellar gas, which generally sample much smaller volumes.

Until recently, radio emission was known from only a handful of symbiotic stars. The earlier work on this subject has been reviewed extensively (Hjellming, 1981,   Kwok, ·1982). More recently, the VLA has made possible more sensitive surveys of these objects, and as a result radio emission has now been detected from about 30 systems to a level of about 0.5 mJy (Seaquist, et al. 1984; Seaquist and Taylor, in preparation).

2. RADIO CHARACTERISTICS

Virtually all stars in the list of Allen (1984) and reachable by the VLA have now been searched for radio emission. Table 1 contains a list

69

J. Mikolajewska et al. (eds.), The Symbiotic Phenomenon, 69–75.
© *1988 by Kluwer Academic Publishers.*

of all known radio emitting symbiotic stars and their approximate 4.9 GHz flux densities.

Table 1 Symbiotic stars detected in the radio
and approximate 4.9 GHz flux density

Name	Flux (mJy)	Ref	Name	Flux (mJy)	Ref	Name	Flux (mJy)	Ref
RX Pup	34v	2	RT Ser	0.5v	1	AS 296	0.4	1
RW Hya	0.2	1	SS 96	2.2	1	He2-390	0.5	1
He2-106	20	3	H1-36	46	3	BF Cyg	2.0	1
AG Dra	0.4	1	W16-312	1.0***	1	CH Cyg	0.4-18v	5
He2-171	1.9	1	RS Oph	<0.3-63v	1,4	HM Sge	15v	3
He2-176	15*	3	V2416 Sgr	1.5	1	V1016 Cyg	45v	3
AS 210	2.0	1	SS 122	1.4	1	RR Tel	28	3
V455 Sco	0.8	1	AS 270	0.3	1	V1329 Cyg	0.9v	1
Hen 1383	2.0v**	1	H2-38	2.9	1	AG Peg	7.5v	1
Th3-7	0.4	1	AS 289	0.6v	1	Z And	0.9v	1
Hen 1410	0.6	1	HD 319167	0.5***	1	R Aqr	12	6

Notes to table

v	Flux variations detected	(1)	Seaquist et al. (1984) and/or
*	Flux at 8.9 GHz		Seaquist and Taylor, in prep.
**	Not in Allen (1984)	(2)	Seaquist and Taylor (1987)
***	ID or detection uncertain	(3)	Purton et al. (1982)
		(4)	Davis (1986)
		(5)	Taylor, Kenyon, and Seaquist, in preparation
		(6)	Hollis et al. (1985)

2.1. Angular Structure

Most of the sources in Table 1 are unresolved by the VLA, so that the angular sizes are smaller than about 1 arcsec. However some of the more intense sources show structure at arcsec or subarcsec resolution. V1016 Syg and HM Sge both exhibit shell-like structures whose diameter is about 0.3-0.5 arcsec, each containing two peaks of emission separated by about 0.1 arcsec (Newell, 1981; Newell and Hjellming, 1981; Hjellming and Bignell, 1982; Kwok et al., 1984). RX Pup has been recently found to have an angular size dependent on frequency, charac-teristic of a stellar wind approximating a $1/r^2$ density profile (Seaquist and Taylor, 1987). In addition, the very extended (2 arcmin) optically emitting structure surrounding R Aqr has now been detected as a source of free-free emission in the radio (Hollis et al., 1987). This material is the remnant of an outburst which occurred in this system 600 years ago.

Recently, 'jet-like' radio emission on arcsec and/or subarcsec scales suggesting collimated ejecta has been discovered in four symbi-otic stars - CH Cyg (Taylor et al. 1986; Taylor, this volume), RS Oph

(Porcas et al, 1986), R Aqr (Kafatos, et al., 1983), and AG Peg (Hjellming, 1985). Such features may be related to the blobs contained within the shells of V1016 Cyg and HM Sge. These discoveries strongly suggest that accretion disks play a role in collimating episodic out-flows in some symbiotic stars.

2.2. Variability

Nearly all of the objects in Table 1 have been observed at least twice over a period of one to five years, and even longer in some cases. The table shows whether or not the radio flux is variable by more than about 30 percent over a time scale of several years. Most of the sources are either quiescent or very slow variables. This signifies the presence of either a stable outflow or an outburst producing radio emitting ejecta at low velocities (<500 km/s). Several are extremely variable in the radio, however, undergoing outbursts usually associated with activity in the optical. Notable examples are V1016 Cyg (Kwok, 1982; Newell and Hjellming 1981), HM Sge (Purton et al., 1983), CH Cyg (Taylor, Seaquist, and Mattei, 1986), RS Oph (Davis, 1986; Hjellming et al., 1986), and RX Pup (Seaquist and Taylor, 1987). The increased radio emission from RX Pup signals a new enhanced mass loss phase for this object. The relationship between radio and optical variations in some of these outbursts suggest nova-like activity or accretion fed outflows, but there is an enormous diversity in this relationship. For example, unlike most cases, the visual brightness of CH Cyg actually declined at the onset of the large radio outburst which began in 1984.

2.3 Radio Spectra

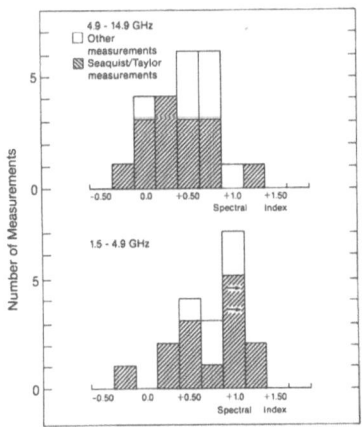

Figure 1 shows the distribution of spectral indices at cm wavelengths from observations at 1.5, 4.9, and 14.9 GHz. Essentially all sources exhibit a positive spectral index characteristic of optically thick emission from an inhomo-geneous ionized region (Seaquist, et al., 1984; Seaquist and Taylor, in preparation). The horizontal arrows in the figure indicate that two of the values plotted are lower bounds. The median spectral index at the lower frequencies in near +1.0, significantly steeper than that expected (+0.6) for a steady

Figure 1. The distributions of spectral indices in two frequency ranges for radio detected symbiotics.

spherically symmetric outflow at constant speed. The median index at
the higher frequencies is lower (flatter spectrum), about +0.5. These
results suggest a more complex geometry, and the flattening of the
spectra at higher frequency indicates a transition from optically thick
to optically thin emission in the cm range.

3. COMPARISON WITH IR DATA

Figure 2 shows plots of the 4.9 GHz flux vs IR magnitude for the K
(2.2μm) and L (3.5μm) bands from data obtained from Allen (1982), and
similar plots for 12μm and 25μm using data from the IRAS point source
catalogue. D-type (dust) and S-type (stellar) IR emitters are distinguished by open and filled symbols. The principal conclusion is
that a significant correlation with IR emission exists, and that it
appears to strengthen with increasing IR wavelength, suggesting that
the radio emission correlates best with the amount of dust in the
envelope of the star. The D-type IR emitters exhibit the largest flux
at both radio and IR wavelengths, and are consequently the most
luminous stars at both bands. Since the D-types are generally Mira
variables possessing the highest mass loss rates, and since the radio
emission is optically thick, this result basically says that the
surface area of the radio emitting photosphere is largest in the
systems with highest mass loss rates.

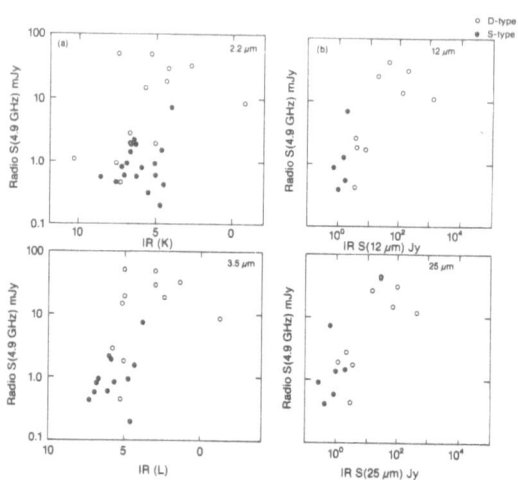

Figure 2. Plots of 4.9 GHz flux vs IR flux for various IR bands.

4. COMPARISON WITH OPTICAL DATA

Seaquist et al. (1984) showed that the radio luminosities are correlated with red giant spectral types, consistent with the aforementioned observation on the correlation with IR emission from dust, since the D-type stars, or Miras, have the latest spectral classes. These authors also showed that the radio luminosity is correlated with the equivalent width of the H beta emission line. The latter correlation is expected since both indices measure in some way the amount of ionized gas in these systems.

However, the radio and H beta emission do not necessarily arise from the same volumes. For S-type stars, the observed (optically thick) radio fluxes are an order of magnitude less than the optically thin radio fluxes predicted from the H beta line (Seaquist and Taylor, in preparation). This factor is somewhat larger than expected, since figure 1 indicates that many sources are nearly optically thin at 14.9 GHz. For D-types, the observed and predicted fluxes are comparable, suggesting that the H beta emission in the D-types emerges primarily from an extended volume comparable to that producing the radio emission. In S-types a significant fraction of the H beta emission must come from smaller volumes not detectable in the radio, such as an accretion disk, a stream between the stars, or chromospheric emission.

5. MODELS FOR THE RADIO EMISSION

A picture compatible with the binary interpretation is the binary model worked out in detail by Seaquist et al. (1984) and Taylor and Seaquist (1984), hereafter the STB model. Note the erratum to these papers in Ap.J. $\underline{317}$, 555, 1987. In this picture, the hot companion to the red giant $\overline{\text{ioni}}$zes a cone-shaped region of the red giant wind, whose geometry is governed by a single parameter $X = \text{const} * a \, L \, (\dot{M}/v)^{-2}$, where a is the binary separation, L is the Lyman continuum luminosity of the hot companion in photons/sec, $\dot{M}$ is the red giant mass loss rate, v is the wind speed. The application of the model to the radio data yield mass loss rates ranging from 10^{-9} to 10^{-7} $M_\odot$ yr^{-1} for S-type and 10^{-7} to 10^{-5} $M_\odot$ yr^{-1} for D-type stars, if v = 10 km/s.

This steady-state model accounts for a large number of the radio emitting characteristics of symbiotic stars. They include for example the steep radio spectra compared with that of spherically symmetric outflows, and the optically thin turnover at cm wavelengths, indicated in figure 1, combined with the relatively quiescent fluxes of many symbiotics.

The model also explains the coexistence of neutral gas (signified by IR emission from dust) with the ionized component, which is a consequence of a source of ionization external to the red giant wind. The model furthermore explains the observation by Kenyon et al. (1986) and Allen (1983) that in D-type systems the hot companion and the emission line region lie outside the region of circumstellar dust, whereas the Mira variable is heavily obscured.

The explanation for the correlations between radio and IR emission and spectral type is not as clear. Higher mass loss rates alone may be insufficient to produce higher radio luminosity because high densities may produce smaller ionized volumes, whereas a larger radio photosphere is required. A likely possibility is that the red giants with the most extensive mass loss reside in the widest binaries, as pointed out, for example by Kenyon (1986). A larger radio photosphere and higher radio luminosity for D-type symbiotics would be the result, as observed.

The diversity in radio morphology and the eruptive nature of symbiotic novae highlight certain limitations of the steady-state binary model, and require some variants of the STB picture. For example, account needs to be taken of the interaction between eruptive mass loss or a fast wind from the hot companion and the pre-existing red giant wind. The interaction between such a wind and mass outflow from the cool star explains the properties of symbiotic novae like HM Sge and V1016 Cyg (Kwok, this volume; Girard and Willson, 1987; Willson et al., 1984; Purton et al., 1983). Other examples include AG Peg (Hjellming, 1985), RX Pup (Seaquist and Taylor, 1987; Allen and Wright, this volume) and R Aqr (Hollis et al., 1985). If episodic outflow from the hot companion dominates the radio emission, it may be possible to recognize this by the spectrum at mm and sub mm wavelengths. The spectrum in this regime may exhibit either no optically thin turnover at all, or a turnover which evolves rapidly, signifying the expansion of an inner boundary of the shell produced when the ejection terminates. In the latter case, the decrease in the optically thin mm flux would be readily detectable even for modest shell expansion velocities (100 km/s).

6. CONCLUSION

The radio emitting properties of symbiotics stars generally support the binary picture of the symbiotic phenomenon. The application of binary models to the radio data promises to yield important constraints on the binary parameters and mass loss from these systems. The diversity of behaviour of symbiotic stars however strain the validity of the simplest form of the binary model in individual cases. Further tests for investigating the geometry of the radio emitting circumbinary nebulae include the use of mm and sub mm wavelengths to probe volumes intermediate in scale between that seen at cm wavelengths and that seen in the IR and optical region.

ACKNOWLEDGEMENTS

This research was supported by an operating grant from the Natural Sciences and Engineering Research Council of Canada. I also wish to thank the staff of the National Radio Astronomy Observatory (NRAO), which operates the VLA, for their generous assistance. The National Radio Observatory is operated by Associated Universities, Inc. under contract with the U.S. National Science Foundation. I am indebted to H. Nussbaumer and M. Vogel for drawing to my attention an error in the published equations describing the STB binary model.

REFERENCES

Allen, D.A.: 1982, in IAU Colloquium No. 70, The Nature of Symbiotic
    Stars, ed. M. Friedjung and R. Viotti (Dordrecht: Reidel), p.209.
Allen, D.A.: 1983, M.N.R.A.S. 204, 113.
Allen, D.A.: 1984, Proc. Ast. Soc. Australia 5, 369.
Becker, R.H. and White, R.: 1985, in Radio Stars, ed. R.H. Hjellming and
    D.M. Gibson (Dordrecht: Reidel), p.139.
Davis, R.J.: 1986, in RS Ophiuchi(1985) and the Recurrent Nova
    Phenomenon, ed. M.F. Bode (Utrecht: VNU Science Press), p.187.
Girard, T. and Wilson, L.A.: 1987, preprint.
Hjellming, R.M.: 1981, in Proceedings of the North American Workshop on
    Symbiotic Stars, ed. R.E. Stencel (Boulder: JILA), p.15.
Hjellming, R.M. and Bignell, R.C.: 1982, Science, 216, 1279.
Hjellming, R.M.: 1985, in Radio Stars, ed. R.M. Hjellming and D.M.
    Gibson (Dordrecht: Reidel), p.97.
Hjellming, R.M., van Gorkom, J.H., Taylor, A.R., Seaquist, E.R., Padin,
    S., Davis, R.J., and Bode, M.F.: 1986, Ap.J. (Letters) 305, L71.
Hollis, J.M., Kafatos, M.K., Michalitsianos, A.G., and McAlister, H.A.:
    1985, Ap.J. 289, 765.
Hollis, J.M., Kafatos, M., Michalitsianos, A.G., Oliversen, R.J., and
    Yusef-Zadeh, F.: 1987, Ap.J. (Letters), in press.
Kenyon, S.J., Fernandez-Castro, T., and Stencel, R.E.: 1986, Astron. J.
    92, 1118.
Kenyon, S.J.: 1986, in The Symbiotic Stars (Cambridge: Cambridge
    University Press).
Kafatos, M., Hollis, J.M., and Michalitsianos, A.G.: 1983, Ap.J.
    (Letters) 267, L103.
Kwok, S., Bignell, R.C., and Purton, C.R.: 1982, Ap.J. 279, 188.
Kwok, S.: 1982, in IAU Colloquium No. 70, The Nature of Symbiotic Stars,
    ed. M. Friedjung and R. Viotti (Dordrecht: Reidel), p.209.
Newell, R.T.: 1981, Ph.D. Thesis, New Mexico Institute of Mining and
    Technology, Socorro, New Mexico.
Newell, R.T. and Hjellming, R.M.: 1981, in Proceedings of the North
    American Workshop on Symbiotic Stars, ed. R.E. Stencel (Boulder:
    JILA) p.16.
Porcas, R.W., Davis, R.J., and Graham, D.A.: 1986, in RS Ophiuchi(1985)
    and the Recurrent Nova Phenomenon, ed. M.F. Bode (Utrecht: VNU
    Science Press), p.203.
Purton, C.R., Feldman, P.A., Marsh, K.A., Allen, D.A., and Wright, A.E.:
    1982, M.N.R.A.S. 198, 321.
Purton, C.R., Kwok, S., and Feldman, P.A.: 1983, Astron. J. 88, 1825.
Seaquist, E.R., Taylor, A.R., and Button, S.: 1984, Ap.J. 284, 202.
Seaquist, E.R. and Taylor, A.R.: 1987, Ap.J. 312, 813.
Spoelstra, T.A.T., Taylor, A.R., Pooley, G., Evans, A. and Albinson,
    J.S.: 1986, in RS Ophiuchi(1985) and the Recurrent Nova Phenomenon,
    ed. M.F. Bode (Utrecht: VNU Science Press), p.193.
Taylor, A.R. and Seaquist, E.R.: 1984, Ap.J. 286, 263.
Taylor, A.R., Seaquist, E.R., and Mattei, J.A.: 1986, Nature, 319, 38.
Willson, L.A., Wallerstein, G., Brugel, E., and Stencel, R.E.: 1984,
    Astron. and Astrophys. 133, 154.

RADIO IMAGING OF SYMBIOTIC STARS

A.R. Taylor
Department of Physics, University of Calgary
Calgary, Alberta, Canada

ABSTRACT. A brief review of radio observations of individual symbiotic stars
is presented, with emphasis on radio imaging of the circumstellar nebulae. The
ionized nebulae are catagorized into two types: outburst ejecta and stellar winds.
Among the ejecta-type there is a strong tendency for bipolar or jet morphology.
In the case of the quiescent, stellar wind emitter, a few very wide binary systems
(eg. H1-36) offer the potential of directly viewing the effects of the binary interaction.

1. INTRODUCTION

The 1980's have seen major advances in radio investigations of symbiotic stars,
primarily due to the combination of high sensitivity and resolving power provided
by the NRAO Very Large Array. The results of an extensive, multi-frequency,
VLA survey of symbiotic stars are reviewed by Seaquist elsewhere in this volume.
A further area of radio investigation, that has lead to significant insights into the
symbiotic phenomenon, is imaging at high resolution, to attempt to map the circumstellar ionized nebulae. This paper presents a brief review of observations of
the radio structure of symbiotic stars, along with some new results.

2. OBSERVATIONS OF RADIO STRUCTURE

It is instructive to carry out a simple calculation that illustrates the limitations
of current instrumentation for imaging of circumstellar gas. In the case of an
optically-thick nebula with brightness temperature $T_B = 10^4$ K (the approximate
equalibrium temperature for a photoionized gas), the radio flux density and angular size are related via

$$S_\nu = 5.7\nu^2\theta^2 \quad mJy, \tag{1}$$

where ν is the observing frequency in GHz and θ is the diameter of either the
source or the telescope beam, whichever is smaller. For symbiotic stars we are interested in regions with dimensions comparable to the binary separation, which

77

J. Mikolajewska et al. (eds.), The Symbiotic Phenomenon, 77–83.

ranges from a few AU for the short period systems (periods of a few years) to perhaps 1000 AU for the very wide systems. A linear dimension of 100 AU at a distance of 1 kpc subtends an angle of 0.1", roughly equal to the resolving power of the VLA at $\lambda = 2$ cm in the highest resolution configuration. From equation (1) the flux density per beam area is then ~10 mJy. Therefore a sensitivity of better than 1 mJy is needed to successfully map emission at 10^4 K on this scale. Note that the required sensitivity increases as the square of the angular resolution. To map emission with a resolution of 0.01" (10 AU at 1 kpc) would require sensitivity of .01 mJy. This combination of resolution and sensitivity is not yet available.

A corrallory of the above calculation is that optically-thick sources with flux density less than a few mJy will have angular sizes less than 0.1" and are not mappable with the present generation of radio telescopes. Unfortunately, this is the case for the majority of radio detected symbiotic stars which are quiescent emitters at flux densities of typically 1 - 2 mJy (Seaquist *et al.* 1984, Seaquist 1988 this volume). However, a few notable exceptions exist. Table 1 lists those symbiotic stars in which the structure of the nebula has been resolved in the radio. The list includes 10 stars. Of these, three, V1016 Cyg, HM Sge and AG Peg are symbiotic novae (see Allen 1980) one, RS Oph, is a recurrent nova and one, CH Cyg, has exhibited outburst behaviour similar to that of novae. This predominance of outburst-type stars in the list is a selection effect, since it is primarily these objects that obtain high enough flux densities to be mappable.

In table 1 the circumstellar nebulae have been catagorized into two types; ejecta and stellar winds. This is an attempt to distinguish between nebulae produced by an impulsive, energetic outburst and those resulting from a steady-state or, perhaps, slowly varying, outflow. Nebulae have been classed as ejecta if they have a clumpy appearance and can be associated with a known outburst of the system. Wind-type nebulae have a smooth, featureless morphology and are signified by an angular size that decreases with frequency. In the idealized case of an isothermal, uniform outflow $\theta \sim \nu^{0.7}$ (see, for example, Wright and Barlow 1975). More generally, Seaquist and Taylor (1987) have shown that for a power law density distribution of the form $\rho \propto r^{-p}$ and a temperature gradient given by $T \propto r^{-n}$, the apparent angular size of the nebulae (ie. the radius corresponding to some value of the optical depth) will vary as $\theta \propto \nu^{-m}$ where

$$m = \frac{2.1}{(2p - 1 - 1.35n)} \tag{2}$$

There is too little space here to discuss all of the stars in table 1 in detail, moreover, excellent discussions of a number are given elsewhere in this volume. Instead, some general comments and results are given below.

2.1 Ejecta

Those stars in table 1 that have undergone a recorded outburst are by far the best studied. The symbiotic novae are distinguished by the long time scale for evolution of the radio outburst relative to classical novae (eg. decades rather than years). The slow evolution is reflected in the expansion rate of the ejecta.

Table 1. Symbiotic Stars Resolved at Radio

Star	Angular Size (")	Type	Radio Variable?	Comment	Refs
V1016 Cyg	0.5	ejecta + wind	Y	symbiotic nova	1
AG Peg	1.5	ejecta?	Y	symbiotic nova	2, 3
R Aqr	6.4	ejecta?	N	jet?	4, 5, 6
HM Sge	0.5	ejecta + wind	Y	symbiotic nova	7
CH Cyg	1.5	ejecta	Y	jet	8, 9
RX Pup	0.25˙	wind	Y		10
RS Oph	0.2	ejecta	Y	recurrent nova	11. 12, 13
H1-36	5.0	complex + wind	N		14
SS 96	0.17˙	wind	N		14
Hen 1383[†]	4.0	ejecta?	Y	jet?	14

˙ measured at $\lambda = 2$ cm

[†] possibly not a symbiotic star

References

1 Newell (1981)
2 Chigo and Cohen (1981)
3 Hjellming (1985)
4 Sopka et al. (1982)
5 Kafatos et al. (1983)
6 Hollis et al. (1985)
7 Kwok et al. (1984)
8 Taylor et al. (1986)
9 Taylor et al. this volume
10 Seaquist and Taylor (1987)
11 Porcas et al. (1986)
12 Hjellming et al. (1986)
13 Taylor et al. (1987)
14 This paper

Figure 1 shows high resolution radio images at $\lambda = 1.3$ cm of the two recent symbiotic novae V1016 Cyg and HM Sge, with outbursts respectively in 1965 and 1975. The morphology of the emission is very similar in the two cases; a central, bipolar source plus a more extended, low surface brightness halo. Lower frequency observations show that the extended emission has the characteristics of a stellar wind (eg. Kwok *et al.* 1984). The angular size of the bipolar features imply upper limits to the expansion velocity of $30D_{kpc}$ km-s^{-1} for V1016 Cyg and $50D_{kpc}$ km-s^{-1} for HM Sge, where D_{kpc} is the distance in kpc. A similar expansion velocity is implied for the third symbiotic nova AG Peg, which experienced an outburst around 1855 and currently has an angular size of 1.5". These low velocities are in marked contrast to the two other symbiotics with recorded radio outbursts, CH Cyg and RS Oph. Figure 2 shows radio images of CH Cyg at two epochs, illustrating the rapid expansion of the radio jet from this system. The expansion velocity for $D = 400$ pc is $1050/\sin i$ km-s^{-1} (Taylor *et al.* 1986). A VLBI image of RS Oph indicates bipolar expansion of a mixture of free-free and non-thermal emitting gas at ~ 3500 km-s^{-1} (Porcas *et al.* 1987, Taylor *et al.* 1987).

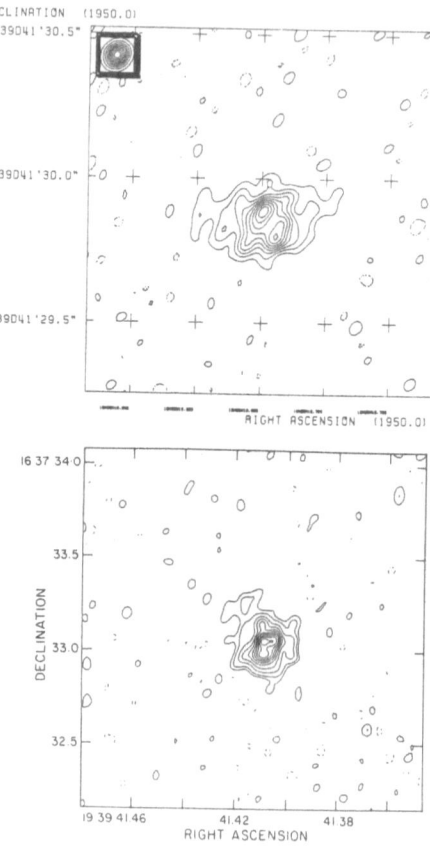

Figure 1. Radio images at $\lambda = 1.3$ of the symbiotic novae V1016 Cyg (*top*) and HM Sge (*bottom*).

The much lower nebular expansion velocities for the symbiotic novae can be explained by the interacting wind model of Kwok and Purton (1979), since in this case the region of high emission measure is a shell of shocked gas with predicted expansion velocity of ~ 100 km-s^{-1} for reasonable parameters for the red giant and hot star winds. The lack of observable proper motion in the R Aqr jet (Hollis *et al.* 1985, Mauron *et al.* 1985) also indicate low outflow velocities for this star (less than a few 100 km-s^{-1}), although an interacting wind interpretation appears unlikely in this case. Solf and Ulrich (1985) have suggested that the R Aqr jet, and the larger scale nebulosity associated with the star, are the result of two, slowly expanding bipolar ejection events occuring 185 and 640 years ago. On the other hand, Kafatos *et al.* (1986) suggests that the jet components are produced

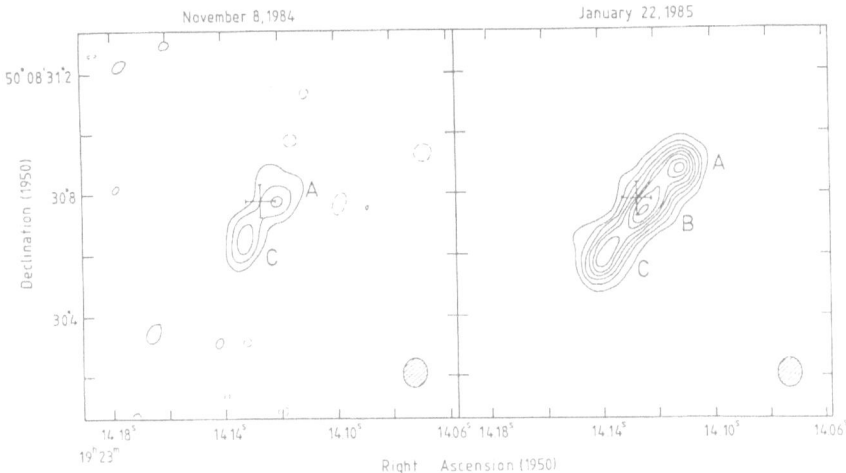

Figure 2. Two Radio Images of CH Cygni at $\lambda = 2$ cm separated by 75 day. The cross marks the position of the star. The angluar expansion rate of the jet is 1.1 "/yr.

by ejection of discrete parcels of material from a geometrically thick accretion disk at intervals of about 50 years.

While there does not appear to be a single ejection mechanism among the symbiotic systems that undergo radio outbursts, one element in common is the striking departure from spherical symmetry. In all cases the ejecta exhibit either bipolar or jet-like morphology. Such axial symmetry might reflect the presence of accretion disks or, particularly for the symbioitic novae, assymmetric mass loss from the red giant. The possibility that red giant winds are non-spherically symmetric is supported by recent MERLIN observations of the distribution of OH and H_2O masers around red supergiants (eg. Chapman and Cohen 1986). Asymmetric mass loss has also been proposed by Manchester (1987) to account for the biannular appearance of many supernova renmants.

2.2 Stellar Winds

Seaquist *et al.* (1984), and Taylor and Seaquist (1984) proposed a binary model for quiescent radio emission from symbiotic stars in which the emission arises from free-free interaction in a portion of the red giant wind ionized by the hot companion. Among other things, this model accounts for the presence of a steady state, high-frequency turnover in the radio continuum spectra of some symbiotic stars. As pointed out by Allen (1983) such a turnover is the result of an inner boundary to the ionized nebula, leaving a cavity of neutral material around the red giant. A prediction of the model is that systems with large binary separation will tend to have a larger neutral region around the giant component, and will thus become optically-thin to free-free absorption at a lower frequency. If the binary separation were large enough, one might expect to observe the structure of the ionized cavity in the radio. While there is some curvature of the radio spectra of symbiotic stars as a class in the 5 - 15 GHz range (Seaquist, this volumme), most symbiotics are optically-thick up to the highest observable VLA frequency. Two notable exceptions are H1-36 and SS 96, which become optically-thin at ∼10 GHz. Model fits to the continuum spectra of these stars yield predicted values for the angular separation of the binary components of 0.5" for H1-36 (Taylor and Seaquist 1984) and 0.4" for SS 96 (unpublished data).

Figure 3 shows a $\lambda = 2$ cm image of H1-36. The map shows a triple radio source with total extent roughly equal to the predicted angular separation of the binary components. Lower frequency observations with lower resolution reveal a more extended nebula with the characteristics of a stellar wind, consistent with the simple wind model. However, the picture on the scale of the binary separation is clearly much more complex. At the suggested separation of ∼10^{16} cm, the red giant is well within its Roche lobe, and the lack of strong variability rules out an outburst/ejection event. The multi-component emission probably reflects the interaction of the hot component with the wind of the red giant. If the hot component is undergoing a nuclear shell flash, a high velocity wind may be present. The complex radio structure might then be a combination of shock-ionized and photo-ionized regions of the red giant wind. The weak eastern component might be chromospheric emission from the red giant itself, as has been observed recently in α Orionis (Newell and Hjellming 1982) and α Sco (Hjellming and Newell 1983).

The radio image of SS 96 shows a single, assymetric, resolved source, without the complexity of the H1-36 image. Fits to the radio continuum spectrum also suggest a much lower luminosity for the hot component relative to H1-36.

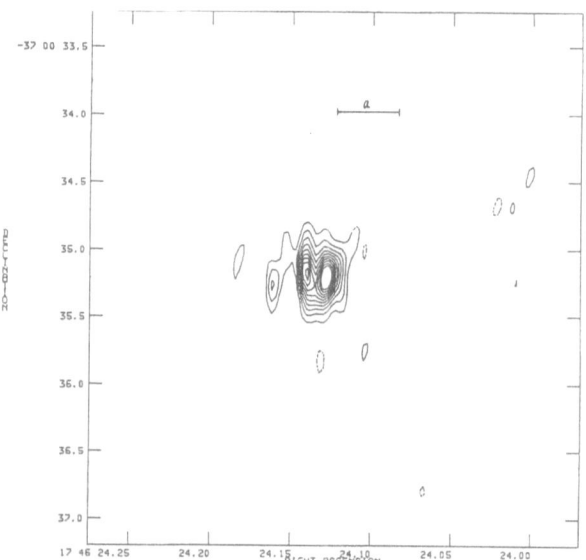

Figure 3. Radio Image of H1-36 at $\lambda = 2$ cm. The angular dimension labeled a is the predicted binary separation based on model fits to the radio continuum spectrum.

CONCLUSIONS

Although radio imaging of the ionized nebulae of symbiotic stars is a rich and important area of investigation, it is a fairly recent endeavour. It has yet to benefit from a sytematic approach. For instance, second epoch images of many of the objects (eg. V1016Cyg, HM Sge and H1-36) remain to be obtained. The main limitation at this time, however, is still instrumental. Significant expansion of the list of radio resolved symbiotics must await instrumental developments. The completion of the Australia Telescope will add a number of southern stars to the list. Planned improvements to the sensitivity of MERLIN and the European VLBI network may allow another step in resolving power.

References

Allen, D.A. 1980, *Mon. Not. Roy. astr. Soc.*, **192**, 521.
Allen, D.A. 1983 *Proc. Astr. Soc. Australia*, **5**, 211.
Chapman, J.M. and Cohen, R.J. 1986, *Mon. Not. Roy. astr. Soc.*, **220**, 513.
Chigo, F.D. and Cohen, N.L. 1981, *Ap. J.*, **245**, 988.
Hjellming, R.M. and Newell, R.T. 1983, *Ap. J.*, **275**, 704.
Hjellming, R.M. 1985 in *Radio Stars*, eds. R.M. Hjellming and D. Gibson, Dordrecht: Reidel, p. 97.
Hjellming, R.M., van Gorkum, J.H., Taylor, A.R., Seaquist, E.R., Padin, S., Davis, R.J. and Bode, M.F. 1986 *Ap. J. Letters*, **305**, L71.
Hollis, J.M., Kafatos, M., Michalitsianos, A.G. and McAlister, H.A. 1985, *Ap. J.*, **289**, 765.
Kafatos, M., Hollis, J.M. and Michalitsianos, A.G. 1983, *Ap. J. Letters*, **267**, L103.
Kafatos, M., Hollis, J.M. and Michalitsianos, A.G. 1986, *Ap. J. Supp.*, **62**, 853.
Kwok, S. and Purton, C.R. 1979, *Ap. J.*, **229**, 187.
Kwok, S., Bignell, R.C. and Purton. C.R. 1984, *Ap. J.*, **279**, 188.
Manchester, R.N. 1987, *A. & A.*, **171**, 205.
Mauron, M., Nieto, J.L., Picat, J.P., Lelievre, G. and Sol, H. 1985, *A. & A. Letters*, **142**, L13.
Newell, R.T. 1981, Ph.D. Thesis, New Mexico Institute for Mining and Technology.
Newell, R.T. and Hjellming, R.M. *Ap. J. Letters*, **263**, L85.
Porcas, R.W., Davis, R.J. and Graham, D.A. 1987, in *RS Ophiuchi (1985) and the Recurrent Nova Phenomenon*, ed. M.F. Bode, VNU Science Press: Utrecht, p. 203.
Seaquist, E.R., Taylor, A.R. and Button, S. 1984 *Ap. J.*, **284**, 202.
Seaquist, E.R. and Taylor, A.R. 1987, *Ap. J.*, **312**, 813.
Solf, J. and Ulrich, H. 1985 *A. & A.*, **148**, 274.
Sopka, R.J., Herbig, G., Kafatos, M. and Michalitsianos, A.G. 1982, *Ap. J. Letters*, **258**, L35.
Taylor, A.R. and Seaquist, E.R. 1984, *Ap. J.*, **286**, 263.
Taylor, A.R., Seaquist, E.R. and Mattei 1986, *Nature*, **319**, 38.
Taylor, A.R., Davis, R.J., Porcas, R.W. and Bode, M.F. 1987, *Mon. Not. Roy. astr. Soc.*, submitted.
Wright, A.E. and Barlow, M.J. 1975, *Mon. Not. Roy. astr. Soc.*, **170**, 41.

OBSERVATIONS OF BIPOLAR MASS FLOW FROM SYMBIOTIC STARS

J. Solf
Max-Planck-Institut für Astronomie
D-6900 Heidelberg, F.R.G.

ABSTRACT. Spectroscopic observations of high spatial and high spectral resolution indicate that mass flow from symbiotic stars generally exhibits a bipolar pattern. Besides the polar features moving at velocities up to several 100 km/s, equatorial structures of much lower expansion rate are present in some cases. Mostly, the high-velocity components appear to be highly collimated and hence can be considered as "jets". The jets probably originate from an accretion disk within a binary star where mass from a late-type giant is accreted by a compact companion.

1. INTRODUCTION

In recent years bipolar mass flow has been discovered to be associated with a large variety of stellar objects in both early or late stages of stellar evolution. Pre-main-sequence stars, such as T Tau or FU Ori stars, are now known as powering sources of highly collimated mass flow observed as jets or Herbig-Haro objects (see e.g. Mundt et al., 1987). Nova shells, if spatially resolvable (e.g. DQ Her or HR Del), exhibit a prolate ellipsoidal shape with polar blobs indicating that a bipolar geometry has been dominating the mass ejection (Solf, 1983b). Among the planetary nebulae most of them are known for strong deviations from an isotropic expansion leading to numerous types of axi-symmetric or bipolar shell structures. The recent discovery of highly collimated flow with a velocity of ~200 km/s in NGC 2392 (Gieseking et al., 1985) shows that the bipolar jet phenomenon is not uncommon even in evolved planetaries. Since in all these cases the bipolar mass flow seems to be related to some sort of mass accretion through a disk in the center region, it will not be surprising if some of the symbiotic stars known for mass ejection do present signatures of bipolarity as well.

2. OBSERVATIONAL METHOD

In order to detect bipolar mass flow from symbiotic stars (or other compact objects) long-slit coudé-spectroscopy of the lines of [O III] or [N II] was carried out at the 2.2-m-telescope on Calar Alto providing high spectral resolution (3-10 km/s) as well as high angular resolution

J. Mikolajewska et al. (eds.), The Symbiotic Phenomenon, 85–88.

(0.7-1.5 arc sec). Usually, the entrance slit was centered on the star and set to various position angles (p.a.). This method allows us to investigate the spatial distribution (along the actual direction of the slit) of spectroscopically resolved components of an emission line. If the components are sufficiently separated in the velocity space ($\Delta v \sim 10$ km/s), their apparent angular separation can be determined far below the seeing limit (down to typically 0.1 arcsec). Combining the results from the various p.a. of the slit, the spatial distribution of the different velocity components on the sky can be reconstructed (for details see Solf, 1984). In the cases of more extended nebulosities (e.g. R Aqr) a fine grid of different slit positions has been used in order to cover the entier nebular structure.

3. RESULTS

In the following, results obtained from R Aqr, V1016 Cyg, HM Sge, and CH Cyg will be summarized, each of them representing a different type of bipolar outflow. R Aqr is associated with extended nebulosities ejected several hundred years ago. V1016 Cyg and HM Sge underwent nova-like outbursts within the last two decades. CH Cyg ejected bipolar jet features in the course of brightening at radio wavelenghts three years ago.

3.1 R Aqr

The deduced velocity field of the nebulosity allows a clear-cut distinction between two (inner and outer) components. Each complex can be described as a bipolar, hour-glass-like, expanding thin shell resulting from a particular geometry of outflow presenting velocities increasing with the latitude angle of the flow vectors (Solf and Ulrich, 1985). Both shells are related to a common bipolar axis of symmetry (inclined by 72° with respect to the line of sight). The deduced ratio between the polar and equatorial expansion rates is about 6 suggesting a rather modest degree of collimation of the gas flow. The pronounced uniformity observed in the kinematic structure of each shell indicates that they must be attributed to no more than two major ejection events, 185 and 640 yr ago. Departures observed in some of the brighter nebular features ("jet") may be explained through local decelerating in the course of an interaction of the ejecta with the ambient material (at rest) rather than postulating additional episodic expulsions of high-velocity material.

3.2 V1016 Cyg

The emission lines of [O III] and [N II] observed in V1016 Cyg present two components separated by 44 and 51 km/s, respectively. The deduced maximum angular separation (p.a.$\sim 80^\circ$) for the [N II] components is 0.4 arc sec (Solf, 1983a). The data indicate a bipolar flow with 120 km/s velocity and a collimation angle of $\sim 20^\circ$ for an adopted distance of 2.2 kpc. (Larger distances would lead to higher velocities and higher collimation). The observations are compatible with the existence of an equatorial ring structure of lower expansion rate.

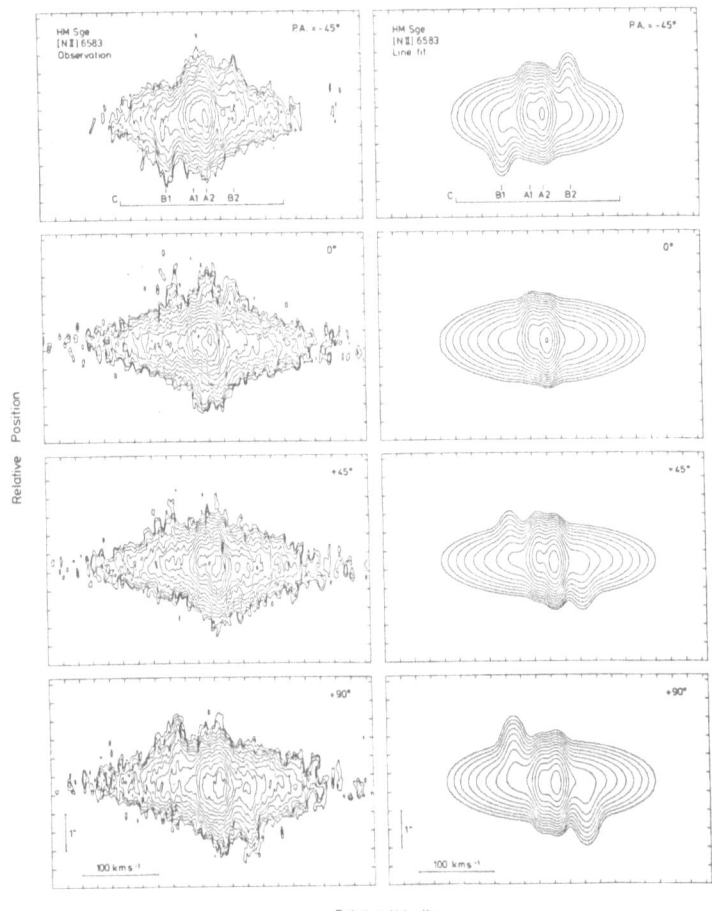

Fig. 1. (Left): Contour diagrams (in a position-velocity representation) of the spatially resolved [N II]λ6583 line of HM Sge observed at various slit position angles. (Right): Two-dimensional gaussian line fits to the observed lines using five distinct components (A1,A2,B1,B2,C)

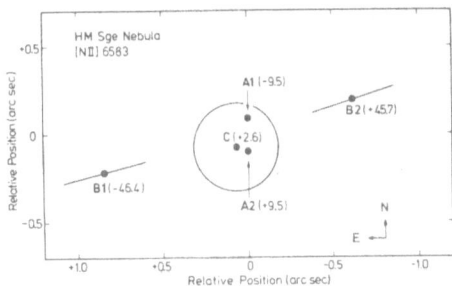

Fig. 2. Map of the relative positions (1983) of the five resolved nebular components of HM Sge as reconstructed from the [N II] line spectra. Relative radial velocities are quoted in parentheses

3.3 HM Sge

The observed [N II] lines (Fig. 1) exhibit four resolved narrow line
components (A1,A2,B1,B2) being superimposed on a broad fifth component
(C). The data show that the inner components A1, A2 are separated by
19 km/s in velocity and 0.2 arc sec along p.a. 0°, whereas the outer
components B1, B2 are separated by 91 km/s and 1.5 arc sec along p.a.
105°, respectively (Solf, 1984). The derived distribution of the five
components on the sky is shown in Fig. 2. The resulting expansion velocities depend on the distance of HM Sge (values beteen 0.4 and 2 kpc are
found in the literature). For d = 0.4 kpc one obtains a bipolar expansion
velocity of 200 km/s for components B1, B2 and a collimation angle of
$\$6^{\circ}$ for the flow. At 2 kpc the corresponding values are 900 km/s and $\sim 1^{\circ}$.
In any case, the bipolar flow is highly collimated and of remarkably high
velocity. Hence components B1, B2 must be considered as jet features
which have been ejected during the 1975 outburst of HM Sge. The low-velo-
city components A1, A2 are likely to represent an equatorial structure,
either slowly expanding (25 km/s) blobs or a rotating (10 km/s) ring.
The broad component C might be due to an unresolved (spheroidal or
ellipsoidal) shell expanding at intermediate velocities (~ 60 km/s).
Both the geometry and kinematics of the nebular components seem to be
less compatible with an single star model of symbiotics, but are in good
agreement with predictions of a binary star model assuming that mass
loss from a giant is accreted through a disk around a compact companion.

3.4 CH Cyg

High-resolution observations of the [O III] 5007 line of CH Cyg in
September 1986 have revealed a compact nebular feature located 1.1 arc
sec northwest of the star (Solf, 1987). The feature has been identified
with a component of the radio jet ejected in 1984 (Taylor et al. 1986).
The data confirm the high tangential velocity (>800 km/s) of the outflow.
The deduced low radial velocity difference beteen the jet feature and the
star suggests that the jet is moving nearly perpendicular to the line of
sight. The observed width of the [O III] line indicates a high collimation of the flow ($<7^{\circ}$). The jet probably originates from an accretion
disk seen nearly edge-on around the hot compact companion of the M6
giant in an eclipsing binary of highly excentric orbit.

4. REFERENCES

Gieseking, F., Becker, I., Solf, J.: 1985, Astrophys. J. Letters 295, L17
Mundt, R., Brugel, E. W., Buehrke, T.: 1987, Astrophys. J. 319, 275
Solf, J.: 1983a, Astrophys. J. Letters 266, L113
Solf, J.: 1983b, Astrophys. J. 273, 647
Solf, J.: 1984, Astron. Astrophys. 139, 296
Solf, J.: 1987, Astron. Astrophys. 180, 207
Solf, J., Ulrich, H.: 1985, Astron. Astrophys. 148, 274
Taylor, A. R., Seaquist, E. R., Mattei, J. A.: 1986, Nature 319, 38

OPTICAL POLARIMETRY OF SYMBIOTIC STARS

Antonio Mário Magalhães
Instituto Astronômico e Geofísico
Universidade de São Paulo
Caixa Postal 30.627
São Paulo 01051
BRAZIL

ABSTRACT. A review of some physical mechanisms that may give rise to detectable optical polarization in symbiotic stars is presented , emphasizing the ability of polarimetry in studying the evolution of dust envelopes, the atmospheric structure of the cool component and the geometry of these systems. A brief summary of the techniques is also given. General polarimetric properties of symbiotics as well as data for a few specific systems are discussed. Proposed models for symbiotics that include the prediction of their polarimetric properties are highly encouraged. Specific suggestions for future work from the ultraviolet through the infrared and which explore the potential of the technique are advanced.

1. INTRODUCTION

For a given object, astrophysical data which have been mostly obtained up to now concern, in a given spectral domain, the intensity as a function of wavelength and time. However,one can typically not afford to throw away information encoded in the other three Stokes parameters, even in the case of symbiotic systems, where the fully application of polarimetric techniques is still in its beginning stages. Symbiotic stars show indeed great potential for such studies due to their gaseous/dusty nebulae and circumstellar environment coupled to asymmetries in the system. Polarimetry can bring information on physical processes and geometry in these systems and provide additional constraints on the models proposed for these objects.

In the next paragraphs, we first outline a few basic polarimetric techniques, followed by an assestment of the physical processes that may originate optical polarization in symbiotic stars. We then move on to the yet unclosed topic of the general polarimetric properties of symbiotic systems and discuss a few objects in some detail. A final section is devoted to suggestions for future work.

Recent, related reviews have been also presented by Aspin and Schwarz (1987) on symbiotics and by Magalhães (1987), Willson (1987)

89

J. Mikolajewska et al. (eds.), The Symbiotic Phenomenon, 89–100.
© 1988 by Kluwer Academic Publishers.

and Schwarz (1986) on luminous late-type giants. Please also refer to the more specific papers by Schwarz et al. (1987), Rodrigues (1987), Kudiakova (1987) and Gershberg and Shakhovskoy (1987) elsewhere in this volume, as well as Piirola (1987). The proceedings of the Vatican Conference on Circumstellar Polarization (Coyne et al. 1987) also carry several other papers of interest in this and other related fields.

Rather than attention to a complete list of observational details, we chose to emphasize here the physical mechanisms that might be present in symbiotics, which type of information, as exemplified by some data, optical polarimetry might bring us and how new observations might lead us into further insight into symbiotic systems.

2. POLARIMETRIC TECHNIQUES

Light can be more generally regarded as being partially elliptically polarized, that is, a mixture of natural (unpolarized) light and completely polarized light. The four parameters, which form the so called Stokes vector (I,Q,U,V), correspond to the need of four quantities (size, shape, orientation and sense of rotation) necessary to fully describe the polarization ellipse. For instance, for unpolarized, circulary polarized and linearly polarized light, we would have $(I,0,0,0)$, $(I,0,0,V)$ and $(I,Q,U,0)$ respectively; fully linearly polarized light, say, would have $(Q^2 + U^2)^{1/2} = I$. In general, $I \geq (Q^2 + U^2 + V^2)^{1/2}$. Also, I, $Q^2 + U^2$ and V are invariant under a rotation of the coordinate system. In fact, $P = (Q^2 + U^2)^{1/2}/I$ and V/I are referred to as the degrees of linear and circular polarization, respectively. For a throuhg discussion of all these concepts, the reader is referred to Clarke and Grainger (1971). Most of this review will concern optical linear polarization.

The basic polarimetric technique is to modulate the beam with optical components which change the state of polarization in a known way. For instance, the light intensity passing through first a half-wave plate oriented at position angle ψ and a double beam analyser at position angle $0°$ is (Serkowski 1974a):

$$I' (\psi) = \frac{I}{2} [1 \pm (Q \cos 4\psi + U \sin 4\psi)],$$

allowing determination of Q and U (say, through least squares) if measurements are made through ψ. The advantages of observing the two orthogonally polarized beams leaving the analyser are the economy of light and possibility of overcoming atmospheric (and other) effects. Measurements with a quarter-wave plate in the beam, in place or together with the half-wave plate, will allow also a determination of the circular polarization. Please refer to papers on instrumentation presented by Coyne et al. (1987) and Magalhães and Velloso (1987).

An important thing to bear in mind is that polarimetry requires typically relatively high precision, by normal photometric or spectrophotometric standards, usually much better than 0.1% and according to one's objects and aims: whether determining wavelength dependence, differences between continuum and spectral features or time variability. Much of this need is of course due to the low values of polarization often found. Nevertheless, many polarimeters have now been built which are basically photon noise limited. Even with such equipment however, information does not come for free and polarimetry typically means long integration times, even for relatively bright objects!

Now, a word about interstellar polarization, which is typically superimposed on the measurements. A good way to separate the effects is via a QxU plot, as discussed, for example, by Serkowski (1970). For instance, the superposition of an intrinsic polarization with an interstellar one give in general an observed wavelength dependent position angle, with the two planes of symmetry more apparent in such a plot. Frequently, in fact, the intrinsic position angle is itself wavelength dependent (section 5). One can estimate the interstellar component at distinct wavelengths by observations of nearby stars and assuming a typical wavelength dependence for the interstellar grains. Of course, if one is interested only in differences in polarization between close spectral features, interstellar polarization will not hide them.

3. SOME PROCESSESS THAT MAY CRIATE POLARIZATION IN SYMBICTICS

Basically, scattering processes are in principle bound to be related to most polarimetric observations in symbiotic systems. Among them, we can cite:
- molecular scattering: cool star's atmosphere (TiO, H_2)
- dust scattering: cool star's envelope, system's envelope
- resonance scattering: cool star, emission lines.
- electron scattering: disk/jet around hot component.

Below, we discuss evidences for each process and lay the background for the remaining sections.

In complex systems like the symbiotics, one may perhaps gain insight into details of some physical processes by studying the individual components, although the interaction among them is of obvious importance. In the case of symbiotics, we can start by looking at the cool companion, which shows at least the first three effects mentioned above. Incidentally, single Luminous Late Type Variables (LLYV's) have been among the first objects to be detected with large and variable polarization. Please refer to Magalhães (1987) for a review of this still evolving field.

Observationally, the optical polarization in LLTV's typically decrease with wavelength, which is normally taken as indicative of Mie (grains) or Rayleigh (small grains, molecules) scattering. Sometimes , temporal correlations between P and brightness exist, as illustrated by V CVn (Coyne and Magalhães 1977), where maximum polarization occurs close to minimum light. Basically, the models that were put forward involve scattering in an non-spherically symmetric circumstellar dust cloud (Kruszewski et al. 1968, Shawl 1975) or photospheric scattering, coupled with an asymmetry across the stellar disk (Harrington 1969).

One of the observational points that would lead more or less naturally to dust scattering models was the rather strong correlation between the average optical polarization and the infrared excess (Dyck et al. 1971a). While it is true that newer observations of highly reddened objects show sometimes large, variable polarization in the near infrared (McCall and Hough 1980) and that multiple dust scattering in such kind of objects is also evidenced by circular polarization measurements (Angel and Martin 1973), Forrest et al. (1975) have not observed significant changes in the [L]-[0] colour while the objects were simultaneously polarimetrically variable. I think is important to stress here that, in the optical, where scattering dominates, one tend to 'see' the effect of smaller grains (say, around a tenth of a micron, if dielectrics) while, since thermal emission is proportional to the mass in the form of dust, in the IR we tend to see the larger grains, presumably further out from the star.

Actually, long term polarimetry may turn out to be very important as a more or less direct probe into the evolution of dust envelopes around LLTV's and, I believe, around symbiotics as well. Data of over a decade for L_2 Puppis (Fig. 1; Magalhães et al. 1986b) show that symmetry of the envelope approached that observed in the IR many years earlier, at which time the wavelength dependence of the polarization indicated increasingly larger grains. For comparison, the variable's period is only 141 days.

Photospheric scattering predicts changes in polarization across the object's spectral features. Higher spectral resolution observations showed that molecular scattering in the photosphere of these late-type stars is indeed one of the ingredients in producing their observed polarization (Landstreet and Angel 1977; McLean and Clarke 1977; Coyne and Magalhães 1977). Interestingly enough, some stars may show enhanced polarization across an observed absorption feature like a TiO band (Coyne and Magalhães 1979; Magalhães et al. 1986a) without position angle rotation. In the framework of Harrington's model, this fact has been interpreted (Magalhães 1981) as a subtle combination of the ratio of absorption to scattering as a function of optical depth coupled with a stronger gradient of the source function at the band depth. As we shall see in section 5, symbiotics may indeed show changes in polariation across TiO bands and other features.

Which asymmetries over stellar disk may give the necessary polarization? Harrington (1969) suggested non-radial pulsations or temperature variations across the stellar surface. Coyne and Magalhães (1979) suggested that even a very slow rotation may induce systematic pole-to-equator temperature differences, due to von Zeippel's theorem. Data for L_2 Puppis, across the CaI 4226 line led Codina and Magalhães (1980) suggest that these star might show spots as seen at that wavelength, causing the polarization to vary across that line. Clarke and Schwarz (1984), Doherty (1986) and Lefevre (1987) also considered a spot model to explain the observations of LLTV's. A fully, consistent treatment of the radiative transport of light leaving a spherical, eventually spotty, photosphere showing limb polarization, with further scattering by a dust envelope, is certainly needed.

Coyne and Magalhães (1977) first proposed, based on their

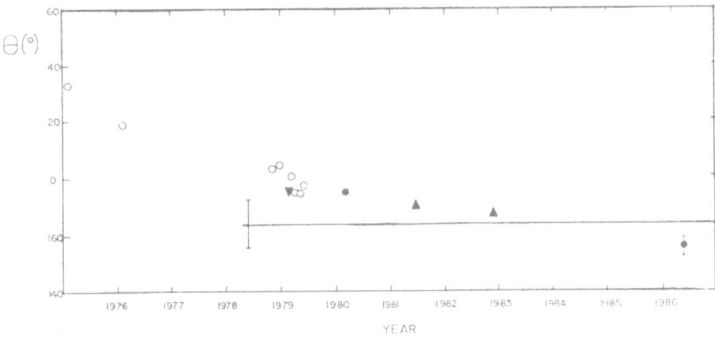

Fig. 1. Visual polarization position angle for the red variable L₂
Pup. The scattering cloud evolves towards the geometry given by
infrared measurements (solid line) (Magalhães et al. 1986).

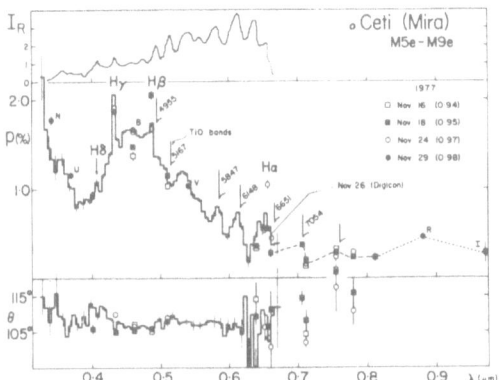
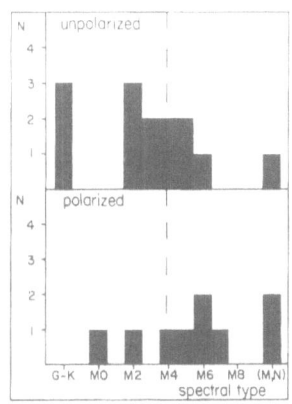

Fig. 2. (left). Spectropolarimetry of Mira near maximum light (McLean
and Coyne 1978), showing polarization changes across spectral features
and an increase into the UV.

Fig. 3. (right). Distribution of polarized and unpolarized symbiotics
according to the spectral type of the cool companion (Schulte-Ladbeck
and Magalhães 1987).

narrow-band observations, that <u>fluorescence scattering</u> (of which
resonance scattering is a particular case) would produce higher
polarization than in the continuum across certain spectral features,
like hydrogen emission lines. In LLTV's these lines result which from
a schock front moving outwards during portions of a late-type
variable's cycle (e.g., Wallerstein 1975). While more quantitative and
observational work are definitely needed, recent models of Willson,
Bowen and collaborators (see Willson 1987 for a review) also predict
this effect. Higher resolution data by McLean and Coyne (1978) of Mira

(o Ceti) may indeed be taken as indicative that this process in indeed operative. Fig. 2, from these authors, is instructive in showing several of the features mentioned above and which are typical of the polarization in late-type variables, including an eventual 'dip' in the UV which might be due to additional molecular opacity (Coyne and Magalhães 1979).

Finally, the literature on electron (Thomson) scattering polarization is far more extensive and will not be reviewed in detail here. The mechanism can perhaps be more easily understood, however, as it results basically from scattering off ionized, non-spherically symmetric (disks, jets, streams) distributed material surrounding the star. Be stars are a classical example (Coyne and Kruszewski 1969), with the wavelength independent mechanism modified by emission and abosrption effects in the plasma. Rudy and Kemp (1978) and Brown, McLean and Emslie (1978) showed how one can derive parameters like the inclination of a binary system from the modulation on the polarization curve as a function of the phase. This effect has been also explored in the context of binaries containing Wolf-Rayet stars (Drissen et al 1987; Meliani et al. 1987).

4. GENERAL POLARIMETRIC PROPERTIES OF SYMBIOTICS

As we have seen, scattering mechanisms and asymmetries present may induce polarization in symbiotics and provide insights into many physical processes. Many basic questions still need answers, however: how common is polarization among symbiotic systems? Which is its wavelength dependence, origin and time variability? Are there correlations with other observed properties (spectra, IR type and so on)? Can we discriminate against competing models for symbiotics? As we shall see below and in the next section, the answers to most of the questions are still being actively pursued.

In the search for the general properties of symbiotics, Schulte-Ladbeck (1985) has presented the results of a survey of northern symbiotics. She found intrinsic polarization to be present in several objects among 18 northern symbiotics and relations possibly existent between the presence of intrinsic polarization and the spectral type of the late-type star and between the presence of intrinsic polarization and IR type. All this would suggest that dust scattering is responsible for the observed polarizations. One should note that a possible lack of variation of the polarization across a particular TiO band would not by itself be indicative of dust (as opposed to photospheric) scattering because, from single variables' studies, this effect is likely to both depend on the particular band and time (Coyne and Magalhães 1979). R Aqr (section 5) seems indeed to confirm this.

A 'southern extension' of such work has recently begun (Schulte-Ladbeck and Magalhães 1987), with 9 out 23 symbiotic stars then showing detectable intrinsic polarization. Polarized symbiotic stars are observed at a slightly higher frequency when the associated cool companion is of spectral type later than M4. This is illustrated in Fig. 3, confirming, to the extent that the available data allow, the earlier finding of Schulte-Ladbeck (1985). Concerning a possible correlation with IR type, of the total sample then available there were 16 S-type objects (with 4 polarized) and 6 D-type (with 5 polarized), the probability of observing such number of intrinsically polarized D-type stars being only about 3%.

The distribution of symbiotic stars in colour-colour diagrams has been also investigated by Schulte-Lacbeck and Magalhães (1987) employing IRAS colours. The IR types were formerly defined by their H-K indices (Allen 1982) but it is illustrative to notice the striking separation that the IRAS data allow, a fact independently noticed by Whitelock (1986). Remarkably, R Aqr stands among the S-type objects in such diagrams. Intrinsically polarized objects are found among in both S- and D-types among the IRAS diagrams.

The survey referred to is being presently undertaken at telescopes at ESO, Hawaii and Brazil (Schwarz et al 1987). In order to cover a great number of objects, a search for a difference between the polarization at the broad R filter and an Hα filter is undertaken. Together, whenever possible, with time variability between distinct epochs, they should give a good indication of how frequent is the presence of intrinsic polarization among symbiotics, allowing statistical studies on a wider and more signficant basis.

5. OBSERVATIONS OF SPECIFIC OBJECTS

With the physical processes discussed above as a background, we now resume some of the observational results on specific symbiotics.

5.1. CH Cyg, CI Cyg and AG Peg

Piirola (1982, 1983) has detected variable linear polarization in these S-type symbiotics (Fig. 4). In 1978, the polarization has increased sharply into the ultraviolet, with very small values in the red and infrared. In early 1978 and May 1980, Piirola notices the appearance of a red component, which leads to a rotation of the position angle with wavelength, with the polarization of this red component increasing into the infrared. In 1981, this component had again disappeared, with the polarization then peaking around the B filter. All these facts may be taken as indicative of changes in the dust size distribution and its geometric distribution, the red/infrared observations suggesting grains with size around 1 μm and a change in size from 0.05 μm to about 0.1 μm to explain the optical/UV data.

Piirola (1983) has also obtained data for CI Cyg and AG Peg. For both objects, a strong rotation of position angle with wavelength is observed, confirming the presence of intrinsic polarization. Moreover, he notes a discontinuity in the θ(λ) curve around the wavelength at which the hot spectrum emerges, evidencing the need for a model where the hot or the cool component dominates, according to wavelength, with the steep increase into the blue for CH Cyg indicating that light from

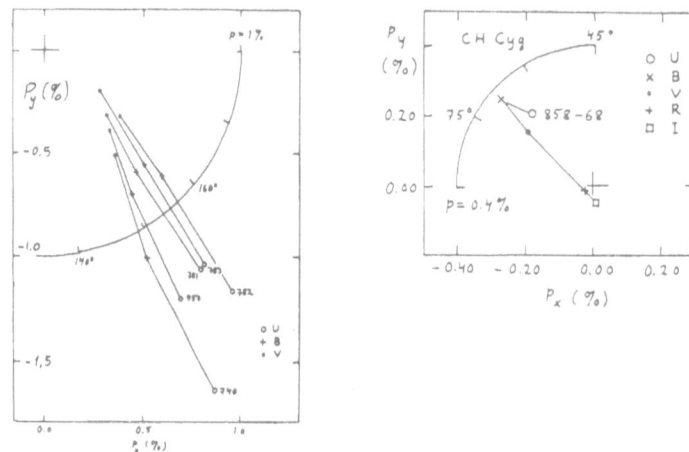

Fig. 4. QxU plots of polarization data for CH Cyg in September 1978 (Piirola 1982, left) and September 1981 (Piirola 1983, right), showing changes in geometry and mean particle sizes as the observed polarization maximum shifts from the U to the B band.

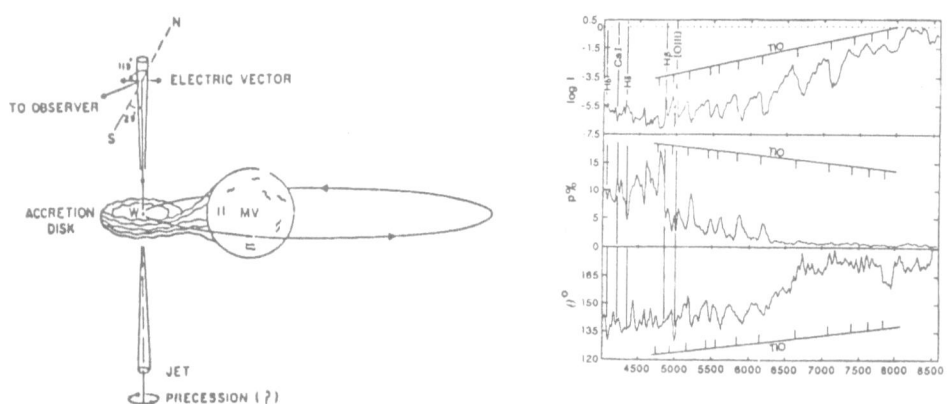

Fig. 5. (left). A model for R Aqr (Deshpande et al. 1987), where the UV polarization arises from light from the subdwarf W scattered off the jets and the RI polarization comes from a scattering shell around the mira variable MV.

Fig. 6 (right). Spectropolarimetry of R Aqr (Aspin et al. 1986), showing changes across spectral features and overall rotation in position angle with wavelength.

the hot source is being scattered in the cool giant's envelope.
Piirola (1987) presents extensive polarimetry of CH Cyg (and other
northern symbiotics) from 1978 through 1986, allowing dust formation
and evolution in the circumstellar envelope to be monitored.

Whenever possible, (quasi) simultaneous near IR polarimetry would
be highly valuable, as the grain size distribution evolves and its
effect is likewise felt also in that wavelength range.

5.2. HM Sge

Data presented for HM Sge by Efimov (1979), gathered in 1977, show
varaibility on a time scale of months. The intrinsic polarization
increased into the red, indicative of relatively large (0.2 μm, if
silicate) grains. Schulte-Ladbeck (1985) has observed similar
wavelength dependence, albeit at a distinct polarization level. She
notes that the strong line emission of the object might dilute the
polarization in the UBV region. As also noted by her, HM Sge is a
prime target for high resolution spectropolarimetric studies, which
would be valuable in deriving the relative geometry between the line
emission and scattering regions.

5.3. R Aqr

This is polarimetrically the best observed (and least
understood?) symbiotic so far. Schulte-Ladbeck (1985), who presents
1983 data and lists references up to that date, finds a very strong
increase of the polarization into the UV, indicative of Rayleigh
scattering. Also, a relatively strong wavelength rotation of the
position angle indicates that the circumstellar material scatters the
light from the two stars.

Deshpande et al (1987) present further, more recent (1984)
observations and clearly deserve merit for advancing a concrete model.
Employing earlier data from Serkowski (1974b) and Schulte-Ladbeck's,
the authors confirm the very large increase in polarization into the
UV, small variation in θ(U) around 120 degrees and large variations in
both polarization and position angle in the VRI filters. They propose
a model in which a white sub-dwarf dominates the ultraviolet light
which is scattered off a precessing jet (Fig. 5), while in the VRI
region the Mira companion dominates. They note that the relative
constancy in the UV position angle does not support the model of
Spergel et al (1983), where discrete clumps that form in the mira's
wind are eventually excited by the UV flux from the hot companion.

The situation may by bound to be more complicated, however. Aspin
et al (1986) present spectropolarimetric data of November 1983, which
show a great number of details (Fig. 6): increases in polarization
across TiO bands and decreases across the [OIII] and Balmer lines,
position angle changes across these features, an overall rotation in
θ(λ) and an increase in polarization across the CaI 4226 line. Their
preliminary interpretation call for two or more mechanisms giving rise
to the observed polarization, with the possible existence of both
spots on the cool stellar companion and resonance scattering at the

CaI line. Time dependent spectropolarimetry of R Aqr should clearly be of immense value.

Schwarz and Aspin (1987) analyse long (17 years) trends in the R Aqr data. They find large variations in the polarization in the blue , as opposed to small ones in the red, with some dilution due to the [OIII] line. They note that the variations in the position angle in UVR are quite strong; I believe this latter point poses the need for some modification to Deshpande et al's picture, which is based on a relatively small variation in $\theta(U)$. Schwarz and Aspin note that, contrary to o Ceti itself, maximum polarization occurs at minimum light. As we saw in section 3, however, this behaviour is similar to other late-type variables.

As a final remark, I would like to mention that, in view of the electronic densities (10^6 cm^{-3}) observed in R Aqr's jet (Schwarz and Aspin 1987; Michalitsianos and Kafatos 1987), an optical depth for Thomson scattering of the order of, say, 0.1 would require a rather large ($\sim$ 0.05 pc) physical path, if the jet is to be responsible for part of the observed polarization in R Aqr.

6. CONCLUDING REMARKS

Symbiotics do offer several sites for scattering of optical radiation as well as the needed asymmetries. Judicious polarimetry can hence prove useful in several respects, probing the evolution of dust envelopes, the atmospheric structure of the cool component and the geometry of particular systems. In this regard, theoreticians and observers alike should try to make an effort in predicting the polarization properties of their models, given the high accuracy of polarimetric measurements nowadays possible; models with such predictions should clearly be at an advantage as compared to others and could be more readily refined.

We have here not discussed magnetic processes such as the ones that can take place near the surface of a magnetic white dwarf. An accretion disk would tend to observationally swamp such effects but these merit a closer look, for instance through circular polarimetry.

Polarimetry may also help discern between competing models for symbiotics, as we have seen. Also, polarimetric data as a function of phase for eclipsing systems like AR Pav and SY Mus, when screening of the hot component by the cool component occurs, should furnish a more direct (and perhaps the only one) evidence for disks, information on geometry and on the density and type of scatterers.

In other instances, spectropolarimetry might prove highly valuable in unravelling the relative positions of the emission line regions of some systems as well as probing into the cool companion's atmosphere through observations across features like TiO bands and resonance lines.

Finally, models of resonance line profiles from accretion flows (Tylenda 1987) have been constructed. My suggestion here is that resonance scattering in emission lines are bound to produce detectable polarization across these features. Observations with the forthcoming

Space Telescope's Faint Object Spectrograph, in its polarimetric mode, should yield a good deal of information and constraints on the geometry and physics of the emission line regions in symbiotic systems.

ACKNOWLEDGEMENTS. It is a pleasure for the author to thank the invitation of the Organizing Committee and the polish hospitality as well as the brazilian Conselho Nacional de Desenvolvimento Científico e Tecnológico who financed his trip to Poland through grant nº 407218/87.4.

REFERENCES

Allen, D.A. 1982, in IAU Coll. 70, The Nature of Symbiotic Stars, eds. M. Friedjung and R. Viotti, Reidel (Dordrecht), p. 27.

Angel, J.R.P. and Martin, P.G. 1973, Astrophys. J. 180, L39.

Aspin, C. and Schwarz, H.E. 1987, in Vatican Conference on Circumstellar Polarization, eds. G. Coyne, D.T. Wickramasinghe, A.M. Magalhães, R. Schulte-Ladbeck, A.F.G. Moffat and S. Tapia, Vatican Observatory Publ., in press.

Aspin, C., Schwarz, H.E., McLean, I.S. and Boyle, R. 1986, Astron. & Astrophys. 142, L21

Brown, J.C., McLean, I.S. and Emslie, A.G. 1978, Astron. Astrophys. 68, 415.

Clarke, D. and Grainger, J.F. 1971, Polarized Light and Optical Measurement, Pergamon Press (Oxford).

Clarke, D. and Schwarz, H.E. 1984, Astron. Astrophys. 132, 375.

Codina-Landaberry, S. and Magalhães, A.M. 1980, Astron. J. 85, 875.

Coyne, G.V. and Kruszewski, A. 1969, Astron. J. 74, 528.

Coyne, G.V. and Magalhães, A.M. 1977, Astron. J. 82, 908.

Coyne, G.V. and Magalhães, A.M. 1979, Astron. J. 84, 1200.

Coyne, G.V., Wickramasinghe, D.T., Magalhães, A.M., Shulte-Ladbeck, R.E. Moffat, A.F.G. and Tapia, S. (eds.) 1987, Vatican Conference on Circumstellar Polarization, Vatican Obs. Publ., in press.

Deshpande, M.R., Joshi, U.C., Kulshrestha, A.K. and Sen, A.K. 1987, Publ. Astron. Soc. Pac. 99, 62.

Doherti, L.R. 1986, Astrophys. J. 307, 261.

Drissen, L., St.-Louis, N., Moffat, A.F.J., Bastein, P. 1987, Astrophys. J., in press.

Dyck, H.M., Forrest, W.J., Gillet, F.C., Stein, W.A., Gehrz, R.D., Wolf, N.J. 1971a, Astrophys. J. 165, 57.

Dyck, H.M., Forbes, F.F. and Shawl, S. 1971b Astron. J. 76, 901.

Efimov, Y.S. 1979, Sov. Astron. Lett. 5, 352.

Forrest, W.J., Gillet, F.C. and Stein, W.A. 1975, Astrophys. J. 195, 423.

Gershberg, R.E. and Shakhovskoy, N.M. 1987, this volume.

Harrington, J.P. 1969, Astrophys. Lett. 3, 165.

Kruszewski, A., Gehrels, T. and Serkowski, K. 1968, Astron. J. 73, 677.

Kudiakova, T.N. 1987, this volume.

Landstreet,J.D. and Angel, J.R.P. 1977, Astrophys. J. 211, 825.
Lefevre, J. 1987, in Vatican Conference on Circumstellar Polarization, op. cit.
Magalhães, A.M. 1981, in Physical Processes in Red Giants, eds. I. Iben and A. Renzini, Reidel (Dordrecht),p. 231.
Magalhães, A.M. 1987, in Vatican Conference on Circumstellar Polarization, op. cit.
Magalhães, A.M. and Velloso, W.F. 1987 in Optical Instrumentation for Ground-Based Telescopes, ed. L. Robinson, Springer-Verlag,in press.
Magalhães, A.M., Coyne, G.V. and Benedetti, E. 1986a, Astron. J. 91, 919.
Magalhães, A.M., Coyne, G.V., Codina-Landaberry, S. and Gneiding, C. 1986b, Astron. Astrophys. 154, 1.
McCall, A. and Hough, J.H. 1980, Astron. Astrophys. Suppl. 42, 141.
McLean, I.S. and Clarke, D. 1977, Mon. Not. Roy. Astron. Soc. 179, 293.
McLean, I.S. and Coyne, G.V. 1978, Astrophys, J. 226, L145.
Meliani, M.T., Velloso, W.F. and Magalhães, A.M. 1987, in Vatican Conference on Circumstellar Polarization, op. cit.
Michalitsianos, A.G. and Kafatos, M. 1987, this volume.
Piirola, V. 1982, in IAU Coll. 70, The Nature of Symbiotic Stars, eds. M. Friedjung and R. Viotti, Reidel (Dordrecht), p. 139.
Piirola, V. 1983, in IAU Coll. 72, Cataclismic Variables and Related Objects, eds. M. Livio and G. Shaviv, Reidel (Dordrecht), p. 211.
Piirola, V. 1987, in Vatican Conference on Circumstellar Polarization, op. cit.
Rodrigues, M.H. 1987, this volume.
Rudy, R. and Kemp, J.C. 1978, Astrophys. J. 221, 200.
Schulte-Ladbeck, R.E. 1985, Astron. Astrophys. 142, 333.
Schulte-Ladbeck, R.E. and Magalhães, A.M. 1987, Astron. Astrophys. 181, 213.
Schwarz, H.E. 1986, Vistas in Astronomy 29, 253.
Schwarz, H.E. and Aspin, C. 1987, IAU Symp. 122, Circumstellar Matter, eds. I. Appenzeller and C. Jordan, Reidel (Dordrecht), p. 471.
Schwarz, H.E., Aspin, C., Magalhães, A.M. and Schulte-Ladbeck, R.E. 1987, this volume.
Serkowski, K. 1970, Astrophys. J. 160, 1083.
Serkowski, K. 1974a, in Methods of Experimental Physics, vol. 12, Part A: Astrophysics, eds. M.L. Meeks and N.P. Carleton,Academic Press (New York), p. 359.
Serkowski, K. 1974b, IAU Circ. nº 2712.
Shawl, S. 1975, Astron. J. 80, 595.
Spergel, D.N., Giuliani, J.L. and Knapp, G.R. 1983, Astrophys. J. 275, 330.
Tylenda, R. 1987, this volume.
Wallerstein, G. 1975, Astrophys. J. Suppl. 29, 375.
Whitelock, P. 1986, in Light on Dark Matter, ed. F.P. Israel, Reidel (Dordrecht), p. 323.
Willson, L.A. 1987, in Vatican Conference on Circumstellar Polarization, op. cit.

POLARIMETRY OF SYMBIOTIC STARS

T.N. Khudyakova

Leningrad University
U.S.S.R.

In an effort to bring out the variations in the polarization parameters of symbiotic stars that result from their binary character, polarimetry in R, I IR bands was initiated at the Byurakan station of the Leningrad University Observatory in 1974. In addition, U, B, V measurements for some stars have been made. Three telescopes have been used, with mirror apertures of 48, 48, 62 cm.

Table I

| Star | l | b | Opt. | B | | I | | E(B-V) | Ref |
	$\circ$	$\circ$	var.	Δq %	Δu %	Δq %	Δu %	mag.	
R Aqr	66.49	-70.32	m	7.3	16.0	7.1	3.7	0.1	Eg
UV Aur	174.22	-02.36	m	1.0	0.4	1.0	0.1	0.2	RG
EG And	121.54	-22.17	m?			<0.3	<0.3		
z And	100.97	-12.09	r			0.9	0.8	0.2-0.4	KW
CH Cyg	81.86	+15.58	r			0.9	0.2	0.1	KW
CI Cyg	70.90	+04.74	r	2.0	1.0	1.3	1.5	0.4	B
V1016Cyg	75.17	+05.67	sm			<0.4	<0.4	0.2-0.3	KW
AG Dra	100.29	+40.97	r			<0.3	<0.3	0.03	KW
AG Peg	69.27	-30.88	s			<0.3	<0.3	0.1-0.2	KW
FR Sct	18.47	+00.35	?	<0.4	<0.5	0.4	1.7	A_v=4.2	L

Key: m - Mira-like; s - slow nova;
       r - classical symbiotic

B - Baratta, G.B. et al: 1982, IAU Colloq. n°70,145.
Eg - Eggen, O.J.:1970, Ap. J. 22, 289.
KW - Kenyon, S.J. and Webbink, R.F.: 1984, Ap. J. 279,252.
L - Lee, T.: 1970, Ap. J. 162, 217.
RG - Reimers, D. and Groot, D.: 1983, Astr. & Ap. 223, 257.

J. Mikolajewska et al. (eds.), The Symbiotic Phenomenon, 101–102.

Table I consists of observed stars list and some
data concerning them. The character of polarization is
complicated: for some stars variations in visual and IR
bands are correlated, for the other ones they are not.

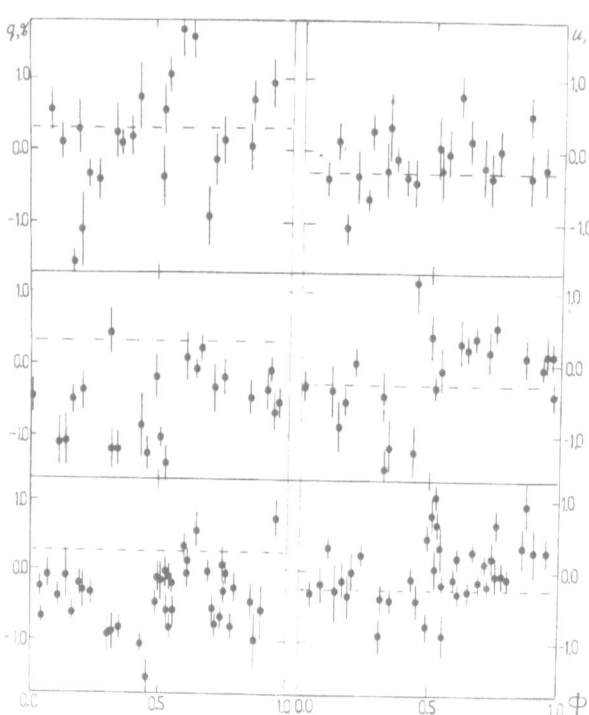

CI Cyg.It is
found that variati-
ons of polarization
are not correlated
with those of stel-
lar brightness. Let
us suppose that the
variable polarizati-
on reflects the or-
bital motion (Khudya
kova 1987). Variati-
ons of normalized
Stokes parameters q,
u in the B, R, I
bands with orbital
phase are shown in
fig.1. Dotted line
indicates the inter-
stellar polarization.
We assume that
the increase of po-
larization (see fig.
1 phase 0.5) is the
result of passage of
the hot object with
accretion disk be-
tween cool star and
observer.

R Aqr. The maximum of degree of polarization in
1976 (Nikitin and Khudyakova 1979) lies near to the eclipse of 1978 (Wallerstein and Greenstein 1980). The period
of the orbital motion is 44 years. In the case R Aqr long-
term variations of polarization combine with intrinsic
ones of the cool companion, the Mira. So for the two stars
we may assume that the passage of the object with a disk
before cool component leads to the increase of the degree
of polarization.

UV Aur.Our data suggest that the polarization presumably accompanying the orbital motion combines with intrinsic one of the cool companion, the Mira variable. The
cyclic behavior of the parameters of polarization implies
an orbital period of 14 years (Khudyakova 1985).

References:

Nikitin, S., Khudyakova, T.: 1979, Pisma Astr. Zh. 5, 611.
Khudyakova, T.: 1985, Sov. Astr. Lett. 11(4), 262.
Khudyakova, T.: 1987, Astr. Tsirc. n°1504.
Wallerstein, G., Greenstein, J.:1980, PASP 92, 547.

THE PRESENT STATE OF SYMBIOTIC POLARIMETRY.

Hugo E. Schwarz,   ESO, Casilla 19001, Santiago 19, Chile
Colin Aspin,   UKIRT, 665 Komohana St., Hilo, Hawaii, USA
A.M. Magalhaes, Instituto Astronomico e Geofisico, Dept. of Astronomy,
                University of Sao Paulo, Brazil
R.E. Schulte-Ladbeck, Washburn Obs.,Madison, WI53706, USA

1. ABSTRACT. We present preliminary statistical results
of an ongoing polarimetric survey of symbiotics. So far,
38% of the known symbiotics have had their polarization
measured at least once. About 40% of S and D types show
some intrinsic polarization while only one of the five
measured D' types is polarized. R Aqr is still the only
symbiotic with strong, variable polarization.

2. AIM, METHOD AND RESULTS.

The aim of this survey is to determine what percentage of
symbiotics   is   intrinsically   polarized and how the
polarization correlates with other parameters.

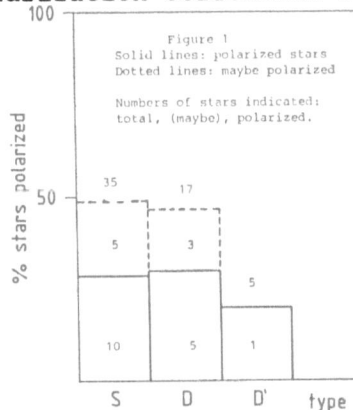

Figure 1
Solid lines: polarized stars
Dotted lines: maybe polarized

Numbers of stars indicated:
total, (maybe), polarized.

By simultaneously measuring with
a narrow Hα filter and in Cousins
R, any dilution by the emission
line will not show up in the
interstellar polarization and
hence a significant difference
between   the   two   measures
indicates   the   presence of
intrinsic polarization.

By combining the present
measurements   with   those of
Schulte-Ladbeck(1985) and S-L and
Magalhães(1987) the result of
Figure 1 has been obtained. This
picture summarizes the present
state of symbiotic polarimetry. In a later paper more
detailed results will be presented.

3. REFERENCES.

Schulte-Ladbeck,R.E. (1985) AA, 142,333.
---------------. ,Magalhães,A.M. (1987) IAU Symp.122, p485.

J. Mikolajewska et al. (eds.), The Symbiotic Phenomenon, 103–104.

Torun - Copernicus monument in front of the gothic city-hall. Nicolaus
Copernicus (1473-1543) was born in a building 30 m away from this place.

"Stars regain equilibrium after coughing"

Mario Livio

Ionization models of symbiotic stars

H. Nussbaumer
Institute of Astronomy
ETH Zentrum
8092 Zürich (Switzerland)

SUMMARY. A model of a symbiotic system is outlined. We present evidence that the main ionization mechanism responsible for the symbiotic nebula is radiative and not collisional. For calculating fractional ionization throughout the emission region we take either a blackbody source or an accreting system. If the ionizing radiation is due to accretion, the boundary layer between accreting star and disc provides the bulk of the ionizing photons. The possibility for distinguishing between different ionization sources are briefly discussed. C/N and O/N abundance ratios, calculated with the outlined ionization model places HM Sge, RS Oph, and T CrB among the novae and not among the symbiotic stars.

1. Introduction

When the first IAU colloquium on symbiotic stars was held at the Haute Provence Observatory in August 1981, the ultraviolet observatory IUE (International Ultraviolet Explorer) had just about finished three years of operation. Already the first results made it obvious that for symbiotic stars IUE had indeed opened an essential part of the spectral range, containing vital information for deriving ionization models. The new observations stimulated a wave of research in symbiotic stars, and the results presented at the 1981 colloquium were impressive. Concerning ionization models, it is now my task to present what has happened in the six years since 1981, and to outline what is required in the near and medium future. – Although the sudden glamorous impact of IUE is certainly responsible for the revival of interest in symbiotic systems (e.g. Nussbaumer and Stencel 1987), the progress in our knowledge would be unthinkable without the infrared (e.g. Feast et al.1983) and radio observations.

When IUE went up, UV spectroscopy of symbiotic stars was taken up without need of much additional line-identifications. The list published by Penston et al.(1983, but worked on and circulated much earlier) on lines identified in RR Tel, was a summary of knowledge that had been accumulated since the times when spectroscopy of the solar transition region was a new field. It was also helped by studies of planetary nebulae and QSO.

107

J. Mikolajewska et al. (eds.), The Symbiotic Phenomenon, 107–118.

2. Spectroscopy at the 1981 colloquium on symbiotic stars

Are all symbiotic stars binary systems? Neither infrared nor visual, nor UV observations on their own can provide a general answer to that question: indeed on the basis of each of those observations one can certainly construct single star models for many symbiotic systems. The planetary nebula type model, as published for example by Nussbaumer and Schild (1981) in order to explain the UV spectrum of V 1016 Cyg, did not need a double star system to match model and observation. And only rarely do periodic wavelengthshifts prove the existence of a double star system, as in the case of HBV 475. The proof of the double star nature of an object like V 1016 Cyg comes mainly from the combination of infrared and UV observations.

The main results of the first phase of UV studies of symbiotic stars was to show that ionization is done by radiation and not by collisions. This then implies the presence of a hot radiation source. How does one prove that symbiotic emission regions are radiatively ionized? We show that the electron temperature, T_e, responsible for the collisional excitation of observed spectral lines, is too low to produce the observed ionization stages by collisional ionization. Suitable lines for a determination of T_e are scarce. C III and Si III can be used. C III] $\lambda 1909$ corresponds to the transition $2s^2\ ^1S_0 - 2s2p\ ^3P^o_1$. We compare it to the strength of the multiplet $2s2p\ ^3P^o - 2p^2\ ^3P$ at $\lambda 1176$. In a collisionally ionized gas C III is seen if $T_e \approx 50\,000$ K. But at that temperature, and $N_e > 10^6$ cm^{-3}, the C III $\lambda 1176$ multiplet should have about one third of the strength of $\lambda 1909$ (Nussbaumer and Schild, 1979). However, in symbiotic stars C III $\lambda 1176$ does not appear with that strength. We may therefore conclude that $T_e < 50\,000$K. We now turn to Si III $3s^2\ ^1S_0 - 3s3p\ ^3P^o_1$ $\lambda 1892$ and the multiplet $3s^2\ ^1S_0 - 3s3p\ ^1P^o_1$ at $\lambda 1206.5$. With the help of these lines Hayes and Nussbaumer (1986) have determined T_e in RR Tel, they find $T_e = (1.3 \pm 0.2) \times 10^4$ K. This rules out collisional ionization for the main emission regions of the nebular lines. It would not rule out that some of the lines – from either higher or lower ionization stages – could originate from a collisionally ionized region. However, the spectral evidence gives no indication for such conditions, with, perhaps, the exception of X–ray observations. Evidently, extension of the observing range down to 912 Å, where many of the resonance lines appear, would greatly facilitate T_e–determinations.

3. Shift on emphasis

As the binary (multiple?) nature of symbiotics has again become the generally accepted opinion, research in symbiotic stars has been shifting to the physical processes taking place within those systems. The fashion for accretion discs has also caught up with symbiotic stars. Models are at present mainly concerned with the following three components:

1) a red giant or supergiant,
2) a hot radiation source,
3) a nebula.

Depending on the authors, there may be coexistence or interaction. Because of geometry and chemical composition the origin of the nebula matters: is it a remnant of what now is the hot radiation source, or is it the stellar wind from the cool star? Nussbaumer et al.(1987) have compared C/N and O/N abundance ratios of red giants and supergiants, carbon stars, planetary nebulae, novae, and symbiotic stars. In Fig.1 I show some of the results. The symbiotic stars lie in the region of the M–giants, and are clearly separated from novae. The authors interprete this result as proof that symbiotic nebulae are composed of the mass lost by the red giant, and that the nebulae are not the product of nova–like nuclear burning, as would be expected in nova-like symbiotic models (e.g. Paczynski and Rudak, 1980); planetary nebulae occupy a yet different field of the plot. It is worth adding that RS Oph and T CrB have been counted as novae.

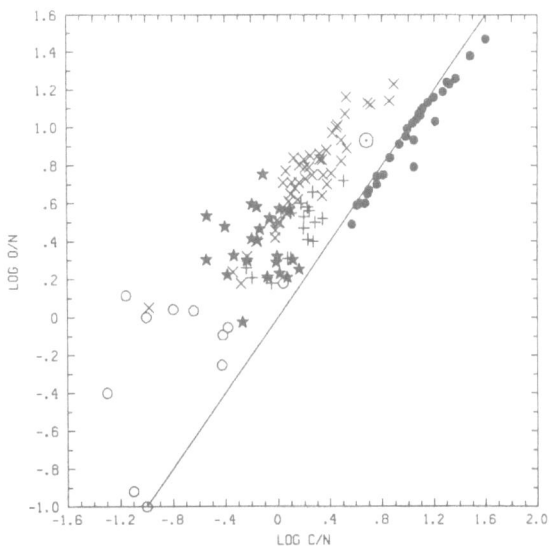

Figure 1. Relative abundances O/N plotted against the relative abundances C/N for red giants, carbon stars, novae, and symbiotic stars (Nussbaumer et al.1987) The relative abundances are given on a logarithmic scale. The symbols stand for the sun: ⊙, G– and K–giants: x, M–giants: +, novae: ○, carbon stars: •, symbiotic stars: *. The symbiotic star among the novae is HM Sge.

One cannot talk about ionization models without specifying a model of the symbiotic system as a whole. I shall proceed on the basis of the symbiotic model of Nussbaumer and Vogel (1987). They suggest that the emission line region observed in symbiotic stars is provided by the wind of the red giant, and that the onset of such a wind may explain the outbreak of symbiotic stars. If that theory is correct, the wind from the cool star is an essential feature in symbiotics. As a source of hot radiation the following candidates are in the running:

(1) normal hot star with blackbody or non-blackbody radiation field

(2) accretion disc and its non-blackbody radiation field.

Because of their basic differences the nature of the source of radiation has – independent of its importance for the ionization structure – become itself a debated object of investigation.

We then assume that the main components and features of a symbiotic system can be described in the following way:

The double star pair consists of a mass-losing red giant or supergiant, and a dwarf, subdwarf or white dwarf, which emits the radiation ionizing the nebula. The dwarf–type star may form an accreting system. The line and continuum radiation at visual, UV and radio wavelengths is mainly due to the nebular emission and to hot stellar or accretion system contribution. The red part of the visual, and the infrared radiation is emitted by the cool mass-loosing star and by dust in its wind. A fraction of the mass lost by the cool star may be captured by the companion, either by accretion from a wind or through Roche–lobe overflow. *The accreted mass may form a disc around the companion. In the latter case the accretion disc could be the main source for the hot radiation field.*

4. Ionization models

4.1 General remarks.

In the ionization model we assume an equilibrium situation. Thus for each volume element the number of ionizations is equal to the number of recombinations:

$$N(X^{+i}) \rightleftharpoons N(X^{+i+1}). \tag{1}$$

We consider that equilibrium is established by the following processes:
  radiative ionization and two-body recombination (including dielectronic recombination)
  collisional ionization and three-body recombination
  charge exchange
The most important of these processes are: radiative ionization with its inverse – two-body recombination – and charge exchange. The radiation field is composed of the stellar radiation, J_ν^*, and the diffuse field, J_ν^d, created in the nebula:

$$J_\nu = J_\nu^* + J_\nu^d. \tag{2}$$

For a source radiating spherically symmetrically, we have

$$4\pi J_\nu^*(r) = \pi F_\nu^*(R^*) \left(\frac{R^*}{r}\right)^2 e^{-\tau_\nu(r)} \tag{3}$$

where the optical depth τ is

$$\tau_\nu(r) = \sum_{X+m} \int_{R^*}^{r} N(X^{+m}) a_\nu(X^{+m}) ds. \tag{4}$$

In principle this sum should be extended over all the continuum and line absorptions active at frequency ν. The difficulty lies in properly treating J^d. Apart from strong spectral lines, which can have an ionizing effect, J^d is mainly created by recombination within the nebula. At a given position $\vec{r}$ in the nebula we should add the contribution to J^d from all other positions $\vec{r}'$, attenuated by optical depth. However, recombination is usually treated in a much simpler fashion, I shall briefly outline how the ionizing field J at a position $\vec{r}$ is obtained in our approach. We follow the ionizing paths of the stellar radiation in straight lines, starting from the ionizing star S2. Along this path J is the result of the attenuated stellar radiation (3) and the diffuse radiation accumulated along the path between S2 and $\vec{r}$. At the position $\vec{r}$ we only regard the recombination radiation arising in a small volume around position $\vec{r}$. We assume that those photons which are capable of ionizing H or He will do so, the others are added to J^d; this includes line emission due to collisional excitation.

Fig.2 shows the model configuration which will be employed in the following chapters. A cool mass–losing star, S1, is accompanied by a source of hot radiation, S2. S2 may be stellar-like, in which case it is assumed to radiate as a blackbody. It may be also be an accreting system, in which case the accretion–disc and the boundary layer are supposed to be the source of the ionizing radiation. We assume the mass–loss of the cool star to be given by

$$\dot{M} = 4\pi r^2 \mu m_H N(r) v_\infty \left(1 - \frac{R}{r}\right). \tag{5}$$

In the examples shown here, the origin of the stellar wind lies at $R = 50R_\odot$, $v_\infty = 80$ km/s, and for the mass–loss we set $\dot{M} = 3.5 \cdot 10^{-6} M_\odot/yr$. For the nebula we thus have a $\approx 1/r^2$ density distribution, with density decreasing towards the source of ionizing radiation. S1 and S2 are separated by $5 \cdot 10^{13}$cm. The model of Figure 2 was originally proposed by Nussbaumer and Schmutz (1983) and further developped by Nussbaumer and Vogel (1987).

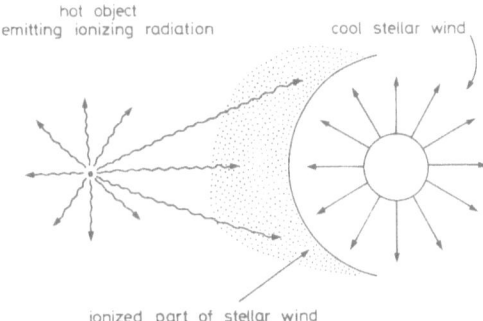

Figure 2. Model of a symbiotic object. A fraction of the wind of a cool, mass–loosing star is ionized by a hot radiation source.

4.2 Black–body radiation sources.

The observed ionization stages in symbiotics mostly indicate high T^*, I shall assume $T^* = 150\,000$ K and $R^* = 9 \cdot 10^8$cm. – It was pointed out by Hayes and Nussbaumer (1986) that with the presently observable spectra we cannot determine T^* with high accuracy if $T^* \gtrsim 150\,000$K. – We first compare the resulting ionization structure with that from a nebula of constant density $N(\mathrm{H})$. The density in the constant-density model was chosen to give the same Stroemgren–radius on the S1-S2 path as the $1/r^2$–model, it resulted in $N(\mathrm{H}) - 8.53 \cdot 10^8$ cm^{-3}; $N(\mathrm{H}) = N(\mathrm{H}^+) + N(\mathrm{H}^0)$. Results along the radiation path S2 ⟩ S1 are given in Figures 3 – 5 for He, C, N, O, Si, and Fe. From Fig. 3 and Fig. 4 we see that in the $1/r^2$ density distribution the higher ionization stages approach the cool mass–loosing star more closely than in the constant density model. However, the two density distributions result in qualitatively similar ionization structures. This is due to the high mass–loss rate; for $\dot{M} = 3.5 \cdot 10^{-6} M_\odot/yr$ the nebula resulting from our configuration appears almost as one of constant density. According to the ionization state of He we distinguish between the He^{+2} and the He$^+$ zones. (For $T^* \gtrsim 40\,000$K the He^0–region coincides with the H^0–region.) As the prominent He–lines in symbiotic systems are recombination lines, the two zones correspond to the He II and the He I emission regions. The He$^+$ zone corresponds to the region where O III and Fe IV are emitted, it is also the region where C III and N III are emitted. There is practically no emission of Fe V and only a narrow Fe VI emission region, corresponding to the O IV region. Fe VII and O V, and Fe VIII and O VI are emitted in about the same regions respectively. In the innermost part of the He^{+2} region oxygen exists as He–like O^{+6}, whereas Fe emits mainly Fe IX, but also some Fe X and Fe XI in tiny, hardly observable amounts.

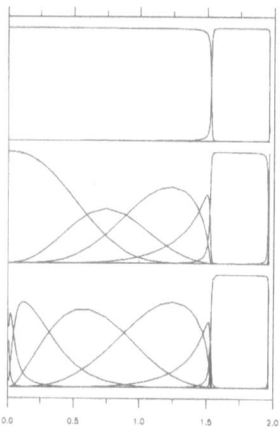 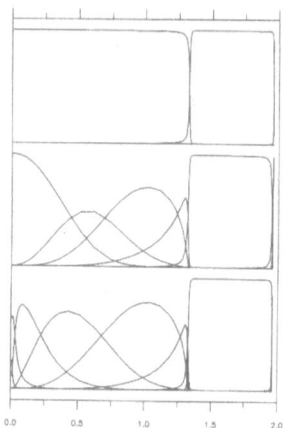

Figure 3 (left). The ionization structure of helium, oxygen, and iron (from top to bottom) calculated with a $N(r) \sim 1/r^2$ density distribution. Given is the fractional ionization $N(X^{+m})/N(X)$. The hot radiation source, with $T^* = 1.5 \cdot 10^5$K, $R^* = 9 \cdot 10^8$cm, is placed at $r = 0$, distances from that point are given in units of 10^{13}cm. The cool mass-loosing star is separated by a distance $p = 5 \cdot 10^{13}$cm from the hot star. The resulting nebula corresponds to the closed model in Fig.1 of Nussbaumer and Vogel (1987). The ionization stages in order of their maxima with growing distance from the radiation source are for He: He^{+2}, He^+, for O: O^{+6}, O^{+5}, O^{+4}, O^{+3}, O^{+2}, O^+ (very narrow), for Fe: Fe^{+10}, Fe^{+9}, Fe^{+8}, Fe^{+7}, Fe^{+6}, Fe^{+5}, Fe^{+4} (practically not visible), Fe^{+3}, and as very narrow peaks at the boundary appear Fe^{+2} and Fe^+.

Figure 4 (right). The ionization structure of helium, oxygen, and iron, as in Fig. 3 but calculated with $N(r)$= constant= 4.03×10^8 cm^{-3}.

Figure 5. The ionization structure of carbon, nitrogen, and silicon, obtained with the same parameters as the curves of Fig. 4. The ionization stages in order of their maxima with growing distance from the radiation source are for C: C^{+4}, C^{+3}, C^{+2}, C^+ (peak at the boundary); for N: N^{+5}, N^{+4}, N^{+3}, N^{+2}, N^+ (peak at the boundary); for Si: Si^{+7}, Si^{+6}, Si^{+5}, Si^{+4} (containing most of Si), Si^{+3}, Si^{+2} and Si^+ (as very narrow peak at the boundary).

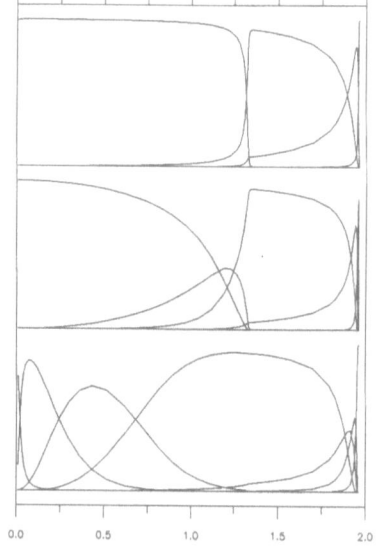

The neat division between O^{+2} and its neighbours, as well as between Fe^{+3} and other ionization stages of Fe, is due to their respective ionization potentials being just slightly higher – less than 1 eV – than that of He^+. As the ionization potentials of C^{+2} and N^{+2} are similar, we find also a fairly neat separation in those elements. Within the He^+ region C, N, O, and also Fe are represented by one or two ionization stages. This is not a general feature, Si being an example to the contrary. The Si ions Si^+, Si^{+2}, Si^{+3} have ionization potentials of 16, 33, and 45 eV. We see that in the He^+ zone C is represented by C III and C IV, N by N III and N IV, O by O III, Fe by Fe IV, however, Si by Si II, Si III, Si IV and Si V. As all four corresponding ionization stages are present, there is a rather large uncertainty about the fractional abundance of each individual stage. This concerns in particular Si III, as Si^{+2} is less abundant than either Si^{+3} or Si^{+4}. The temptation, to use the prominent Si III $\lambda 1892$ line as representative for Si is therefore dangerous.

As long as densities are sufficiently low to render collisional de-excitation ineffective, differences between the two density distributions do not effect relative line strengths. The reason is the relation between the radiation source (T^*, R^*) and the mass–loss. The present model, with $\dot{M} = 3.5 \cdot 10^{-6} M_\odot / yr$ results in a closed nebula (but see also the figures in Nussbaumer and Vogel, 1987). It is of interest to compare relative fluxes of ions with ionization energies above and below the He^+ ionization energy. Thus O V and O IV originate mainly in the He^{+2} region, whereas O III originates in the He^+ region. As an example we compare the ratios O III $\lambda 1660$/O IV $\lambda 1400$/O V $\lambda 630$ for of the total nebular fluxes for the two models. Normalising the multiplet O III $\lambda 1663$ in the $N(r) \sim 1/r^2$ model to 100, we find:

O V $\lambda 630$/ O IV $\lambda 1400$ / O III $\lambda 1663 = 6/39/100$ for the $N(r) \sim 1/r^2$ model
and
O V $\lambda 630$/ O IV $\lambda 1400$ / O III $\lambda 1663 = 7/37/102$ for the $N(r) =$ constant model.

4.3 Accretion discs.

We shall now turn to accretion discs. The luminosity in an accretion disc is obtained through conversion of kinetic into radiative energy when particles on their orbits drift closer to the accreting star. Our model for the emitted spectrum is mainly based on the work of Shakura and Sunyaev (1973), Lynden–Bell and Pringle (1974), and Pringle (1981 and ref.cit.). At any given time the particles are supposed to be on Keplerian orbits, except for the inner boundary where the Keplerian orbits have been stopped. The inward drifting is due to viscosity. According to the virial theorem, half of the energy available to a particle is radiated away within the trajectory through the disc, and the other half on the arrival at the boundary layer. The main uncertainties are viscosity and the structure of the boundary layer.

For calculating the flux, F_ν^{disc}, emitted by the accretion disc, a temperature $T(r)$ can be defined at any point in the accretion disc. This function has the form

$$T(x) = T_* \left(x^{-3} - x^{-\frac{7}{2}} \right)^{\frac{1}{4}}, \tag{6}$$

where

$$x = \frac{r}{R_*}, \quad \text{and} \quad T_* = \left(\frac{3GM\dot{M}}{8\pi\sigma R_*^3} \right)^{1/4}. \tag{7}$$

R_* is the radius of the boundary layer. We further find $T_{max} = 0.488 \times T_*$. In Figure 6 we show T(x) for $1 \leq x \leq 10$.

Lynden–Bell and Pringle (1974) and Pringle (1977) have investigated the properties of the inner boundary. The temperature, T^{bl}, calculated for the boundary layer depends on the physical assumptions concerning the structure of the boundary layer. The work of Lynden-Bell and Pringle gives a choice of three temperatures. Lynden-Bell und Pringle (1974) calculate the thickness of the boundary layer from the Navier-Stokes relations, neglecting pressure. Pringle (1977) includes pressure. A different T^{bl} results when opacity due to electron scattering is included. All three results have the form

$$T^{bl} \sim \left(\dot{M} \right)^{\alpha} (M)^{\beta} \left(\frac{1}{R_*} \right)^{\gamma}. \tag{8}$$

The values for α and β vary between 2/9 and 4/7, those for γ between 3/4 and 18/19. The total flux emitted from the accretion disc and the boundary layer is

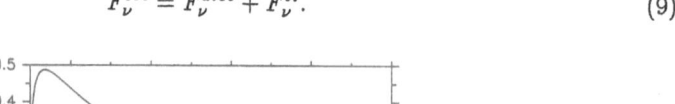

$$F_{\nu}^{tot} = F_{\nu}^{disc} + F_{\nu}^{bl}. \tag{9}$$

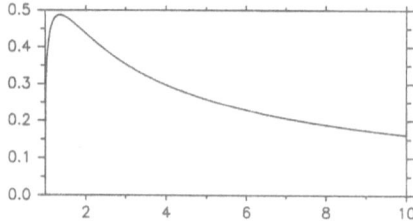

Figure 6. The radiation temperature of the accretion disc as a function of the distance from the inner boundary of radius R_*. The temperature is given in units of T_*, the distance is given as $x = r/R_*$.

The separation of the accretion system in disc and boundary layer is artificial. This is obvious from the drop in $T(x)$ to zero for $x = 1$, as shown in Fig.6. The work of Regev (1983) is an example for the attempt to match disc and boundary layer.

Vogel (1988) will report in some detail on the differences expected between ionization from accretion discs or blackbodies. In the next section I shall summarily mention his main results.

5. Accretion

Is accretion an essential feature in symbiotic systems? Based on the work of Livio et al.(1986), who place severe limits on the accretion cross section, Nussbaumer and Vogel (1987) conclude that if the hot companion moves in a spherically symmetrical wind emitted by the cool component, there is little likelyhood of accretion disc formation in symbiotic systems. The case is different, if the giant fills the Roche–lobe.

Kenyon and Webbink (1984) have set themselves the task of finding the most pronounced observable spectral differences longwards of 1000 Å among the various ionization models, including accretion systems. However, to avoid dealing with the many free parameters of the nebular emission, they restrict their efforts to the continuum flux from the stars or discs. They find the following configurations to be the most likely models for hot sources:

(a) Main sequence stars accreting $\approx 10^{-5} M_\odot/yr$. Such large rates require the giants of these systems (E Aud, CI Cyg, YY Her, AR Pav) to fill their Roche lobes. The nature of the giants could not be properly determined.

(b) The rest of the 15 systems observed they explain by the presence of a hot stellar source $25\,000\,\text{K} < T^* < 120\,000\,\text{K}$; but they are subdwarfs rather than white dwarfs.

They think that the two types are different in their outburst mechanism. The main sequence accretor outbursts are thought to be dwarf-nova-type outbursts, whereas the hot star outbursts are thought to be rather of the nova-type outburst.

But can we confidently distinguish, from presently available spectral features, whether the source of hot radiation is a blackbody stellar object or an accretion disc? Vogel (1988) discusses the details of this problem. The results he found do not agree on all points with earlier investigation by Kenyon and Webbink (1984). Vogel finds that on the basis of emission line spectroscopy it is almost impossible to distinguish between a blackbody source and a main sequence accretor. Uncertainties in the theoretical treatment of the boundary layer are partly responsible for this state.

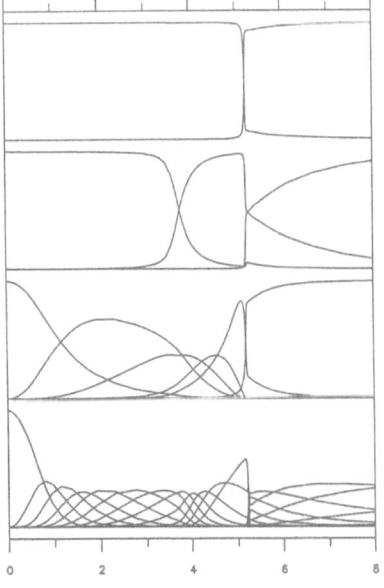

Figure 7. The ionization structure of hydrogen, helium, carbon, and iron, obtained in the radiation field of an accreting white dwarf. The ionization stages in order of their maxima with growing distance from the radiation source are for H: H^+, H^0, for He: He^{+2}, He^+, He^0, for C: C^{+6}, C^{+5}, C^{+4}, C^{+3}, C^{+2}, C^+, for Fe: Fe^{+15} and higher, then follow Fe^{+14} and lower stages, reaching a peak for Fe^{+3}. For the region covered by H^0 see text. Distances are given in units of 10^{15}cm.

We can easily spot the presence of accretion onto a white dwarf. As shown by the formulae for T^{bl}, small values of R_* produce high radiation temperatures; for white dwarfs they easily reach $T^{bl} \approx 10^6$K. Fe is then ionized to Fe^{+9} and higher, and we expect [Fe X] $\lambda 6374$ or [Fe XIV] $\lambda 5303$ to be prominent lines in the spectra. However, neither [Fe X] nor [Fe XIV] are prominent in symbiotic spectra. In agreement with Kenyon and Webbink (1984), who for different reasons arrive at the same conclusion, we can probably rule out accretion on white dwarfs as relevant for symbiotic systems.

In Figure 7 we show the fractional ionizations resulting from the radiation field of an accreting white dwarf with $M = 1 M_\odot$, $R^* = 0.009 R_\odot$, and an accretion rate of $10^{-6} M_\odot/yr$.

Within the disc we find $T_{max}(disc) = 250\,000$ K. The ionizing field is dominated by the boundary layer with $T = 1.2 \cdot 10^6$K. The ionization structure is qualitatively different from those shown in Fig. 3 – 5. Helium does not completely recombine at the border of the H^+-zone, nor does Fe. This behaviour has the following reason: The combined disc and boundary–layer continuum is relatively flat, with an important fraction of hard photons having energies far beyond the He^+ ionization limit. Even in the He^+-zone, there is still a sufficient quantity of hard photons to keep a few % of He doubly ionized. When the bulk of the photons capable of easily ionizing H have disappeared, hydrogen will become mainly neutral. But with the disappearance of about 90% of the e^-, recombination for other elements is more difficult. Thus, the remaining hard photons, peaking in our case at about 10 Å, are able to keep He and other elements ionized. We thus have an outer part of the nebula, where hydrogen is neutral but many other elements are partially ionized.

6. Outlook

In recent years the models have evolved away from full spherical symmetry in radiation and mass–loss. For the interpretation of radio observations Taylor and Seaquist (1984) have modelled the symbiotic nebula as consisting of hydrogen only. They also simplified the treatment of ionization and recombination. Thus, they are able to calculate analytically the extent of the ionization region. Their basic model is the same as the one shown in Figure 2. By comparing the analytical results with much more extended model calculations, Nussbaumer and Vogel (1987, this publication also contains corrections to the formulae given by Taylor and Seaquist) show that the analytical treatment can serve as a very useful first approximation.

To account for the observed X–ray emission from HM Sge, Willson et al.(1984) propose a model, where the head-on collision of two stellar winds is supposed to produce a collisionally ionized hot region as a source of free–free radiation. However, the full implications for line and continuum spectra have not been calculated. But, as they need only $\approx 10^{-11} M_\odot$ to explain the X-ray emission, which is a tiny fraction of the total nebula, we obviously deal with fringe effects – this remark does not imply an invitation to disregard fringe–happenings. X–ray emission has also been detected from the jet of R Aquari. In connection with the model proposed by Kafatos et al.(1986), Viotti et al.(1987) feel that the observed X-ray emission of R Aqr may originate either in a radiatively or a collisionally ionized part of the jet. We see, that in spite of the fact that the normal spectrum of symbiotic stars indicates radiative ionization, collisional ionization may after all be active in some parts.

I now want to trace the steps the ionization models have followed, or are likely to follow. (I design the center of mass–loss as C_w and the origin of the ionizing flux as C_r.) Published results:

1) Spherically symmetrical radiation field centered on C_r, spherically symmetrical mass–loss centered on C_w. The two centers coincide. The ionized nebula is spherically symmetrical around $C_r = C_w$. The computational problem is one-dimensional. Often this approximation is further simplified to a one–point model for the nebula.

2) Spherically symmetrical radiation field, spherically symmetrical mass–loss. The two centers C_r and C_w do not coincide. The resulting ionzed nebula is axially symmetrical around the line $C_r - C_w$. The computational problem is two-dimensional.

Still to be done:

3) Disclike radiation field, spherically symmetrical mass–loss. The two centers C_r and C_w do not coincide. The only symmetry of the resulting ionzed nebula is relative to the plane containing the disc, and to the plane at a right angle, containing the line C_r – C_w. The computational problem is three-dimensional with a certain facilitation due to the properties of symmetry

4) Disclike radiation field, mass–loss through one point. This corresponds to an accretion disc as radiation source and mass–loss via Roche–lobe overflow.

5) And further. In order to copy nature perfectly, we would certainly have to abandon any symmetry in mass–loss and radiation field.

We have seen that it is rather difficult to distinguish on purely spectroscopical grounds between blackbody radiation sources and accretion disc radiation sources. We also have to ask, how far we want to push modelling, so as not to introduce many more free parameters than can be fixed by observations.

In the ionization models employed to calculate symbiotic systems optical depth effects are usually taken into account only in a first approximation (see section 4.1). These methods will have to be extended to account properly for radiative transfer. Although it is useful to pay attention to what is done in other fields of astrophysics, where the nebular spectrum also contains major clues about the physical environment, we must always take into account the particular situation of symbiotic nebulae. Thus, in comparison with planetary nebulae, our nebular densities are higher by several powers of ten.

Up to now I have not mentioned the presence of dust. On this point I should like to point to Allens' (1982) comments. He considers the presence of dust as certain for D–type symbiotics, with some dust present in S–type symbiotics. Dust found in the interstellar medium is thought to be created in the winds of red giants and supergiants. But as shown by observations of planetary nebulae, dust is not necessarily the same everywhere. – A review of work up to 1982 on dust in planetary nebulae has been given by Barlow(1983). – The presence of dust, and whether observationally determined element abundances in symbiotic stars are influenced by condensation into dust grains, is in itself of high interest. In the model adopted, the symbiotic nebula is linked directly to the wind of the red giant. At a distance of $2 \cdot 10^{13}$cm (see Fig.3 – 5) the not attenuated radiation field of a $T^* = 150\,000$K, $R^* - 9 \cdot 10^8$cm star would result in dust temperature of ≈ 700K. At the same distance a giant with $T^* = 3\,000$K, $R^* = 7 \cdot 10^{12}$cm gives $T^{dust} = 1250$K, if calculated on the basis of energy balance. To deal properly with dust, its presence will have to be included in the ionization model. There it will also have to be considered as an agent altering the ionizing radiation field.

As concluding remark I want to stress the importance of backing up speculative ideas about symbiotic systems, by matching them against properly calculated line and continuum fluxes. This requires ionization models, which abandon many of the simplifications employed up to now.

Acknowledgments: I thank H.M. Schmid and M. Vogel for discussions and help.

118

References:

Allen, D.A.: 1982, *The Nature of Symbiotic Stars*, IAU Coll. No.30, ed M. Friedjung and R. Viotti, D. Reidel Dordrecht, p.27

Barlow, M.J.: 1983, IUE Symp.103, *Planetary Nebulae*, ed. D.R. Flower, p 105, Reidel (Dordrecht)

Feast, M.W., Whitelock, P.A., Catchpole R.M., Roberts, G., Carter, B.S.: 1983, *Monthly Notices Roy.Astron.Soc.* **202**, 951

Hayes, M.A., Nussbaumer, H.: 1986, *Astron.Astrophys.* **161**, 287

Kafatos, M., Michalitsianos, A.G., Hollis, J.M.: 1986, *Astrophys.J.Suppl.* **62**, 853

Kenyon, S.J., Webbink, R.F.: 1984, *Astrophys.J.* **279**, 252

Livio, M., Soker, N., de Kool, M., Savonije, G.J.: 1986, *Monthly Notices Roy.Astron.Soc.* **222**, 235

Lynden–Bell, D., Pringle, J.E.: 1974, *Monthly Notices Roy.Astron.Soc.* **168**, 603

Nussbaumer, H., Schild, H.: 1979, *Astron.Astrophys.* **75**, L19

Nussbaumer, H., Schild, H.: 1981, *Astron.Astrophys.* **101**, 118

Nussbaumer, H., Schild, H., Schmid, H.M., Vogel, M.: 1987, *Astron.Astrophys.* (to be submitted)

Nussbaumer, H., Schmutz, W.: 1983, *Astron.Astrophys.* **126**, 59

Nussbaumer, H., Stencel, R.E.: 1987, *Exploring the Universe with IUE*, ed. Y. Kondo et al., Reidel Dordrecht, p.203

Nussbaumer, H., Vogel, M.: 1987, *Astron.Astrophys.* **182**, 51

Paczynski, B., Rudak, B.: 1980, *Astron.Astrophys.* **82**, 349

Penston, M.V., Benvenuti, P., Cassatella, A., Heck, A., Selvelli, P., Machetto, F., Ponz, D., Jordan, C., Cramer, N., Rufener, F., Manfroid, J.: 1983, *Monthly Notices Roy.Astron.Soc.* **202**, 833

Pringle, J.E.: 1977, *Monthly Notices Roy.Astron.Soc.* **178**, 195

Pringle, J.E.: 1981, *Ann.Rev.Astron.Astrophys.* **19**, 137

Regev, O.: 1983, *Astron.Astrophys.* **126**, 146

Shakura,, N.I., Sunyaev, R.A.: 1973, *Astron.Astrophys.* **24**, 337

Taylor, A.R., Seaquist, E.R.: 1984, *Astrophys.J.* **286**, 263

Viotti, R., Piro, L., Friedjung, M., Cassatella, A.: 1984, *Astrophys.J.* **319**, L7

Vogel, M.: 1988, *this volume*

Willson, L.A., Wallerstein, G., Brugel, E.W., Stencel, R.E.: 1984, *Astron.Astrophys.* **133**, 154

Photoionization models with accretion discs

M. Vogel
Institute of Astronomy
ETH Zentrum
8092 Zürich (Switzerland)

ABSTRACT. The diagnositic possibilities for identifying the ionizing source in symbiotic systems are explored. As possible sources we consider hot blackbodies and accretion discs. It turns out that main sequence accretors and hot blackbodies may have the same appearance in both, emission line and continuum flux distribution. However, UV continuum indices of models containing an accretion disc around a white dwarf are confined to a very small region, separated from main sequence accretors and blackbodies. Furthermore, if symbiotic systems containing a white dwarf accretor exist, they might be recognizable by strong emission in Fe X $\lambda6374$.

INTRODUCTION, ASSUMPTIONS AND CONCLUSIONS

We investigate the possibilities of nebular diagnostics as a tool for distinguishing blackbody sources from accretion discs. For this purpose we compare calculated nebular spectra (e.g. Nussbaumer and Vogel 1987), where the ionizing sources are blackbodies of various temperatures, and accretion discs simplified in the following way :

(i) The contributions from the hot spot and from the accreting star can be neglected.
(ii) The accretion disc is treated according to the standard disc theory as summarized for example by Shakura and Sunyaev (1973) and Pringle (1981). Thus, we assume a steady, geometrically thin and optically thick disc, where every surface element radiates as a blackbody with a temperature given by the energy dissipation rate.
(iii) Since up to now no rigorous theory for the boundary layer is available, we introduce a free parameter f for defining a boundary layer blackbody temperature T_{BL}.

$$L_{BL} = L_{disc} = \frac{1}{2}L_{acc} = \frac{GM\dot{M}}{2R^*} = 2\,(2\pi R^* b)\,\sigma T_{BL}^4 = (4\pi R^{*2})\,f\,\sigma T_{BL}^4 \quad,$$

where M and R^* are the mass and the radius of the accretor and $\dot{M}$ is the mass accretion rate. The parameter f represents the ratio of the boundary layer aera to the surface of the accreting star. All other parameters adopted for our calculations are the same as used by Kenyon and Webbink (1984) in their study on the nature of the hot component in symbiotic stars.

J. Mikolajewska et al. (eds.), The Symbiotic Phenomenon, 119–121.

Since the nebular density distribution around an accreting object is unknown, we disregard the angular dependence of the ionizing radiation and assume for simplicity that all photons from the disc and the boundary layer are emitted spherically symmetrically and that they cross a nebula of constant density with $N(H) = 10^6 \mathrm{cm}^{-3}$ and cosmic abundances.

Due to the energetic photons emitted from the boundary layer, accretion discs around white dwarfs can be distinguished from hot stellar sources by increased fluxes from the highly ionized ions relative to the low and medium ionized ones. Unfortunately, for most of these lines the fluxes are either not strong enough for an easy detection or they lie shortward of 912 Å and are affected by nebular absorption. The multiplets O VI $\lambda\lambda1032,1038$ and Ne VI $\lambda\lambda988,1001$, which would reveal the presence of a hot boundary layer, are not accessible to IUE. From the observable strong emission lines the O IV multiplet at $\lambda1400$ and the red coronal line Fe X $\lambda6374$ are the most suitable lines for detecting boundary layer photons. As an example we show in Fig.1 the flux of Fe X $\lambda6374$ and Fe XIV $\lambda5303$ relative to H_β.

In the case of an accretion disc around a main sequence star, most of the ionizing photons are provided by the boundary layer. The disc photons contribute an additional flux to the UV continuum, and can therefore best be distinguished from hot stellar sources by the slope of the UV continuum. Kenyon and Webbink (1984) propose reddening-free colour indices C_1 and C_2 for this task.

Figure 1.
Logarithmic line flux ratios
Fe X $\lambda6374$ / H_β (a) and
Fe XIV $\lambda5303$ / H_β (b) as
functions of the accretion
rate for white dwarf (heavy
lines) and main sequence
accretors (dashed lines),
and as a function of the
effective temperature for
hot blackbodies (thin solid
line). The trajectories of
the blackbodies and the
$f = 0.001$ main sequence
accretors are only close to
each other due to the
scaling of T^* on the $\dot{M}$
axis.

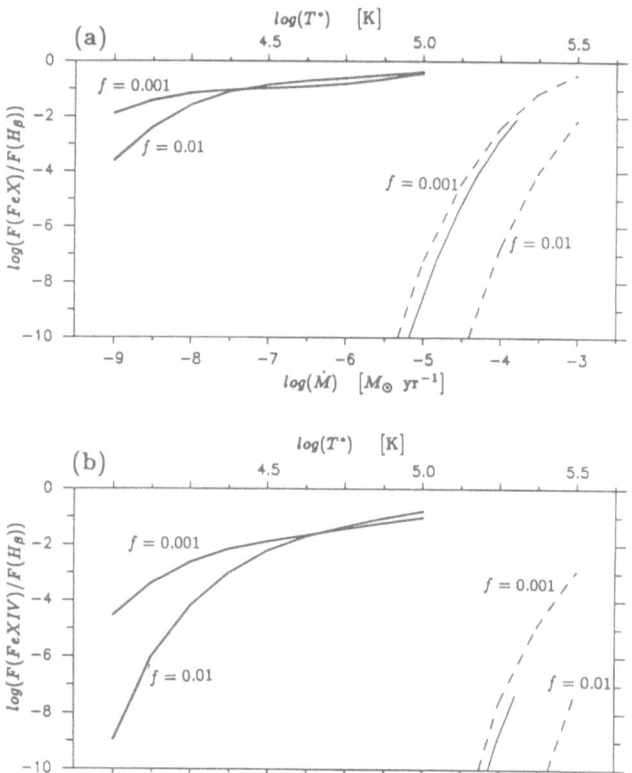

Figure 2.
Colour indices C_1 and C_2 for white dwarf (heavy line) and main sequence accretors (dashed lines), and for hot blackbodies (thin solid line).

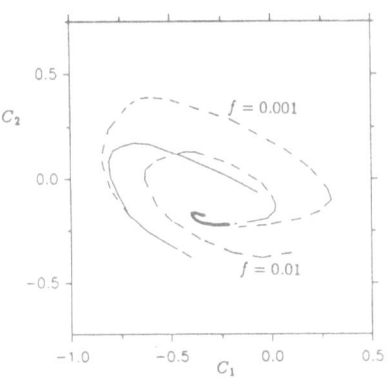

For main sequence accretors and for the hot stellar objects we have reproduced the colour indices of Kenyon and Webbink (1984). For the white dwarf accretors however, our results are different. This is particularly interesting because they found no correlation between their theoretical colour indices for white dwarf accretors and observations of symbiotic stars. Kenyon and Webbink (1984) pointed out that the reddening–free colour indices are not very sensitive to the adopted parameters. This is true for hot stellar sources and white dwarf accretors. However, for main sequence accretors the colour indices depend on the assumed boundary layer parameter f. Therefore, some symbiotics, identified as hot stellar sources, might nevertheless belong to the group of main sequence accretors. In Fig.2 we show our trajectories in the reddening–free colour–colour plot. Kenyon and Webbink used an estimate by Lynden–Bell and Pringle (1974) for the boundary layer temperature which is equivalent to $f = 0.01$, for all M, R^* and $\dot{M}$.

Although a main sequence accretor has relatively more soft UV continuum than that of a blackbody or a white dwarf accretor, distinguishing them observationally may still be impossible because of interstellar reddening. Another reddening independent criterion would be the Balmer jump measurement as shown and defined in Fig.3, where F^- and F^+ is the continuum flux short– and longwards of the Balmer jump. For some symbiotics, the practical determination of $\mathcal{B}$ may not be easy. Furthermore, in some systems the cool component may already contribute some radiation to that wavelength region.

Figure 3.
The behaviour of the Balmer jump index $\mathcal{B}$ as a function of the accretion rate for white dwarf (heavy lines) and main sequence accretors (dashed lines), and as a function of the effective temperature for hot blackbodies (thin solid line).

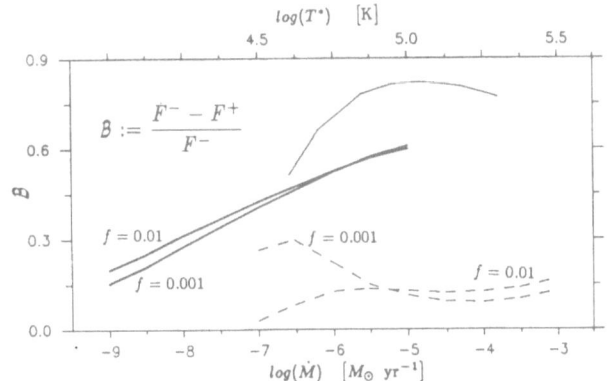

REFERENCES

Kenyon, S.J., Webbink, R.F.: 1984, Astrophys.J. **279**, 252
Lynden–Bell, D., Pringle, J.E.: 1974, Monthly Notices Roy. astron. Soc. **168**, 603
Nussbaumer, H., Vogel, M.: 1987, Astron. Astrophys. **182**, 51
Pringle, J.E.: 1981, Ann. Rev. Astron. Astrophys. **19**, 137
Shakura, N.I., Sunyaev, R.A.: 1973, Astron. Astrophys. **24**, 337

EMISSION LINE RATIO CLASSIFICATION OF SYMBIOTIC STARS.

Hugo E. Schwarz
European Southern Observatory
Casilla 19001
Santiago 19
Chile

ABSTRACT. A 2-dimensional classification system for symbiotic stars based on the 5007/4861 and 3727/5007 line ratios adapted from the method of Baldwin et al.(1981) is presented. It is shown that a simple measurement suffices to classify stars and no reddening has to be taken into account. Symbiotics fall into 2 distinct classes in the line ratio plot. The dusty and S types. Neither type coincides with either the planetary nebulae or the HII region locus. The highest excitation is found for the dusty types. The two dimensional distribution can be interpreted as evolutionary, the more evolved Mira containing symbiotics lying closer to the PN locus. The fact that there are few D, D' type symbiotics would indicate that the phase between Mira and PN is short.

1. INTRODUCTION

Symbiotic stars are generally considered to consist of a cool (super)giant primary and a white dwarf or main sequence star companion. Both stars are embedded in a common envelope of hot, partially ionised gas and or dust. Their spectra are dominated by strong emission lines (both permitted and forbidden) of H, He, O, N and other elements and usually show a red continuum with TiO absorption bands. Webster and Allen(1975) have classsified symbiotics into three types on the basis of IR photometry. S types show a normal photospheric continuum while D and D' types show respectively a hot and cool dust excess in the IR. A cataloque of about 140 symbiotics has been published by Allen(1984) which also contains the optical spectra of about 100 of these stars. Our knowledge of these stars is far from complete and any classification scheme which can shed some light on the symbiotic phenomenon is useful. Here such a scheme is presented.

J. Mikolajewska et al. (eds.), The Symbiotic Phenomenon, 123–125.

2. EMISSION LINE CLASSIFICATION.

A two dimensional classification of emission line objects has been published by Baldwin, Phillips and Terlevich(1981) (BPT). In their plot, the PN and HII regions occupy two well defined and distinct loci.

In a recent paper, Gutierrez-Moreno et al.(1986)(GMMC) apply the scheme of BPT to two symbiotics. The stars fall between the HII region and PN loci. Since a number of PN are inside the HII region locus, the significance of the position of the two stars is not clear. A larger sample of symbiotics would indicate if these stars occupy a special place in the BPT diagram and if different types of symbiotic star occupy different loci.

In this paper a large sample of symbiotics is used in a simplyfied version of the BPT scheme and it is shown that the emission line classification separates the dusty (D, D') types from the S types.

3. DATA AND RESULTS.

The spectra used for this work are mainly selected from those published by Allen(1984), the two from GMMC and for some 15 stars IDS spectra were obtained by the author at the 1.5m ESO telescope on La Silla, Chile.

Since for most of the stars only the 3727, 4861 and 5007 line fluxes were available, the full BPT scheme could not be used since it needs the 6300, 6563 and 6584 lines too. For a larger proportion of the stars, the 6300 and 6584 lines were available and on the basis of these ratios, it was determined that the 5007/4861 ratio is the most sensitive parameter. It was therefore decided to use this ratio in the BPT diagram in addition to the usual 3 ratio plot which uses the 5007/4861, the 6300/6563 and the 6584/6563 ratios.

Reddening has not been taken into account since the sensitivity to its effects is very low. The effect of de-reddening would be to shift stars to the top right hand corner of the diagram.

In Figure 1 the adapted BPT diagram for 92 stars is shown. Of the 16 D and D' types 1 lies in, 11 lie close to and 4 lie well away from the PN region. There is a clear separation into S types and D, D' types. The 2 stars of GMM are included without their reddening correction.

The full BPT diagram shows the same trend with 13 stars of which the 8 D, D' types are all near the PN locus.

The separation of D, D' types from S types towards the PN locus could indicate an evolutionary trend in the diagram. This would mean that S types evolve into D, D' types and eventually become PN. The scarcity of dusty types indicates a rapid stage of evolution, with D' types perhaps as a "late" D type.

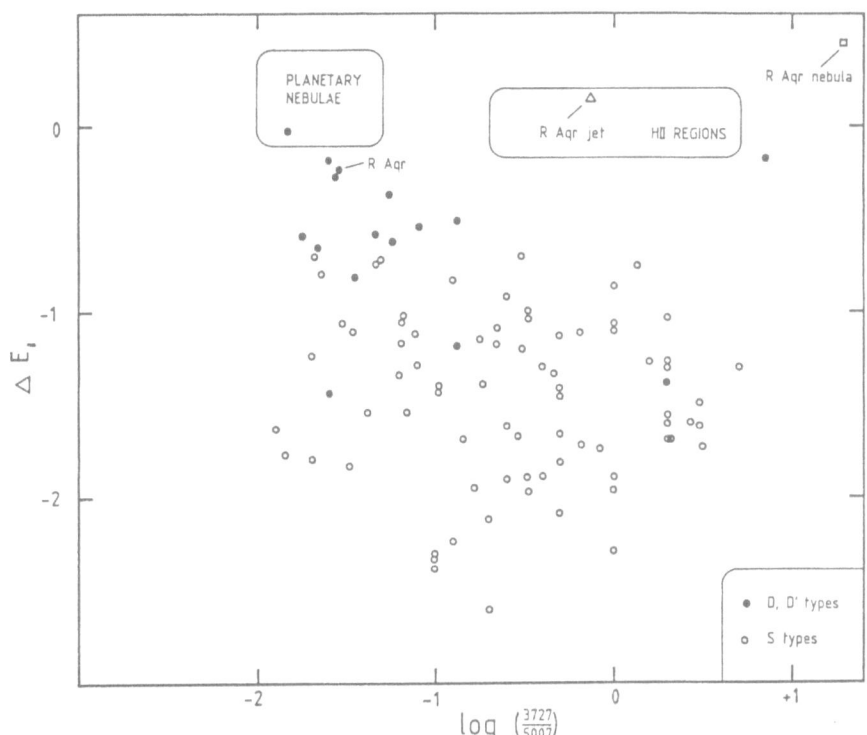

Figure 1   Plot of ΔE_1 (see BPT) against log 3727/5007. Clearly, the
S and D types are separated with the D types nearer the PN
locus.

This scheme would be the symbiotic equivalent of that for
"normal" stars: through a Mira and a short OH/IR stage
stars on the upper AGB become PN.   A search for OH/IR
symbiotics might prove interesting. Also, are there such
things as symbiotic PN?

ACKNOWLEDGEMENTS.

My thanks to J.W.Pel,   W.Seitter   and   D.A.Allen   for
interesting discussions and to Prof.L.Woltjer for his
support.

REFERENCES.

Allen,D.A. (1984) Proc.ASA, 5,369
Baldwin,J.A.,Phillips,M.M.,Terlevich,R. (1981) PASP, 93,5
Gutierrez-Moreno,A.,Moreno,H.,Cortes,C. (1986) AA, 166,143
Webster,B.L.,Allen,D.A. (1975) MN, 171,171

RESONANCE LINE PROFILES FROM RADIAL ACCRETION FLOWS

R. Tylenda
Laboratory for Astrophysics
Copernicus Astronomical Centre
Chopina 12/18
87-100 Torun, Poland

A compact star in a detached binary system can accrete the matter from the stellar wind of the companion. In this case a more or less radial accretion flow is formed. By analogy to stellar winds from early type stars it is usually believed that the accretion of this sort should produce inverse P-Cygni profiles in resonance lines. However, there are physical differences between the wind and the radial accretion which can alter the outgoing profile significantly.

The matter outflowing from an early type star cools off very fast due to the adiabatic expansion. However, it remains highly ionized as the quickly decreasing density does not allow it to recombine. Therefore the principal mechanism for the resonance line formation here is the scattering of the stellar continuum photons in the wind.

The situation in the accretion flow is qualitatively different. The free-falling gas is compressed which should lead to its heating. More importantly, just above the surface of the accreting star a stand-off shock is formed which heats up the matter to very high temperatures. The UV and X radiation emitted by the shocked gas is then absorbed in the free-falling matter causing its further heating and ionization. Consequently it is expected that the accretion flow itself is a strong source of the resonance-line photons due to collisional excitation of corresponding transitions. Subsequent scattering of these photons in the accretion flow determines the outgoing line profile. The stellar continuum scattering is here less important. The calculations on which we briefly report in this paper show that in these conditions the outgoing profile is mainly constituted of a strong emission component. The absorption feature, if present, is very shallow.

The transfer of resonance-line photons in spherically symmetric accretion flow has been studied using a Monte Carlo scattering code. The principles of the method are the same as those used for similar problems in stellar winds (e.g. Natta and Beckwith, 1986). In an accretion flow the velocity varies as $(1/r)^{1/2}$ while the density is proportional to $(1/r)^{3/2}$. The interdependance between the scattering of the stellar continuum photons and the transfer of the photons originating in the accretion flow has been specified by a parameter P_{env}. It is defined as

$$P_{env} = F_1/(F_1 + F_\odot)$$

J. Mikolajewska et al. (eds.), The Symbiotic Phenomenon, 127–128.
© *1988 by Kluwer Academic Publishers.*

128

where F_1 is the photon flux emitted in the accretion envelope while F_0
denotes the flux of the stellar continuum photons which can interact
with the accretion flow. The local emissivity of the line photons in the
accretion flow is proportional to the square of the density.

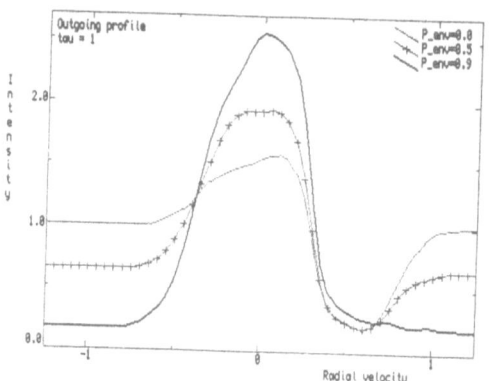

Fig. 1. The resonance line profiles from radial accretion flows with
different values of P_{env}. The radial velocity on the abscissa is
expressed in terms of the free-fall velocity at the stellar surface.

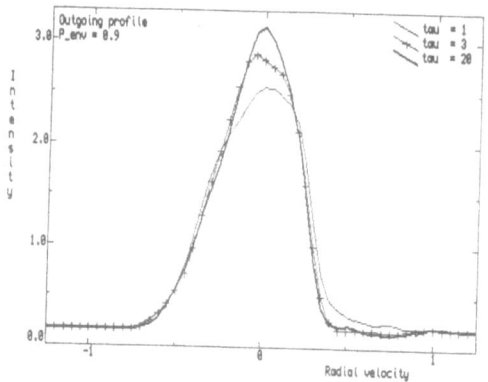

Fig. 2. The line profiles from flows of different optical thickness.

Fig. 1 compares the outgoing line profiles for different values of
P_{env}. Note increase of the emission component and disappearance of the
absorption feature as P_{env} approaches 1, i.e. when the photon emission
from the accretion flow more and more dominates the stellar continuum
scattering. As explained above the latter is expected to occur in real
radial accretion flows. In this case the absorption feature does not
appear even for large optical thicknesses as illustrated in Fig. 2.

REFERENCES

Natta, A., Beckwith, S., 1986, Astron. Astrophys. 158, 310.

COLLIDING WINDS IN SYMBIOTIC SYSTEMS

Sun Kwok
Department of Physics
University of Calgary
Calgary, Alberta
Canada T2N 1N4

ABSTRACT. The physics of colliding winds in symbiotic systems is reviewed. The theoretical predictions are compared with observational data of symbiotic novae, in particular the recently erupted system HM Sge. It is suggested that the spectral behaviour of HM Sge from X-ray to radio can be explained by the colliding winds process.

1. INTRODUCTION

One of the most interesting aspects of symbiotic stars is the extreme difference in spectral types between the two components of the system and symbiotic novae represent the most extreme case among symbiotic stars. It is now generally believed that a symbiotic nova system consists of a Mira Variable of effective temperature (T_e) <3000K and a degenerate white dwarf (probably a C-O white dwarf) of T_e>100000 K. While the combination of the large extent of the atmosphere of the cool component and the strong gravity of the hot component suggests the possibility of interaction, in fact the symbiotic novae systems have very large binary separations and very long periods and belong to an evolution scenario more extreme than the Case C class of Paczyński (1980). If the outbursts observed in these systems are the result of binary interaction, then stellar winds must play an important role.
    Some symbiotic stars are known to contain Mira Variables, which as a class is believed to have large mass loss rates of >10^{-6} $M_\odot$ yr^{-1}. Other symbiotic stars have outbursts similar to that of classical novae which undergo ejections with velocities >1000 km s^{-1}. If both of these physical processes take place within a single system, then the interaction between the two ejecta is inevitable, resulting in physical consequences which manifest in many parts of the electromagnetic spectrum. In this review, we will examine the physics of wind interactions and compare the theoretical predictions with observations.

2. SYMBIOTIC NOVAE

The class of objects called symbiotic novae is defined by Allen (1980)

J. Mikolajewska et al. (eds.), The Symbiotic Phenomenon, 129–136.
© 1988 by Kluwer Academic Publishers.

Among the seven (or eight if PU Vul is included) objects in this class, HM Sge is the one example which has the largest amount of multi-frequency data available. HM Sge has been detected in almost every spectral band from X-ray to radio (Kwok 1982) and has been monitored extensively since its outburst in 1975. Although one can argue that every symbiotic object is unique and it is unfair to concentrate on one object as an representative of the whole class, yet the wealth of data for HM Sge makes it attractive to use this object as a proto-type for the physical phenomenon of colliding winds.

2.1 The cool component

The detection of photospheric absorption features of CO and H_2O suggests that the cool component in HM Sge is a Mira Variable (Puetter *et al.* 1978), and this is confirmed by the Mira-like variations due to atmospheric pulsation (Ipatov, Taranova and Yudin 1985). A period of 540 days has been determined by Whitelock (this volume), suggesting that it is a very evolved star on the asymptotic giant branch (AGB). Figure 1 shows the 8-100 μm infrared spectrum of HM Sge as observed by the *IRAS* satellite. The strong emission features at 10 and 18 μm are identical

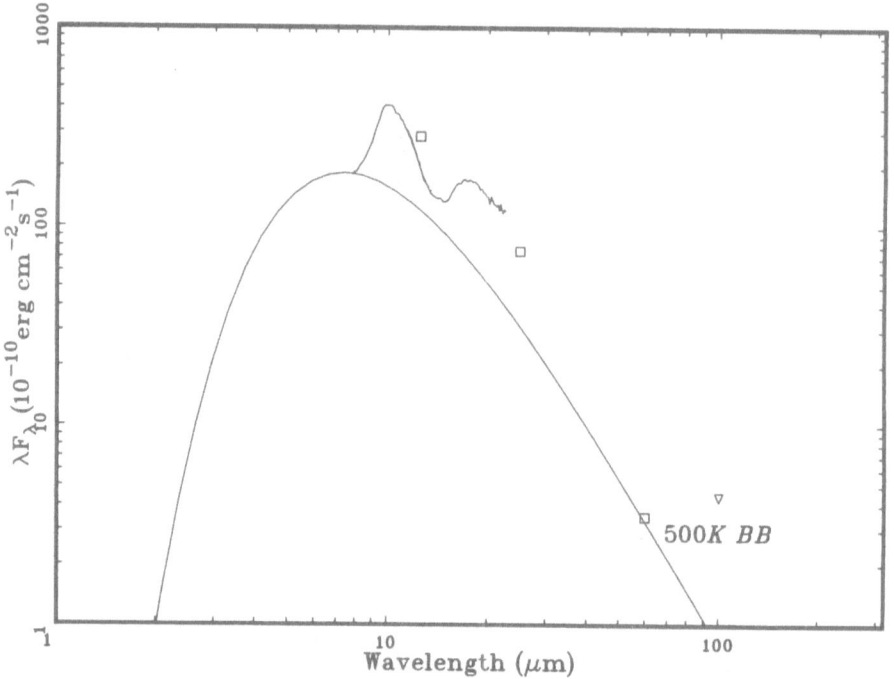

Fig. 1 The IRAS spectrum of HM Sge. The 8-23 μm spectrum is obtained by the Low Resolution Spectrometer and has been normalized to the 12 μm photometry point. The four photometric measurements have been corrected for colour. A 500 K blackbody is also shown for comparison.

to the circumstellar silicate dust features observed in >2000 AGB stars observed by *IRAS* (Volk and Kwok 1987). This suggests that HM Sge must have a very extended circumstellar envelope which has not been severely disturbed by the outburst event in 1975.

2.2 The hot component

Even in the ultraviolet, the continuum emission is dominated by the nebular component and the existence of the hot component is only inferred from the effect of its ionizing radiation on the circumstellar region. Analysis of the emission lines implies a hot-star temperature of 50,000K after the outburst and a temperature of ~160,000 K now (Stauffer 1984; Mueller and Nussbaumer 1985). The most interesting aspect of the hot star is the detection of a Wolf-Rayet feature suggesting an expansion velocity of >2000 km s^{-1}. The presence of the WR feature is similar to those observed in central stars of planetary nebulae, which are now commonly observed by *IUE* to have high-velocity mass outflows. Unfortunately the WR phase in HM Sge ended before *IUE* spectra could be obtained.

2.3 The nebular component

The detection of thermal radio emission implies that the ionized nebula extends over a radius >10^{15} cm. The radio spectrum is optically thick to at least 22 GHz, and as in the case of the optically thick phase of classical novae, the flux level has been steadily increasing at all wavelengths. In contrast to classical novae, which optically thick phase only lasts a few months, HM Sge has remained optically thick for over 10 years! If we approximate this behaviour by an expanding blackbody then the expansion velocity of the radio-emitting region can be estimated to be ~56 (D/kpc) km s^{-1} (Kwok, Bignell and Purton 1984). Figure 2 shows the radio light curves of HM Sge as the result of monitoring programs at the Algonquin Radio Observatory and the Very Large Array since 1977. Wallerstein (1978) was the first to note the existence of multi-emission components in HM Sge.

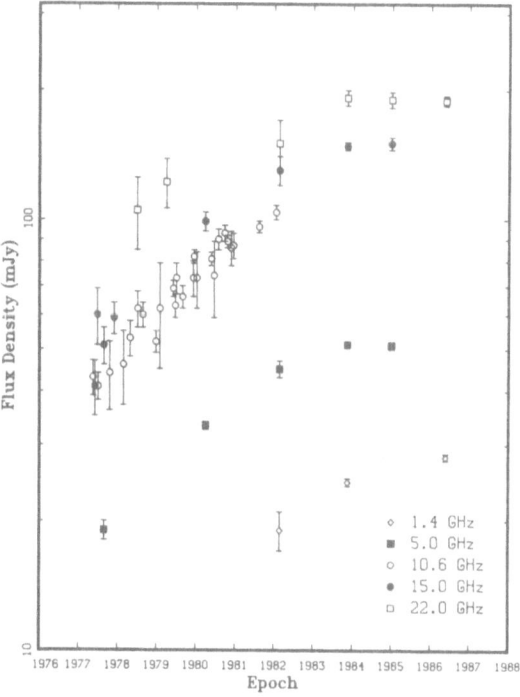

Fig. 2 Radio light curves of HM Sge.

He found that while permitted lines like Hα and HeI have broad wings indicative of expansion velocities of 1700 km s⁻¹, [OI] and [SIII] show velocities of 75 km s⁻¹ and [NII] is very narrow with a velocity of 20 km s⁻¹.

3. THE INTERACTING WINDS MODEL

Wallerstein's observations led to the proposal of the interacting stellar winds model for HM Sge (Kwok and Purton 1979). In this model, the permitted lines (Hα and HeI) are suggested to arise from a high-speed wind from the hot star and low-critical-density lines ([NII]) are emitted from the wind of the Mira. If the hot-star wind began at the time of the outburst, then the interaction of this wind with the pre-existing Mira wind will lead to the formation of a high-density shell from which the intermediate-critical-density lines (e.g. [SIII]) are emitted. Mass of the shell will increase with time as more of the Mira wind material is swept up by the hot-star wind.

3.1 The dynamical equations

Assuming that the shell is made up of mostly swept-up cool-star material, mass loss from the hot star is steady, and the interaction between the hot-star wind and the shell is adiabatic, then the equations of continuity, motion and the conservation of energy can be solved by similarity analysis yielding the following solutions (Kwok 1986):

$$R_s = V_s t \tag{1}$$

$$M_s = \dot{M} (V_s/V - 1)t \tag{2}$$

$$P = \frac{\tfrac{1}{2} \dot{m} v^2}{6 \pi V_s^3} t^{-2} \tag{3}$$

where M_s and V_s are the mass and velocity of the shell, $\dot{M}$ and V are the mass loss rate and velocity of the Mira wind, $\dot{m}$ and v are the mass loss rate and velocity of the hot-star wind, and P is the pressure in the shocked region which is responsible for pushing the shell. The expansion velocity V_s is given by the root of the following equation:

$$(\dot{M}/V)V_s^3 - 2\dot{M}V_s^2 + \dot{M}VV_s = \tfrac{1}{3} \dot{m}v^2 \tag{4}$$

3.2 Thermal structure of the wind interaction region

Applying the jump conditions for a strong adiabatic shock, one finds that the location of the inner shock (R_{in}) is given by:

$$R_{in} = 1.5 (V_s/v)^{\tfrac{1}{2}} R_s \tag{5}$$

For $V_s \sim 100$ km s⁻¹ and $v \sim 2000$ km s⁻¹, $R_{in} \sim 0.34 R_s$, or approximately 96% of the volume inside R_s is shocked. Using the ideal gas law and assuming that the shocked region is uniform in density we have

$$T = \frac{\mu \ m_H \ v^2 \epsilon}{9 \ k} \tag{6}$$

where μ (~0.6) is the mean atomic weight, k is Boltzmann constant, and ϵ is the filling factor. With the above parameters, the shocked region is found to have a temperature of ~10^7K (Kwok and Leahy 1984).

X-ray emission is expected at such high temperatures and HM Sge is indeed found to be an X-ray source by the *Einstein* Satellite (Allen 1981). Willson *et al.* (1984) suggest that the X-ray emission originates from the head-on collision region of the two winds. If the interaction between the two winds is indeed adiabatic, then the analysis above suggests that the high-temperature shocked-region is not confined to a small volume but instead occupies a volume much larger than the binary separation. The analysis of the *Einstein* data by Kwok and Leahy (1984) finds that the observed X-ray flux is consistent with the expected emission from a shocked hot-star wind of mass loss rate of $\dot{m}$~10^{-7} $M_\odot$ yr^{-1}.

4. OBSERVATIONAL TESTS OF THE INTERACTING WINDS MODEL

4.1 Radio

The most obvious test of the predicted wind-shell structure of HM Sge is by direct imaging. Very high (~0.08 arc sec) resolution radio observations (Kwok, Bignell and Purton 1984) show a diffuse halo surrounding a central core of 0".15 in size in qualitative agreement with the model prediction. A more quantitative test would be to fit the multi-frequency radio light curves of Fig. 1 since measurements at different frequencies probe into different depths of the source and the optical depths are evolving with time. Using Wallerstein's (1978) velocities as input parameters, Purton, Kwok and Feldman (1983) were able to obtain reasonable fits to the light curves with the interacting winds model.

4.2 Infrared

While the presence of a circumstellar envelope is evident in the infrared spectrum (Fig. 1), it would be desirable to directly detect the extended far infrared emission. Figure 3 shows the 50 μm map of HM Sge

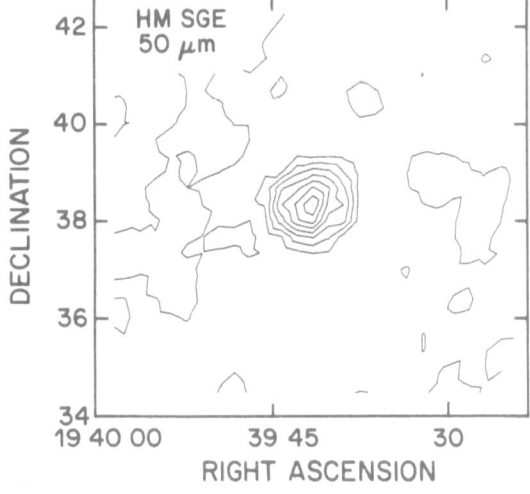

Fig. 3   IRAS CPC map of HM Sge.

obtained by the Chopped Photometric Channel instrument on the *IRAS* satellite. The infrared source is not resolved at the resolution of ~90 arc sec.

4.3 Optical

High-resolution optical spectroscopic observations by Stauffer (1984) have found double-peak profiles in the emission lines [SIII], [NII], [OIII], etc. which suggest that these emissions originate in a shell expanding at ~42 km s^{-1} (32 km s^{-1} in V1016 Cygni). The simultaneous presence of a narrow component (e.g. in [OI]) is interpreted as arising from the Mira wind. It is interesting to note that features with stronger narrow components have lower critical densities. In the interacting winds picture, one can then associate a densities of $10^{6.7}$ for the shell and 10^5 for the halo.

4.4 Profile asymmetries

Asymmetric emission line profiles have been noted in V1016 Cygni and HM Sge by Solf (1983; 1984), Stauffer (1984) and Wallerstein *et al.* (1984). A biconical model is proposed by Solf who finds that the ejections in both systems to be well collimated, with opening angles of 6° and 35° for HM Sge and V1016 Cygni respectively. Wallerstein *et al.* offer a "sphere minus cone" geometry, which could arise naturally from the collision of the two winds (Girard and Willson 1987).

5. ENERGY SOURCES

Figure 4 shows the energy distribution of HM Sge. It should be noted that these plotted observations were not taken at the same time, and in view of the variable nature of the object, only represent an approximate picture of the overall spectrum. Also not corrected for are the effects due to interstellar extinction. The *IUE* measurements are the continuum level as estimated by Feibelman (1982) and do not include the contribution from emission lines. The solid line represents the photospheric contribution as measured by Puetter *et al.* (1978). It is clear that most of the flux from HM Sge is emitted in the infrared, but the possibility of significant amount of flux escaping in the far ultraviolet cannot be excluded.

The distance and luminosity of HM Sge is still in controversy (Solf 1984, Kwok 1986). While the measured angular size and expansion velocity of the shell suggest a distance of ~400 pc, the total observed flux and the inferred luminosity of the Mira component would put its distance >1.4 kpc. We note that the LRS spectrum of HM Sge is among the best quality obtained by *IRAS* which also argues for a small distance.

6. DURATION OF THE HOT-STAR EJECTION

One of the assumptions used in §3 is that the system is in a steady state. However, there is evidence that the WR feature attributed to the hot-star wind has disappeared after 1980 (Feibelman 1982; Stauffer 1984)

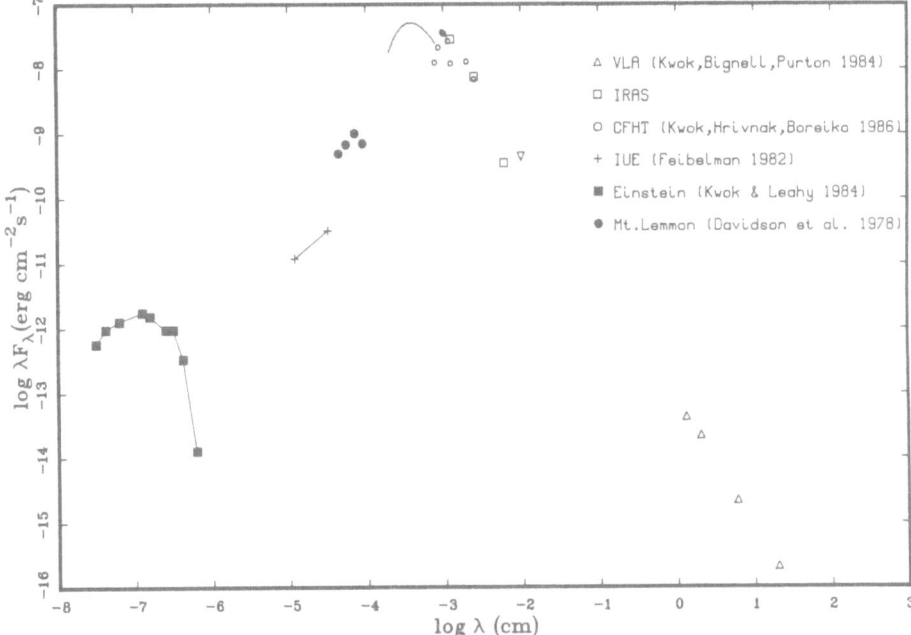

Fig. 4   Energy distribution of HM Sge.

and this may imply that the wind has been weakening with time and ceased completely in 1980.   This event was accompanied by the rise in HeII 4686 line and the increase in ultraviolet fluxes of CIV, NIV, NV, etc. (Mueller and Nussbaumer 1985).   If the ejection from the hot star has indeed ceased, then the high-density shell must be decelerating and such deceleration should be detectable by continued monitoring of the radio light curves as well as by observing the profile changes of the emission lines.

The reduction in ejection rate would also cause the shrinking of the pseudo-photosphere with the effect of increasing effective tempera-ture of the hot star.   There is evidence that the excitation state of the nebula has increased after 1980, with an inferred temperature of the hot star changing from ~50,000 K to 160,000 K (Stauffer 1984).

7.   EVOLUTIONARY SCENARIO

We can postulate the following evolutionary scenario for symbiotic novae.   The two stars in the system evolve independently as if they are single stars.- Sometime after the primary has gone through the planetary nebula phase and has become a white dwarf, the secondary star evolves up the AGB and enters the Mira phase.   Stellar wind from the secondary (now the more massive of the two) is accreted by the white dwarf.   After enough material is accumulated, H-shell burning is ignited and a fast wind begins to flow from the white dwarf.   The circumstellar material of the Mira acts an a barrier to this new wind and the interaction of the two results in the formation of a dense shell and a high-temperature bubble. The expansion of the shell leads to the observed radio brightening and the gradually diluting bubble is responsible for X-ray emission. It is likely that the visible brightening observed in HM Sge is not dominated by an increase in the level of continuum emission, but is the result of the dissipation of circumstellar dust and the subsequent ionization of the circumstellar envelope.

7. CONCLUSIONS

It has become apparent that the wind from the Mira component in symbiotic stars is responsible for the presence of nebular emission lines in many D-type symbiotics. Nova-like ejections in the hot component, even for a short period of time, introduce the possibility of colliding winds which can lead to many interesting observational consequences, including slow light curves and X-ray emissions. One may even speculate that some symbiotic stars (e.g. H1-36, AG Peg) had a history of colliding winds but have now become quiescent. This process therefore may be more common than presently realized and deserves more investigation in our quest for the understanding of the symbiotic phenomenon.

REFERENCES

Allen, D.A. 1980, *Mon. Not. Roy. Astron. Soc.*, **912**, 521.
Allen, D.A. 1981, *Mon. Not. Roy. Astron. Soc.*, **197**, 739.
Feibelman, W. A. 1982, *Astrophys. J.*, **238**, 548.
Girard, T., and Willson, L.A. 1987, *Astron. Astrophys.*, in press.
Ipatov, A.P., Taranova, O.G., and Yudin, B.F. 1985, *Astron. Astrophys.*, **142**, 85.
Kwok, S. 1982, in *The Nature of Symbiotic Stars*, eds. M. Friedjung and R. Viotti, p. 209.
Kwok, S. 1986, in *Highlights of Astronomy*, ed. J.P. Swings, p. 189.
Kwok, S. and Purton, C.R. 1979, *Astrophys. J.*, **229**, 187.
Kwok, S. and Leahy, D.A. 1984, *Astrophys. J.*, **283**, 675.
Kwok, S., Bignell, R.C., and Purton, C.R. 1984, *Astrophys. J.*, **279**, 188.
Kwok, S., Hrivnak, B.J., and Boreiko, R.T. 1986, unpublished data.
Mueller, B.E.A., and Nussbaumer, H. 1985, *Astron. Astrophys.*, **145**, 144.
Paczyński, B. 1980, *Highlights of Astronomy*, ed. P.A. Wayman, **5**, 27.
Puetter, R.C., Russell, R.W., Soifer, B.T., and Willner, S.P. 1978, *Astrophys. J. (Letters)*, **223**, L93.
Purton, C.R., Kwok, S., and Feldman, P.A. 1983, *Astron. J.*, **88**, 1825.
Solf, J. 1983, *Astrophys. J. (Letters)*, **266**, L113.
Solf, J. 1984, *Astron. Astrophys.*, **139**, 296.
Stauffer, J.R. 1984, *Astrophys. J.*, **280**, 695.
Volk, K., and Kwok, S. 1987, *Astrophys. J.*, **315**, 654.
Wallerstein, G. 1978, *Publ. Astron. Soc. Pacific*, **90**, 36.
Wallerstein, G., Willson, L.A., Salzer, J. and Brugel, E. 1984, *Astron. Astrophys.*, **133**, 137.
Willson, L.A., Wallerstein, G., Brugel, E.W., and Stencel, R.E. 1984, *Astron. Astrophys.*, **133**, 154.

ACCRETION DISKS IN SYMBIOTIC STARS.

Wolfgang J. DUSCHL,
Max-Planck-Institute for Astrophysics, Karl-Schwarzschild-Strasse 1,
D-8046 Garching bei München, West Germany.

ABSTRACT.

We give an overview over the theory of geometrically thin α-accretion disks; further we introduce the two different proposed mechanisms that can cause outbursts of accretion disks; and finally we compare the results of these models applied to symbiotic stars (=SS).

1. INTRODUCTION.

Accretion disks were *invented* almost a quarter of a millennium ago: In 1755 Immanuel Kant explained the origin of the solar system by a model that we today would call an accretion disk model. The main principles behind accretion disks, namely the inward transport of matter with outward transport of angular momentum, were formulated by von Weizsäcker (1943) and, especially, Lüst (1952). It took another two decades until accretion disk physics made the next major step forward: in 1973 Shakura and Sunyaev introduced the so-called α-accretion disks. We shall follow the line of this model and its approximations in presenting the basic physics of accretion disks in Sect. 2.

While this ansatz was successful in describing the overall behaviour of accretion disks, especially in dwarf nova systems, there remained the problem of the outbursts or active phases of these systems. As early as 1972 and 1974, Bath, and Osaki, resp., proposed two different models. While in Bath's description the outburst takes place in the envelope of the companion star (an instability there leads to a strong increase in the mass transfer rate; what we observe is this material being processed through the disk), in Osaki's picture an essentially constant mass transfer feeds an unstable accretion disk (matter first stored in the outer regions of the disk, is released by a – then unknown – instability and processed through the disk). Meyer and Meyer-Hofmeister (1981) found an instability with the properties required by Osaki's model. In Sect. 3 the different possible outburst mechanisms are discussed.

The application of accretion disks to and their outburst behaviour in SS was analyzed by several authors (Bath, 1977; Bath and Pringle, 1982; Plavec, 1982; Kenyon and Webbink, 1984; Kenyon, 1985, 1986; Duschl, 1985a, 1986a, 1986b; among others). We are discussing SS with a characteristic behaviour like CI Cyg, Z And, or AR Pav. In Sect. 4 we shall describe the present situation and compare the different models for SS. An independent indication that accretion disks exist in SS comes from the observations of for instance bipolar outflow (Solf, 1984).

2. ACCRETION DISKS ...

In this Section we give a concise overview over accretion disk physics; for a more detailed treat-

J. Mikolajewska et al. (eds.), The Symbiotic Phenomenon, 137–148.

138

ment, and for original references we refer to Frank *et al.* (1985).

We treat the accretion disk in a cylindrical coordinate system $\{s, z, \varphi\}$, and introduce a radial distance $r = \sqrt{s^2 + z^2}$. The structure and evolution is described by the continuity equation, the Navier-Stokes-equations, and equations for the energy transport. In the following we first present these basic equations and introduce the relevant quantities (2.1.); then we discuss the commonly used approximations (*thin α-disk theory*) (2.2.); in the next paragraph (2.3.) we shall describe the set of resulting equations that are actually solved; and finally the results for stationary disks are presented (2.4.). If not stated otherwise, we use cgs-units.

2.1. Basic equations.

2.1.1. The continuity equation.

$$\frac{d\rho}{dt} + \rho \cdot \nabla \underline{v} = 0 ,$$ (1)

where ρ is the matter density, t the time, and $\underline{v}$ the velocity vector.

2.1.2. Navier-Stokes-equation.

$$\frac{d\underline{v}}{dt} = -\nabla \Phi - \frac{1}{\rho} \cdot \left(\nabla P + \nabla \tilde{\Theta} \right) ,$$ (2a)

P being the pressure, and $\tilde{\Theta}$ the viscous stress tensor (see e.g. Landau and Lifshitz, 1959). The gravitational potential Φ is that of a point mass, M:

$$\Phi = -\frac{G \cdot M}{r} .$$ (2b)

We assume the gas to be an ideal one (P_g: gas pressure), and take into account radiation (P_r). With the temperature T, and the molecular weight μ we get ($\Re$ is the gas constant.)

$$P = P_g + P_r = \frac{\rho \cdot \Re \cdot T}{\mu} + \frac{a}{3} \cdot T^4.$$ (2c)

2.1.3. The Energy Transport Equation.

$$\frac{d\varepsilon}{dt} = -\frac{P}{\rho} \cdot \nabla \underline{v} - \frac{1}{\rho} \cdot \left(\left(\tilde{\Theta} \nabla \right) \underline{v} - \nabla \underline{q} \right) .$$ (3a)

ε is the specific internal energy, and $\underline{q}$ the heat flux vector. After Baker and Kippenhahn (1962) we take ε to be

$$\varepsilon = \frac{3}{2} \cdot \frac{P}{\rho} \cdot (2 - \beta) ,$$ (3b)

with β being the gas pressure in units of the total pressure:

$$\beta = \frac{P_g}{P} .$$ (3c)

The heat flux , $\underline{q}$, consists of a radiative ($\underline{q}_r$) and a convective ($\underline{q}_c$) part:

$$\underline{q} = \underline{q}_r + \underline{q}_c .$$ (3d)

For the radiative part we use the diffusion approximation, and for the convective contribution the mixing length formalism as described by Kippenhahn *et al.* (1967).

2.2. Approximations.

2.2.1. Symmetry.
We assume the accretion disk to be symmetric in azimuth, i.e. in the φ-direction, and in the z-direction.

2.2.2. Mass of the disk.
The mass of the accretion disk shall be negligible compared to the central mass (M); only this justifies a) taking (2b) for the potential, and b) neglecting selfgravitation in the vertical direction.

2.2.3. Velocities in the disk.
Velocities in the vertical, i.e. z-direction are neglected; which is equivalent to assuming that equilibrium in this direction is always reached on time scales shorter than the ones we are interested in.

The influence of pressure on the horizontal disk structure shall be very small compared to that of gravity:

$$\frac{\partial P}{\partial s} \ll \rho \cdot \frac{\partial \Phi}{\partial s} \; ; \tag{4a}$$

this means $v_s \ll v_\varphi$. We further assume that the velocities do not vary in the z-direction, i.e.

$$\frac{\partial v_s}{\partial z} = \frac{\partial v_\varphi}{\partial z} = 0 \; . \tag{4b}$$

2.2.4. Surface density, vertical scale height.
The time dependent evolution of the disk will be evaluated only for the equations integrated in vertical direction, and a surface density, Σ, is introduced:

$$\Sigma = \int_{-\infty}^{\infty} \rho \, dz \; . \tag{5a}$$

Futher a vertical scale height, h, is defined, where $\bar{\rho}$ is the density averaged in z-direction:

$$h = \frac{\Sigma}{2 \cdot \bar{\rho}} \; . \tag{5b}$$

2.2.5. Time scales.
Lightman (1974) has shown that in geometrically thin accretion disks the longest time scale is that over which variations due to viscosity occur. Only processes which run on this time scale are taken into account; this means that all other processes are regarded as reaching equilibrium instantaneously. During outbursts also processes on shorter time scales are included, if important.

2.3. Resulting Equations.

Applying the approximations described in Paragraph 2.2. to the equations introduced in Paragraph 2.1. gives the set of equations to describe the stationary structure and time dependent evolution of α-accretion disks.

2.3.1. Stationary accretion disks.
For a (constant) mass flow rate, $\dot{M}$, one gets from the continuity and Navier-Stokes-equations (subscripts s, z, and φ denote the components of the vectors/tensors in the respective directions; in

brackets we give the results for an even simpler approximation where integrations and differentiations with respect to z are replaced by multiplications with, and divisions by, h, resp.):

$$\dot{M} = -2 \cdot \pi \cdot s \cdot v_s \cdot \int_{-\infty}^{\infty} \rho \, dz \quad (= -2 \cdot \pi \cdot s \cdot \Sigma \cdot v_s) \ ,$$

$$v_\varphi = \sqrt{\frac{G \cdot M}{s}} \ ,$$

$$\dot{M} \cdot s \cdot v_\varphi = 2 \cdot \pi \cdot s^2 \cdot \int_{-\infty}^{\infty} \Theta_{s\varphi} \, dz + \dot{I} \quad \left(\approx 2 \cdot \pi \cdot s^2 \cdot \Theta_{s\varphi} \cdot (2 \cdot h) + \dot{I} \right) \ ,$$

$$\frac{\partial P}{\partial z} = -\frac{G \cdot M}{s^2} \cdot \frac{z}{s} \cdot \rho \quad \left(\Rightarrow P \approx \frac{\Sigma \cdot h \cdot \Omega^2}{2} \right) \ .$$

(6a)

$\dot{I}$ is the net flux of angular momentum; $\dot{M}$ and $\dot{I}$ are determined by boundary conditions. In all models we give $\dot{M}$ as the mass inflow rate at the disk's outer boundary, and assume no net flux of angular momentum, i.e. $\dot{I} = 0$.

In geometrically thin accretion disks heat flows predominantly in the vertical direction, i.e.

$$\underline{q} \approx q_z \cdot \underline{e}_z \ ,$$

(6b)

where $\underline{e}_z$ is a vector in the z-direction of unity length. From Eqs. (3) we get:

$$\frac{\partial q_z}{\partial z} = -\Theta_{s\varphi} \cdot s \cdot \frac{d\Omega}{ds},$$

$$\frac{\partial T}{\partial z} = \begin{cases} -\frac{3 \cdot \kappa \cdot \rho}{4 \cdot a \cdot c \cdot T^3} \cdot q_z \ , & \text{for radiative energy transport,} \\ \frac{G \cdot M}{s^2} \cdot \frac{z}{s} \cdot \rho \cdot \frac{T}{P} \cdot \nabla_{conv} \ , & \text{for convective energy transport,} \end{cases}$$

(6c)

where ∇_{conv} is the convective gradient, κ the opacity, and $\Omega = v_\varphi/s$ the angular velocity.

2.3.2. Time dependent evolution of accretion disks.

The evolution of the disk is described – under these approximations – by a diffusion-type equation:

$$\frac{\partial \Sigma}{dt} - \frac{3}{s} \cdot \frac{\partial}{\partial s} \left(\sqrt{s} \cdot \frac{\partial}{\partial s} \left(\sqrt{s} \cdot f \right) \right) = 0 \ ,$$

(7a)

with the normalized stress tensor element integrated in vertical direction,

$$f = \frac{2}{3 \cdot \Omega} \cdot \int_{-\infty}^{\infty} \Theta_{s\varphi} \, dz \ .$$

(7b)

2.4. Stationary accretion disks.

2.4.1. Material functions.

In order to solve Eqs. (6) we have to define a set of material functions; for our models we use:

$$\Theta_{s\varphi} = \alpha \cdot P \ ,$$

$$\alpha = \min \left(0.05, \ 5 \cdot \left(\frac{h}{s} \right)^{1.5} \right) \ .$$

(8)

We choose α following Meyer and Meyer-Hofmeister (1983), but introduce also a satuaration value (Duschl, 1986a). The exact values in the α-prescription were chosen to give the best fit to the observed time scales.

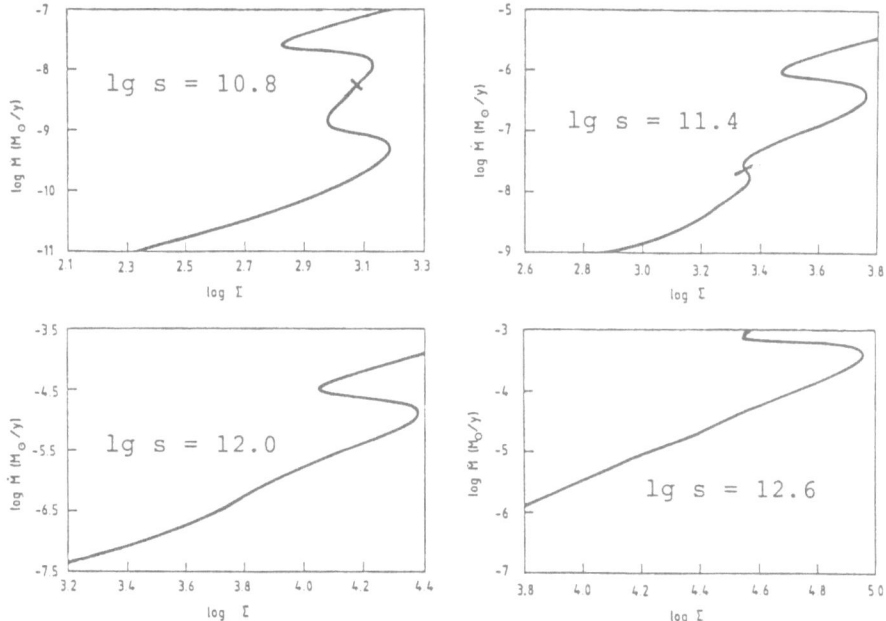

Figure 1: *The mass transfer rate as function of the surface density for different distances to the accretor.*

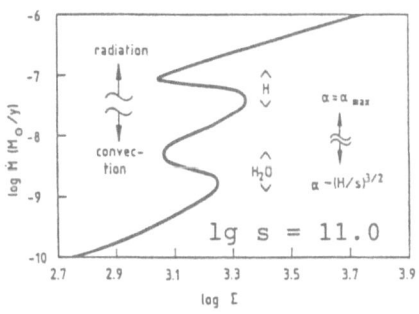

Figure 2: *The different domains of energy transport, viscosity parameter, and main contributors to the opacity, for a typical case ("H" indicates the region of ionization/recombination of hydrogen, "H_2O" the domain of disintegration/formation of water).*

We take κ from Cox and Stewart (1969) for $\log T \geq 4$, and Alexander (1975) for $\log T \leq 3.8$; in between we interpolate between the two sets of tables; the chemical composition is taken to be $X = 0.739, Y = 0.240$. The specific heat is calculated as described by Kippenhahn et al. (1967) for the regime of the Cox and Stewart-opacities, and taken from tables by Sharp (1981) for the lower temperatures. For the molecular weight we follow Kippenhahn et al. (1967), and Alexander (1975), resp.. The central accreting mass is assumed to be one solar mass.

Figs. 1 show the surface densities for different radii and mass transfer rates. Tickmarks on the $\log \Sigma - \log \dot{M}$-curve mark the points where the saturation value of α is reached; for higher mass transfer rates α is constant; at the two largest radii (10^{12}, and $10^{12.6}$cm) the entire curve shown has a constant viscosity parameter.

In Fig. 2 the different domains are indicated for a characteristic radius, 10^{11} cm. The saturation of the viscosity was already discussed above. The two distinct S-shaped features in the curve result from strong changes of the opacity within a comparatively small temperature range: at about $\dot{M} = 10^{-7.2}$ M$\odot$ /yr it is due to the ionization/recombination of hydrogen, while at around $10^{-8.6}$ M$\odot$ /yr it is due to the formation/disintegration of the water molecule. These S-shaped features indicate the regions of the disk instabilities.

The resulting accretion disks a) are geometrically thin, and b) have negligible mass compared to M, so that the approximations introduced above are justified. In contrast to the results for dwarf novae the accretion disks in SS are always optically thick.

3. ... AND THEIR OUTBURSTS ...

There are basically two different reasons why accretion disks may show outbursts: Either the mass inflow into the disk changes strongly, or — at an essentially constant inflow rate — there exists an intrinsic instability in the disk.

Integrating the energy equation (6c) in the vertical direction and introducing an effective temperature, T_{eff}, one obtains locally a relation between T_{eff} and $\dot{M}$:

$$\sigma \cdot T_{eff}^4 = \frac{3}{8 \cdot \pi} \cdot \frac{G \cdot M \cdot \dot{M}}{s^3} , \qquad (9)$$

i.e. the higher the local mass flux, the higher is the corresponding temperature. This shows that an outburst has to be associated with an increased mass flow — at least somewhere in the disk.

The remaining, but crucial question is which physical reason stands behind the increase in $\dot{M}$.

3.1. The mass transfer instability (=MTI).

Bath proposed in 1972 that actually the instability is situated in the atmosphere of the companion; because of such an instability the mass overflow to the accretion disk may vary quite strongly, and cause thus the outbursts. Although there seem to exist some problems (Gilliland, 1985) the question whether such an instability exists or not is not settled as yet (Bath, 1975; Edwards, 1985; among others).

In the following we shall not deal with the question how such an increase of $\dot{M}$ can originate, but take for the time being the optimistic point of view that some suitable mechanism exists.

Eq. (7a) allows us to define an accretion time scale, τ_{accr}, that describes the time within which variations of the mass flow at some radius — because of whatever reason — reach the inner edge of the disk due to viscous transport:

$$\tau_{accr} = \frac{s \cdot v_\varphi}{\alpha \cdot c_s^2} , \qquad (10a)$$

where we used the fact that according to Eqs. (6a) (integrated in the z-direction)

$$\frac{h}{s} = \frac{c_s}{v_\varphi} . \qquad (10b)$$

Comparing the fastest changes in the observed light curves with τ_{accr}, one obtains a lower limit for the viscosity parameter α, as variations on time scales shorter than τ_{accr} cannot be caused by a viscous process that involves large parts of the disk. Further the definition of α gives an upper limit for it as one expects turbulence to be sub-sonic (assuming that the length scale of the turbulence is of the order of the smallest length scale in the disk, the vertical scale height h):

$$\alpha \not> 1 . \qquad (10c)$$

Bath and Pringle (1982) find that the MTI model needs $\alpha \not< 0.2$ to be capable of reproducing the light curves of SS. For smaller α, τ_{accr} becomes so large that not even a step-like change of $\dot{M}$ at the outer edge suffices to reproduce fast modulations of the luminosity.

The results of model calculations for SS in the framework of the MTI will be presented and discussed in Sect. 4.

3.2. The disk instability (=DI).

Meyer and Meyer-Hofmeister (1981) were the first to calculate the stationary vertical disk structure with a realistic description of the energy transport a n d the opacity. They realized that an instability as proposed by Osaki (1974) may exist since in some regions Lightman's (1974) stability criterion

$$\frac{\partial \dot{M}}{\partial \Sigma} \geq 0 \tag{11}$$

is violated. (The exact form of this condition depends on the choice of $\dot{I}$ (see Eq. (6a)); for $\dot{I}$ as chosen here, Eq. (11) is exact, for other $\dot{I}$ the principles of this reasoning do not change.)

The calculations presented in Figs. 1 and 2 show that for a certain range of $\dot{M}$ and s condition (11) is violated. The resulting behaviour of the disk is shown in Fig. 3.

Figure 3: *The limit cycle ABCD in an unstable disk (the Figure is taken from Meyer and Meyer-Hofmeister's (1981) models for dwarf novae; in addition to mass flow rate and surface density, it also shows the viscosity integral (Eq. (7b)).*

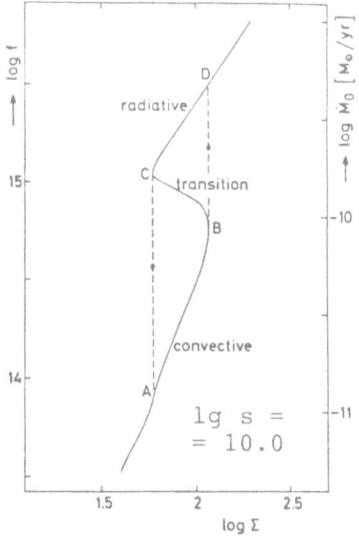

If at some radius the stationary solution lies on an unstable branch, we shall instead have a limit cycle type evolution of the disk. This will influence neighbouring regions thus leading to a *cooling* or *heating* front moving through the disk. "Cooling (heating) front" means that the transition from the stable branch with high (low) local mass flow, i.e. effective temperature, to the other stable branch is initiated. Such a front can move through the disk until it reaches zones where for the involved range of $\dot{M}$ only stable solutions exist. Meyer (1984) has discussed their behaviour in detail; other authors used slightly different physical and numerical approximations to describe this phenomenon (see e.g. Faulkner *et al.*, 1983; Mineshige and Osaki, 1983, 1985; Papaloizou *et al.*, 1983; and others).

Such fronts move through the disk with a velocity that is of the order of the sound velocity of the front's final state times the viscosity parameter α. Thus one can define a time scale, $\tau_{f,a}$, that describes within which timespan a front can change the energy output of an accretion disk markedly (a stands for c or h, which again stand for cooling, or heating front, resp.):

$$\tau_{f,a} = \frac{s}{\alpha \cdot c_s(T_a)} \cdot \tag{12a}$$

The (relatively) thinner an accretion disk, the shorter is τ_f compared to τ_{accr}:

$$\tau_f = \frac{h}{s} \cdot \tau_{accr} ; \tag{12b}$$

the type of the front determines h.

So in the DI model one can reproduce variations on shorter time scales than is possible with the MTI model, i.e. has a larger allowed range of values of α, although $\dot{M}$ is kept constant. On the other hand the rise and the decay time are related to each other (Eq. (12a)). With τ_{f+} being the rise, and τ_{f-} the decay time scale, we get:

$$\frac{\tau_{f+}}{\tau_{f-}} = \frac{\alpha_c \cdot \sqrt{T_c}}{\alpha_h \cdot \sqrt{T_h}} \, . \tag{12c}$$

Here we introduced the possibilty that the hot and the cool state may have different values of the viscosity parameter as is indicated from dwarf nova models (e.g. Meyer and Meyer-Hofmeister, 1983; Mineshige and Osaki, 1983; Smak, 1984). A further restriction arises from the timescale within which outbursts follow each other. One expects the disk to be refilled after an outburst within the accretion time scale of the cool state, so the typical repetition time scale of outbursts, τ_{out}, is

$$\tau_{out} = \frac{s \cdot v_\varphi}{\alpha_c \cdot c_{s,c}^2}. \tag{12d}$$

All these arguments are only of the *order of magnitude* type but are shown to be reasonably accurate by numerical models that will be discussed later.

There are obviously stronger restrictions on the parameters in the DI model than there are in the MTI model. The reason is that in the latter one is free to choose any suitable form for the mass overflow rate as one cannot give a selfconsistent physical model for the actual form as yet. In contrast to that, in the DI model these restrictions come from the physical processes that determine the outburst; this means that the DI model has a higher degree of selfconsistency compared to the MTI model. In this sense the higher degree of freedom in the latter ansatz actually results from the partial lack of a physical description, rather than being model inherent.

4. ... IN SYMBIOTIC STARS.

In the following we shall discuss the two outburst models and their application to SS; for the details we refer to the papers by Bath and Pringle (1982; MTI), and Duschl (1985a, 1986b; DI).

4.1. The accretor: main sequence star vs. white dwarf.

The first question is whether one expects a main sequence star, or a white dwarf to be the accreting object. Integrating Eq. (9) one gets for a stationary accretion disk a luminosity, L, of

$$L = \frac{G \cdot M \cdot \dot{M}}{s_*} \, , \tag{13}$$

where s_* is the radius of the accreting star. One can deduce from this for both cases some average accretion rates; in the case of a white dwarf these would be smaller by a factor of the order 10^2. According to Eq. (10a) we have $\tau_{accr} \propto T_{eff}^{-1}$, estimating the relation between temperature and mass inflow at the (here relevant) outer radius of the disk (that does not depend on the central object) we find that $T_{eff} \propto \dot{M}^{1/4}$. This would be equivalent to a higher accretion time scale for a white dwarf accretor. We give as an example the following numbers (s_0: outer radius of the disk): $M = M\odot$; $s_* = R_\odot$; $s_0 = 10^{12} \ cm$; $L = 10^{36} \ erg/s$, and $\alpha = 1$, and get an accretion time scale of the order one year, while for a white dwarf we would reach already several years, which is definitely too long. As we have natural upper limits for α and M (white dwarf), and observed values for L, the only way out would be a much smaller s_0. From the observed orbital periods of SS, and from the relative sizes of accretion disks in dwarf novae (e.g. Smak, 1982), we regard that as very unlikely. This argument only uses τ_{accr} and thus applies strictly only to the MTI. But similar arguments are true for τ_{out}. So both models point strongly towards a main sequence accretor.

This is in agreement with the analysis of spectra of SS (e.g. Mikołajewska and Mikołajewski, 1983; Kenyon and Webbink, 1984; Kenyon, 1985, 1986; for a more detailed compilation see Kenyon, 1986).

4.2. The MTI ansatz.

Figure 4: *The evolution of the mass transfer rate for CI Cyg in the model calculations by Bath and Pringle (1982); the Figure is taken from their paper.*

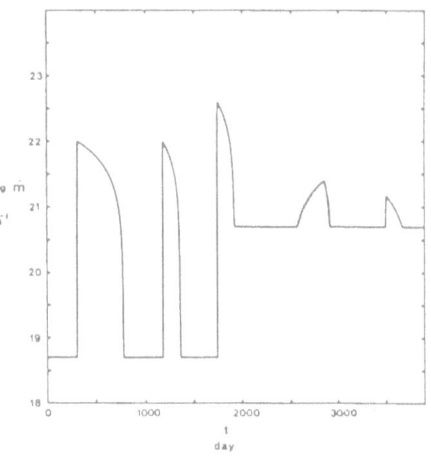

Bath (1977), and Bath and Pringle (1982) investigated how far the MTI model is capable of reproducing the lightcurves of SS. They chose *CI Cyg* as their example. In Figure 4 we show the evolution of the mass transfer rate they needed for their best fit of the lightcurve. Their model dimensions were the following: $M=M\odot$; $s_* = R_\odot$; $s_0 = 8.5 \cdot 10^{12}$ cm; $\alpha = 1$. They find good agreement with the observed lightcurves. But there remains a problem with the evolution of the mass transfer rate which this model needs to reproduce the lightcurves: a) The minimum mass overflow rate between outbursts has to vary itself by more than order of magnitude within few orbital periods; this could be attributed to long term variations in the companion's atmosphere, so that it does not seem to be a serious problem, although it is a remarkable feature in the evolution of the transfer rate; b) The structure of the bursts of $\dot{M}$ itself varies; while most bursts show a very sharp increase over several orders of magnitude, followed by a much shallower decline, at least for some cases the opposite configuration is necessary; this points towards two different types of instabilities being needed in the companion star's envelope (which remains to be explained).

The basic assumption of the disk theory presented here is that the disks are geometrically thin; as stated by Bath and Pringle, during the maxima of the outbursts this becomes a poor approximation as the disk's thickness reaches values of a third of the distance to the accretor, i.e. the neglected quantities (which are of relative order $(h/s)^2$) are less than an order of magnitude smaller then those taken into account, so one is at the limit of the applicability of the theory. This latter statement is also true for the DI ansatz.

4.3. The DI ansatz.

Duschl (1985a, 1986a, 1986b) and Mineshige (1986) have applied the DI model to SS. Since in this type of model the degrees of freedom are far fewer one cannot reasonably try to reproduce an individual system as this would need *fine tuning* of all parameters to an exactness that is far beyond what one can reach observationally; since one on the other hand does not expect a unique parameter set to fit all the observations, it is also not suitable as a part of observational diagnostics. One therefore attempts to model an average system and analyze which parameter range is allowed; these results are then compared with observations.

We also assume a one solar mass main sequence star as accretor; here we take the α-prescription as defined in Eq. (8). The outer radii are varied from $10^{11.6}$ to $10^{12.8}$ cm, the − constant − mass inflow rate between 10^{-6} and 10^{-3} M$\odot$ /yr. What one wants to reproduce are lightcurves with typical variations of the order of several month to years; the increase in luminosity is sharper than the decrease; one observes only minima that last for about the same timespan as − or less than −

the outbursts (in contrast to dwarf novae where the minimum state is far longer than the outburst). One can mark a region in the $s_0 - \dot{M}$-diagram where the parameter combinations lie that are capable of producing light curves like the ones observed in SS (Fig. 5). The limitations are the following: A: (dashed line) The disks are no longer geometrically thin, i.e. the approximations break down (this might very well be a technical rather than a physical limit); B: τ_{out} becomes too long as the disks become too large; C: τ_{out} becomes too long as $\dot{M}$ ($\Rightarrow T_{eff} \Rightarrow c_s$) becomes too small; D: The maximum brightness becomes so small that it is no longer consistent with observations; E: There are no bursts at all as the entire disk is stable.

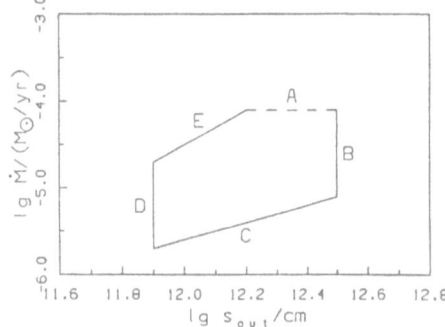

Figure 5: *The domain where the models for SS according to the DI model can be found.*

One might wonder how changes of this α-prescription would change these results. While the value of α determines the overall timescale of the evolution, it has almost no influence on the relation between τ_{out} and τ_f as h/s is only weakly dependent on α. So a change in the viscosity parameter would not leave unaffected the good absolute and relative agreement of time scales as well as the luminosities, that is achieved in the models. Changes in α would either change the frequency of outbursts or the luminosity; either are undesired effects thus making the possible range for an α-law rather small; this means that the results shown in Fig. 5 are not strongly dependent on α.

4.4. Comparison to dwarf nova models.

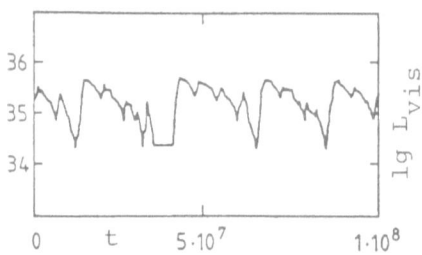

Figure 6: *A typical light curve for a SS in the DI model; the parameters for this model are:* $\dot{M} = 10^{21}$ *g/s, and* $\lg s_0 = 10^{12}$ *cm.*

The MTI as well as the DI model were both originally developed to model dwarf nova outbursts (e.g. Bath and Pringle, 1981; Meyer and Meyer-Hofmeister, 1981). While the behaviour of accretion disks in the MTI model is essentially the same as in the case of dwarf novae – only scaled to the dimensions of SS – in the DI model new types of behaviour of the fronts show up. In dwarf novae one has usually only either a heating or a cooling front moving through the disk, or – in the minimum state – no fronts at all; in the models for SS one almost always finds fronts moving through the disk; often there are several fronts of both types moving through the disk; and these fronts can interact. Because of the different timescales (see Eq. (12c)) two types of interaction are possible: a) A slower moving cooling front is *overtaken* by a faster heating front, and the two fronts cancel each other out; b) Two fronts of the same type move towards each other, also cancelling themselves out. These different interactions are the reason for the typically not very smooth light curves as shown in Fig. 6. Duschl (1986b) gave a color figure showing the evolution of the fronts in an accretion disk of a SS; this may be compared to the case of dwarf novae (Duschl, 1985b; there the color bar was reversed during the printing process!)

4.5. Comparison of the two outburst models.

Both models are capable of reproducing the observed light curves reasonably well. The DI model has a higher degree of physical selfconsistency. In the field of dwarf novae where the comparison of both models with observations is in a much more detailed state than in that of SS, there are strong indications that the DI model is in better agreement with the observations although this question is far from being settled.

In both models one has T_{eff} during the maxima of the outbursts of several 10^4 K; in addition one expects a boundary layer between the disk and the star where the characteristic temperatures during the outbursts reach values of the same order, i.e. sensible temperatures for SS.

5. SUMMARY.

There exist many − mutually incompatible − models for SS, and hopefully nobody will expect one model to be applicable to all SS. For a subgroup of the *type-B*-SS accretion disks seem to be the main reason for the behaviour we observe. Members of this subgroup are stars like CI Cyg, Z And, and AR Pav.

ACKNOWLEDGEMENTS.

I thank Prof. R. Kippenhahn, and Drs. E. and F. Meyer for all I have learnt from them about accretion disks in many, very valuable discussions; I also thank Drs. H.-C. Thomas and S. White who read this manuscript and helped by their suggestions to improve it, and Drs. G.T. Bath, E. and F. Meyer, and J.E. Pringle, for allowing me to use figures from their papers in this review.

REFERENCES.

Alexander, D.R.: 1975, *AstrophysJSuppl* **29**,363
Baker, N., Kippenhahn, R.: 1962, *ZeitschrAstrophys* **54**,114
Bath, G.T.: 1972, *AstrophysJ* **173**,121
Bath, G.T.: 1975, *MonthlyNoticesRoyAstronSoc* **171**,311
Bath, G.T.: 1977, *MonthlyNoticesRoyAstronSoc* **178**,203
Bath, G.T., Pringle, J.E.: 1981, *MonthlyNoticesRoyAstronSoc* **194**,967
Bath, G.T., Pringle, J.E.: 1982, *MonthlyNoticesRoyAstronSoc* **201**,345
Cox, A.N., Stewart, J.N.: 1969, *Scientific Informations of the Astronomical Council, USSR Academy of Sciences,* **15**
Duschl, W.J.: 1985a, *Ludwig-Maximilians-Universität, München,* Thesis
Duschl, W.J.: 1985b, in: *Recent Results On Cataclysmic Variables,* ESA, Paris, France, **SP-236**, 91
Duschl, W.J.: 1986a, *AstronAstrophys* **163**,56
Duschl, W.J.: 1986b, *AstronAstrophys* **163**,61
Edwards, D.A.: 1985, *MonthlyNoticesRoyAstronSoc* **212**,623
Faulkner, J., Lin, D.N.C., Papaloizou, J.: 1983, *MonthlyNoticesRoyAstronSoc* **205**,359
Frank, J., King, A.R., Raine, D.J.: 1985, *Accretion Power In Astrophysics,* Cambridge University Press, Cambridge, U.K.
Gilliland, R.L.: 1985, *AstrophysJ* **292**,522
Kant, I.: 1755, *Allgemeine Naturgeschichte und Theorie des Himmels.*
Kenyon, S.J.: 1985, in: *Cataclysmic Variables And Low-mass X-ray Binaries,* Eds.: Lamb, D.Q., Patterson, J., Reidel Publ. Corp., Dordrecht, Netherlands, p.417
Kenyon, S.J.: 1986, *The Symbiotic Stars,* Cambridge University Press, Cambridge, U.K.
Kenyon, S.J., Webbink, R.F.: 1984, *AstrophysJ* **279**,252
Kippenhahn, R., Weigert, A., Hofmeister, E.: 1967, *Methods In Computational Physics* **7**,129
Landau, L.D., Lifshitz, E.M.: 1959, *Course Of Theoretical Physics, Vol. 6: Fluid Mechanics,* Pergamon Press, London, U.K.
Lightman, A.P.: 1974, *AstrophysJ* **194**,419

Lüst, R.: 1952, *ZeitschrNaturforschung* **7a**,87

Meyer, F.: 1984, *AstronAstrophys* **131**,303

Meyer, F., Meyer-Hofmeister, E.: 1981, *AstronAstrophys* **104**, L10

Meyer, F., Meyer-Hofmeister, E.: 1983, *AstronAstrophys* **128**, 420

Mikołajewska, J., Mikołajewski, M.: 1983, *ActaAstron* **33**,403

Mineshige, S.: 1986, *Second Japan-China Workshop On Stellar Activities And Observational Techniques*, Eds.: Sadakane K., Yamasaki, A., Department of Astronmy, University of Kyoto, Kyoto, Japan, p.51

Mineshige, S., Osaki, Y.: 1983, *PublAstronSocJapan* **35**,377

Mineshige, S., Osaki, Y.: 1985, *PublAstronSocJapan* **37**,1

Osaki, Y.: 1974, *PublAstronSocJapan* **26**,429

Papaloizou, J., Faulkner, J., Lin, D.N.C.: 1983, *MonthlyNoticesRoyAstronSoc* **205**,487

Plavec, M.J.: 1982, in: *The Nature Of Symbiotic Stars*, IAU-Coll. 70, Eds.: Friegjung, M., Viotti, R., Reidel, Dordrecht, p.231

Shakura, N.I., Sunyaev, R.A.: 1973, *AstronAstrophys* **24**,337

Sharp, C.: 1981, *University of St.Andrews*, Thesis

Smak, J.: 1982, *ActaAstron* **32**,213

Smak, J.: 1984, *ActaAstron* **34**,161

Solf, J.: 1984, *AstronAstrophys* **139**,296

von Weizsäcker, C.F.: 1943, *ZeitschrAstrophys* **22**,319

ACCRETION FROM STELLAR WINDS

Mario Livio
Dept. of Astronomy, Univ. of Illinois, Urbana, IL 61801
and Dept. of Physics, Technion, Haifa 32000, Israel

ABSTRACT. Recent calculations have demonstrated that accretion from a stellar wind is very probably unsteady. The average rate of accretion of angular momentum is lower by about a factor 5 than the rate at which angular momentum is deposited into the Bondi-Hoyle accretion cylinder. This makes disk formation from wind accretion very difficult, in particular in the case of massive x-ray binaries. A combination of x-ray, uv and optical observations of symbiotic and related systems, as well as spin-up information on x-ray binaries, can be used to determine whether an accretion disk does form. Such observations can provide us with valuable information on the process of accretion from an inhomogeneous medium.

1. INTRODUCTION

The problem of accretion from an infinite medium by a gravitating object, is important for neutron stars accreting from stellar winds of early type companions, white dwarfs and main sequence stars accreting from winds of cool giants and galaxies moving in the intergalactic medium.

In the original Hoyle and Lyttleton (1939) picture, particles in the medium were assumed to follow their free Keplerian paths far from the accreting object, and interact inelastically only on the downstream axis. This interaction was assumed to lead to a complete cancellation of transverse (to the accretion line) momentum, leaving the momentum component parallel to the accretion line unchanged. Based on the Hoyle-Lyttleton assumption, it is possible to calculate the maximal impact parameter (the "accretion radius") for which, following the interaction, the matter will have a velocity lower than the escape velocity. All the material entering a cylinder with a radius equal to the accretion radius R_A, was therefore assumed to be accreted, giving an accretion rate

$$\dot{M}_{HL} = \pi R_A^2 \, \rho_\infty \, V = \frac{4\pi (GM)^2 \rho_\infty}{V^3} \quad , \tag{1}$$

149

where M is the mass of the accreting object, ρ_∞ and V are the density and velocity of the gas at infinity. Bondi and Hoyle (1944) replaced the unphysical, infinite density, accretion line in the Hoyle-Lyttleton picture by a finite density accretion cone. They found an accretion rate (still for a velocity dominated flow) of

$$\dot{M}_{BH} = \frac{4\pi \ \alpha(GM)^2 \rho_\infty}{V^3} \tag{2}$$

where $\frac{1}{2} \lesssim \alpha \lesssim 1$ is a parameter that is indeterminate by the theory (dependent on initial conditions). A different limiting case has been treated by Bondi (1952) who considered the problem of spherical accretion from a stationary medium onto a gravitating object. He found for this (pressure dominated) case an accretion rate of

$$\dot{M}_B = 4\pi \ \lambda(\gamma)R_B^2 \ \rho_\infty \ C_\infty = \frac{4\pi \ \lambda(\gamma) \ (GM)^2 \rho_\infty}{C_\infty^3} \tag{3}$$

where C_∞ is the speed of sound (at infinity) and $\lambda(\gamma)$ is a parameter (depending on the specific heat ratio) assuming values in the range 0.25 - 1.12. Combining the two limiting cases represented by eqs. (2) and (3), an interpolation formula for the accretion rate can be written as (Bondi 1952)

$$\dot{M}_B \simeq \frac{4\pi \ (GM)^2 \rho_\infty}{(V^2 + C_\infty^2)^{3/2}} \tag{4}$$

The problem of accretion from an infinite medium has gained renewed interest, in particular with the realization of its relevance to compact objects accreting from stellar winds and to common envelope evolution. Much of the more recent work has concentrated on the use of multi-dimensional numerical hydrodynamics (e.g. Hunt 1971, 1975, 1979, Livio, Shara and Shaviv 1979, Okuda 1983, Shima et al. 1985). The results obtained in these calculations were generally consistent with the Bondi-Hoyle theory, as expressed by eq. 4. The most recent fluid dynamical calculations were performed by Shima et al. (1985) who used a two dimensional, second-order accurate, Osher scheme (e.g. Chakravarthy and Osher 1983). A pseudo-particle description of the hydrodynamics was used by Livio et al. (1986a, b) and by de Kool and Savonije (1987). The results of these works show that the accretion rate (in the homogeneous case) can be expressed as

$$\dot{M} = F(\gamma,\mu) \cdot \frac{4\pi \ (GM)^2 \rho}{V^3} \tag{5}$$

with average values of $F(\gamma,\mu)$ of order 0.75 for $\gamma = 5/3$ and 1.0 for $\gamma \lesssim 4/3$ (μ is the Mach number).

2. ACCRETION FROM STELLAR WINDS - ACCRETION OF ANGULAR MOMENTUM

The basic problem associated with accretion from a stellar wind is the fact that the medium can no longer be considered homogeneous. Indeed, if the accretion cylinder remains unchanged, then due to the fact that there exists a gradient in the density (and possibly the velocity) of the stellar wind, material entering on one side of the accretion cylinder is denser (and may have a different velocity) than on the other side. This of course leads to a net deposition of angular momentum into the symmetric (about the accreting object) Bondi-Hoyle cylinder. As a consequence of the lack of a basic theory for the inhomogeneous case (despite some early attempts by Gethig (1951) and Dodd and McCrea (1952)), it has been assumed by a number of authors (e.g. Illarionov and Sunyaev 1975; Shapiro and Lightman 1976, Wang 1981) that all the angular momentum deposited into the symmetric cylinder is actually accreted. This leads to a rate of accretion of angular momentum of

$$\dot{L}_{BH} = \frac{1}{2} \eta R_A^2 \Omega_{orb} \dot{M} \tag{6}$$

where η is a parameter of order unity, detailed expressions for which (in terms of the radial and azimuthal density and velocity gradients) were obtained by Wang (1981).

The first to note that eq. (6) may significantly overestimate the rate of accretion of angular momentum were Davies and Pringle (1980). They pointed out that in the context of the original Hoyle-Lyttleton picture, in order to get accreted, particles from both sides of the accreting object have to cancel their momentum component transverse to the accretion line. They suggested therefore, that a similar situation should occur in the inhomogeneous case, with the accretion line somewhat displaced towards the lower density side. This should of course result in no accretion of angular momentum. Davies and Pringle admitted that their highly simplified, two dimensional treatment may not be valid in the real case which requires a fully three-dimensional calculation.

The first (and to date the only) three dimensional numerical calculations of accretion from a medium containing a density gradient, were performed by Livio et al. (1986a, b) and Soker et al. (1986). They found that when pressure effects were neglected (corresponding to a hypersonic flow), the accretion cylinder (representing the material that was eventually accreted), was displaced towards the lower density side in such a way that a displaced accretion cone formed (Fig. 1). The average rate of accretion of angular momentum was found to be only about 8% of that represented by eq. (6) (with $\eta = 1$). The calculations were then extended to include pressure effects, first in the isothermal case (Soker et al. 1986) and then for different values of the specific heat ratio γ. All cases resulted in a displacement of the accretion cylinder (and a corresponding one of the accretion cone, e.g. Fig. 2 for the isothermal case). The average rate of accretion of angular momentum, never exceeded ~ 23% of the rate at which angular momentum was deposited into the Bondi-

Hoyle symmetric accretion cylinder (represented by eq. (6)). Very similar results were obtained by Anzer, Börner and Monaghan (1987), who performed, however, a two dimensional calculation (which does not represent accurately the real flow in the inhomogeneous case).

The following important point should be emphasized. Due to the relatively coarse grid (imposed by memory limitations) in a three dimensional calculation, the values quoted above for the rate of accretion of angular momentum are average values only. Also, the 3D calculations could not resolve the flow structure over small scales (in particular, the scale of the density gradient was always chosen to be lrger than the accretion radius). When the finest possible grid (allowed on the CRAY X-MP/48 supercomputer) has been used (Livio and Soker 1986), no steady state was found. A similar result was obtained in the two dimensional calculations of Matsuda, Inoue and Sawada (1987) and of Fryxell and Taam (1987) who found the flow pattern to change its structure, from a displaced cone to a disk. The disk itself was found to reverse its rotational direction on a dynamical timescale. It should be noted however, that these fluctuations in the flow pattern are very probably enhanced in a two dimensional calculation, which does not allow the extra smoothing that a flow in the third dimension can provide. If the variations found by Matsuda et al. (1987) are real, then they should lead to oscillations in both the mass accretion rate and (to an even larger extent) the angular momentum accretion rate (which can even change sign).

3. GENERAL IMPLICATIONS FOR SYMBIOTIC SYSTEMS, X-RAY BINARIES AND RELATED OBJECTS

The determination of spectral types and luminosity classes for the cool components of a number of symbiotic systems (Kenyon and Fernandez-Castro 1987), as well as the determination of orbital periods from radial-velocity variations in the primary (Garcia 1986), suggest a division into semi-detached systems and detached systems. The detached systems include AG Dra ($P \simeq 554$ days, Meinunger 1981), AG Peg ($P \simeq 830$ days Hutchings et al. 1975), SY Mus ($P \simeq 627$ days Kenyon et al. 1985), Mira variable systems such as R Aqr and V1016 Cyg and possibly V1329 Cyg ($P \simeq 950$ days, Taranova and Yudin 1986), EG And ($P \simeq 470$ days, Oliversen et al. 1985) BX Mon, BF Cyg, TX CVn and HM Sge (Garcia 1986 and references therein). In the detached systems, the hot component must accrete mass from the giant's stellar wind. The total accretion luminosity that is obtained, based on the results presented in Section 1, is

$$L_{acc} = \frac{GM_{WD} \dot{M}_{acc}}{R_{WD}} = 4.0 \times 10^{36} \ [\frac{F(\gamma,\mu)}{0.9}] (\frac{M_{WD}}{M_{WD}+M_G})^{2/3} (\frac{M_{WD}}{M_\odot})^{2/3} (\frac{R_{WD}}{6\times10^8 \text{cm}})^{-1}$$

$$(\frac{V_W}{2\times10^6 \text{cm/s}})^{-1} (\frac{V_{rel}}{5\times10^6 \text{cm/s}})^{-3} (\frac{P_{orb}}{1 \text{ yr}})^{-4/3} (\frac{\dot{M}_W}{10^{-6} M_\odot/\text{yr}}) \text{erg s}^{-1} \qquad (7)$$

where M_{WD} and R_{WD} are the mass and radius of the accreting component (presumably a white dwarf), P_{orb} is the orbital period, V_W is the wind velocity, $\dot{M}_W$ is the rate of mass loss from the giant and V_{rel} is the relative velocity between the white dwarf and the wind $V_{rel} = (V^2_{orb} + V^2_W)^{\frac{1}{2}}$.

If an accretion disk does not form (see below), then as the accreted matter flows radially onto the white dwarf, a strong standoff shock forms, at a distance above the star, allowing the post-shock material to cool. If the dominant cooling mechanism is Bremsstrahlung, then the distance of the shock from the stellar surface is roughly (e.g. Hoshi 1973, Aizu 1973, Fabian, Pringle, and Rees 1976)

$$d = \frac{L_{acc}}{f \cdot 4\pi R^2_{WD} \, \epsilon_{ff}} \simeq 2.6 \times 10^8 \; f(\frac{\dot{M}_{acc}}{10^{19} gs^{-1}})^{-1} (\frac{M_{WD}}{M_\odot})^{3/2} (\frac{R_{WD}}{6\times 10^8 cm})^{\frac{1}{2}} cm \qquad (8)$$

where f is the fraction of the white dwarf's surface over which accretion takes place. The x-ray and uv luminosity produced by the accretion is characterized (for non magnetic white dwarfs) by three components (e.g. Lamb 1983): (1) A hard x-ray component, resulting from bremsstrahlung emission is the post-shock region, with a characteristic shock temperature

$$T_s = \frac{3}{8} \frac{GM_{WD} \, m_p \, \mu_m}{kR_{WD}} \simeq 6.2 \times 10^8 \; (\frac{M_{WD}}{M_\odot})(\frac{\mu_m}{0.615})(\frac{R_{WD}}{6\times 10^8 cm})^{-1} K \qquad (9)$$

where μ_m is the mean molecular weight, m_p is the proton mass and k is Boltzmann's constant. The maximal luminosity in this component is about one half of the accretion luminosity (eq. 7). (2) A soft x-ray blackbody component, produced by bremsstrahlung radiation emitted towards the white dwarf, absorbed, and re-radiated from the star's surface. The characteristic temperature of this component is

$$T_{bb} = (\frac{L_{acc}}{f \cdot 4\pi R^2_{WD} \sigma})^{\frac{1}{4}} = 3 \times 10^5 \; f^{-\frac{1}{4}} (\frac{\dot{M}_{acc}}{10^{19} gs^{-1}})^{\frac{1}{4}} (\frac{M_{WD}}{M_\odot})^{\frac{1}{4}} (\frac{R_{WD}}{6\times 10^8 cm})^{-3/4} k. \qquad (10)$$

Detailed calculations (e.g. Kylafis and Lamb 1982) show that this soft x-ray component is always present in non magnetic white dwarfs. (3) Secondary radiation from Compton heated pre-shock material. If the white dwarf is magnetized, a fourth, uv cyclotron component is produced in the hot emission region and component (2) above is partly produced by cyclotron photons. In general the relative strength of the different components depends sensitively on the accretion rate and on the magnetic field strength (e.g. Lamb and Masters 1979). In particular, white dwarfs with fields of the order of 10^7 gauss or more should appear as intense uv sources with an x-ray luminosity (and optical luminosity) amounting to not more than a few percent of

L_{acc}. Another process that can have a significant effect on the observed spectrum is steady nuclear burning on the white dwarf surface. The energy produced by nuclear burning results in a blackbody flux of soft x-rays, which is capable of cooling the hard x-ray emission region by inverse Compton scattering. This results in a reduction in the hard x-ray luminosity by about an order of magnitude (Weast et al. 1981).

The situation can be quite different if accretion is mediated by an accretion disk. In this case, half of the accretion energy is radiated as black body radiation from the disk. In the context of standard, steady state, optically thick disks, this produces a $\nu^{1/3}$ power law spectrum in the optical and uv (e.g. Shakara and Sunyaev 1973, Pringle 1981). The other half of the accretion energy is emitted in the boundary layer between the disk and the white dwarf. The boundary layer can emit soft x-rays (as a black body if optically thick, Pringle 1977) and perhaps hard x-rays (by shocks or turbulent viscosity, Pringle and Savonije 1979, Tylenda 1981) if optically thin. Unfortunately, the exact processes occurring within the boundary layer are still unknown (see e.g. Livio and Truran 1987), but many cataclysmic variables have been identified as x-ray sources (mostly showing a hard component, e.g. Cordova and Mason 1983).

The differences between radial and disk accretion, coupled with the fact that in many symbiotic systems, accretion takes place via a stellar wind, <u>demonstrate the importance of being able to determine whether disks can form from wind accretion</u>. Furthermore, in massive x-ray binaries (in which neutron stars accrete from winds of early type companions), the question of accretion of angular momentum has direct consequences for the spin-up of neutron stars.

In the case of a non magnetic white dwarf accreting from a stellar wind, an accretion disk can form, if the specific angular momentum of the accreted matter is sufficient to allow it to enter a Keplerian orbit around the white dwarf. This can be expressed by the condition

$$\ell_{acc} > (GM_{WD} \, R_{WD})^{\frac{1}{2}} \, , \tag{11}$$

where ℓ_{acc} is the specific angular momentum of the material captured from the wind. Using the results presented in Section 2 for the <u>average</u> rate of accretion of angular momentum, eq. (11) translates into the following condition for the formation of a (quasi steady) accretion disk

$$V_{rel} \lesssim 8.4 \times 10^6 (\frac{\zeta \eta}{0.2})^{\frac{1}{4}} (\frac{P_{orb}}{1 \text{ yr}})^{-\frac{1}{4}} (\frac{M_{WD}}{M_\odot})^{3/8} (\frac{R_{WD}}{6 \times 10^8 \text{cm}})^{-1/8} \text{cm s}^{-1} \tag{12}$$

where η is the factor appearing in eq. 6 and ζ represents the average fraction of the angular momentum deposited into the symmetric accretion cylinder that is actually accreted (according to the 3D calculations). The value of ζ was found to be of order 0.1 in the isothermal case and of order 0.2-0.25 for γ in the range 4/3 to 5/3.

If the accreting object is magnetized, as is often the case for the neutron stars in massive x-ray binaries, then for a disk to form, its radius must be larger than the magnetospheric radius. This results in the condition (where values appropriate for x-ray binaries were used)

$$V_{rel} \lesssim 4.0 \times 10^7 (\frac{\zeta\eta}{0.2})^{\frac{1}{4}} \mu_{30}^{-\frac{1}{4}} (\frac{M_x}{M_\odot})^{5/14} (\frac{P_{orb}}{1day})^{-\frac{1}{4}} (\frac{R_x}{10^6 cm})^{1/28}$$

$$\times (\frac{L_x}{10^{37} ergs^{-1}})^{1/28} \quad cm \ s^{-1} \qquad (13)$$

where μ_{30} is the neutron star's magnetic moment (in 10^{30} gauss cm^3). An additional piece of information in the binary x-ray sources case is provided by observations of spin-up of the neutron star. Equating the average rate of accretion of angular momentum to the one implied by the observed spin-up, gives

$$\dot{L}_{acc} = |\frac{\dot{P}}{P_s}|_{obs} (\frac{2\pi}{P_s}) \ I_x \quad , \qquad (14)$$

where P_s is the spin period of the neutron star and I_x its moment of inertia. This provides another constraint on the relative velocity

$$V_{rel} = 8.9 \times 10^7 \ (\frac{\zeta\eta}{0.2})^{\frac{1}{4}} \ (\frac{|\dot{P}/P_s|}{10^{-11} ss^{-1}})^{-\frac{1}{4}} (\frac{L_x}{10^{37} ergs^{-1}})^{\frac{1}{4}} \ (\frac{M_x}{M_\odot})^{\frac{1}{4}}$$

$$\times \ (\frac{P_{orb}}{1d})^{-\frac{1}{4}} (\frac{R_x}{10^6 cm})^{\frac{1}{4}} (\frac{I_x}{10^{45} gcm^2})^{-\frac{1}{4}} (\frac{P_s}{100s})^{\frac{1}{4}} cm \ s^{-1} \qquad (15)$$

We are now in the position to apply these results to a few specific systems.

4. DISCUSSION OF SOME INDIVIDUAL SYSTEMS

We shall now use the general results of the preceding sections, for some symbiotic and related systems.

AG Dra

The orbital period of this system is P_{orb} = 554 days (Meinunger 1979, Oliversen and Anderson 1982, Garcia 1986). If we adopt $M_G \approx 1.5 \ M_\odot$, $M_{WD} \approx M_\odot$, and a wind velocity $V_W \approx 100$ km/sec (Garcia 1986), then from eq. (7) we obtain an accretion luminosity of

$$L_{acc} \simeq 2.6 \times 10^{32} \ (\frac{\dot{M}_W}{10^{-8} M_\odot/yr}) \ erg \ s^{-1} \ . \tag{16}$$

AG Dra has been detected as a soft x-ray source with a luminosity of $L_x = 2 \times 10^{32}$ erg s^{-1} (Anderson, Cassinelli, and Sanders 1981). For a (quasi steady) disk to form, the relative velocity between the white dwarf and the wind must satisfy $V_{rel} \lesssim 7.6 \times 10^6$ cm, implying $V_W \lesssim 67$ km s^{-1}. This is lower than the value we have used, $V_W \sim 100$ km s^{-1}, but in view of the uncertainties in both the observations and the theoretical predictions, a wind fed disk can probably form in this system. It should be noted that once a disk starts to form, it can spread by viscous transport of angular momentum. The absence of a harder x-ray component is somewhat puzzling since at the relatively low accretion rates implied for this system (for any reasonable wind mass loss rate, $\dot{M}_W \lesssim 10^{-7}$ $M_\odot$/yr), a hard component would perhaps be expected both for radial and disk accretion. However, since the source of the hard x-ray emission in cataclysmic variables is not really entirely clear, the absence of hard x-rays favors probably the existence of a disk.

Mira A + B

While the Mira AB system itself can be considered only "mildly" symbiotic (e.g. Whitelock 1987), it is closely related to symbiotic systems which contain Mira variables (e.g. R Aqr and V1016 Cyg). Mira B is thought to be a white dwarf accreting from the wind of Mira A (Warner 1972, Livio and Warner 1984). The situation here is somewhat complicated by the fact that no reliable orbital period is known. Fernie and Brooker (1961) considered 261 yr as the most plausible of their possible solutions, which included 59 and 169 years. Hopmann (1969) found 139 and 842 years. Baize (1980) found 400 years, but Walker (1985) concluded that all the existing orbits are bad. We shall use $P_{orb} = 400$ yrs, this being close to the average of all the existing periods. While the distance of 77 pc (Jenkins 1952) is often quoted, this has been recently questioned by Whitelock (1987), who uses d = 120 pc. If we still use the former value, a wind velocity of $V_W = 5.6 \times 10^5$ cm s^{-1} (Wannier et al. 1980) and a wind mass loss rate of $\dot{M}_W \simeq 10^{-7}$ $M_\odot$/yr (Reimers and Cassatella 1985), we obtain an accretion luminosity (for an average mass white dwarf, $M_{WD} = 0.6$ $M_\odot$), of $L_{acc} = 7.9 \times 10^{33}$ erg s^{-1}.

The condition for disk formation (eq. 12) gives $V_{rel} \lesssim 1.4 \times 10^6$ cm s^{-1}, which is satisfied for our assumed parameters. Thus, disk formation is possible for the Mira AB system. This is consistent with the observations of Reimers and Cassatella (1985) and Cassatella et al. (1985), which implied the existence of a disk around Mira B. Mira has been detected in soft x-ray (0.15-2.5 kev), with a luminosity (for an assumed distance of 114 pc) of $L_x \simeq 2.3 \pm 0.8 \times 10^{29}$ ergs s^{-1} (Jura and Helfand 1984). These authors tried to argue that the low value of the x-ray luminosity suggests that Mira B is a main

sequence star. However, in view of the many uncertainties, the
accretion luminosity we obtained should be regarded as consistent
with the lower limit on the Mira B luminosity, $L_B > 3.3 \times 10^{32}$ ergs
s^{-1} (Reimers and Cassatella 1985). The fact that a wind fed disk can
form in the system, can explain the low x-ray luminosity.

GX 1 + 4

This known bright galactic binary x-ray source is often mentioned as
a neutron star powered by wind accretion (e.g. Garcia 1986). It has
a spin period of 122 sec and a spin-up rate of $|\dot{P}/P_s| = 7.0 \times 10^{-10}$
s^{-1} (Elsner et al. 1985). The value of the relative velocity
required to produce such a spin-up rate (eq. 15) $V_{rel} = 4.6 \times 10^7$
$(P_{orb}/1d)^{-\frac{3}{4}}$ cm s^{-1} appears too low.
    Also, the extremely smooth spin-up behaviour (Fig. 3) is in
direct conflict with the non-steady behaviour expected for wind
accretion. A fluctuating spin-up is expected both because of the non
steady nature of the flow, as described in section 2, and because of
the ionization feedback of the x-rays on the wind, as found by Ho and
Arons (1987). We thus conclude that GX 1+4 is very probably powered
by disk accretion resulting from Roche lobe overflow.

5. CONCLUSIONS

Recent calculations have shown that accretion from a stellar wind is
very probably unsteady, leading to fluctuations in both the mass and
angular momentum accretion rates. While the _average_ mass accretion
rate is described quite adequately by the Bondi-Hoyle theory, the
average rate of accretion of angular momentum is considerably lower
(by about a factor 5) than the rate at which angular momentum is
deposited into the symmetric accretion cylinder. This makes (quasi-
steady) disk formation from wind accretion very difficult. A combination of x-ray, uv, and optical observations of symbiotics, x-ray
binaries, and related systems can be used to determine the existence
of such a disk, thus providing a better understanding of the
accretion process.

ACKNOWLEDGEMENT

This work has been supported in part by NSF Grant AST 86-11500 at the
University of Illinois. I would like to thank the LOC and in
particular J. Mikolajewska for their support.

REFERENCES

Aizu, K. 1973, _Prog. Theoret. Phys._, 49, 1184.
Anderson, C.M., Cassinelli, J.P., and Sanders, W.T. 1981, _Ap. J._
    J. Lett., 247, L127.

158

Anzer, U., Borner, G., and Monaghan, J.J. 1987, <u>Astron. Ap.</u>, 176, 235.
Baize, P. 1980, <u>Astron. Ap. Suppl.</u>, 39, 83.
Bondi, H. 1952, <u>M.N.R.A.S.</u>, 114, 195.
Bondi, H. and Hoyle, F. 1944, <u>M.N.R.A.S.</u>, 104, 273.
Cassatella, A., Holm, A., Reimers, D., Ake, T., and Stickland, D.J. 1985, <u>M.N.R.A.S.</u>, 217, 589.
Chakravarthy, S.R. and Osher, S. 1983, <u>A.I.A.A.Jl.</u>, 29, 1241.
Cordova, F.A. and Mason, K.O. 1983, in <u>Accretion Driven Stellar X-Ray Sources</u>, eds. W.H.G. Lewin and E.P.J. van den Heuvel (Cambridge: Cambridge University Press).
Davies, R.E. and Pringle, J.E. 1980, <u>M.N.R.A.S.</u>, 191, 599.
de Kool, M. and Savonije, G.J. 1987, in M. de Kool's Ph.D. Thesis, University of Amsterdam.
Dodd, K.N. and McCrea, W.H. 1952, <u>M.N.R.A.S.</u>, 112, 205.
Elsner, R.F., Weisskopf, M.C., Apparao, K.M.V., Darbro, W., Ramsej, B.D., Williams, A.C., Grindlay, J.E., and Sutherland, P.G. 1985, <u>Ap. J.</u>, 297, 288.
Fabian, A.C., Pringle, J.E., and Rees, M.J. 1976, <u>M.N.R.A.S.</u>, 173, 43.
Fernie, J.D. and Brooker, A.A. 1961, <u>Ap. J.</u>, 133, 1088.
Fryxell, B.A. and Taam, R.E. 1987, private communication.
Garcia, M.R. 1986, <u>Astron. J.</u>, 91, 1400.
Gething, J.P.D. 1951, <u>M.N.R.A.S.</u>, 111, 468.
Ho, C. and Arons, J. 1987, preprint.
Hopmann, J. 1964, <u>Ann. Univ. Sternw. Wien</u>, 26, 1.
Hoshi, R. 1973, <u>Prog. Theoret. Phys.</u>, 49, 776.
Hoyle, F. and Lyttleton, R.A. 1939, <u>Proc. Camb. Phil. Soc.</u>, 35, 405.
Hunt, R. 1971, <u>M.N.R.A.S.</u>, 154, 541.
Hunt, R. 1975, <u>M.N.R.A.S.</u>, 173, 465.
Hunt, R. 1979, <u>M.N.R.A.S.</u>, 188, 83.
Hutchings, J.B., Cowley, A.P., and Redman, R.D. 1975, <u>Ap. J.</u>, 201, 404.
Illarionov, A.F. and Sunyaev, R.A. 1975, <u>Astron. Ap.</u>, 39, 185.
Jenkins, L.F. 1952, <u>General Catalogue of Stellar Parallaxes</u>, Yale University, New Haven.
Jura, M. and Helfand, D.J. 1984, <u>Ap. J.</u>, 287, 785.
Kenyon, S.J., Michalitsianos, A.G., Lutz, J.H., and Kafatos, M. 1985, <u>P.A.S.P.</u>, 97, 268.
Kylafis, N.D. and Lamb, D.Q. 1982, <u>Ap. J. Supp.</u>, 48, 239.
Lamb, D.Q. 1983, in <u>Cataclysmic Variables and Related Objects</u>, eds. M. Livio and G. Shaviv (Dordrecht: D. Reidel), p. 299.
Livio, M., Shara, M.M., and Shaviv, G. 1979, <u>Ap. J.</u>, 233, 704.
Livio, M. and Warner, B. 1984, <u>Observatory</u>, 104, 152.
Livio, M., Soker, N., de Kool, M., and Savonije, G.J. 1986, <u>M.N.R.A.S.</u>, 218, 593.
Livio, M., Soker, N., de Kool, M., and Savonije, G.J. 1986, <u>M.N.R.A.S.</u>, 222, 235.
Livio, M. and Soker, N. 1986, unpublished results.
Livio, M. and Truran, J.W. 1987, <u>Ap. J.</u>, 318, 316.

Matsuka, T., Inone, M., and Sawada, K. 1987, preprint.
Meinunger, L. 1979, Inf. Bull. Var. Stars, No. 1611.
Meinunger, L. 1981, Inf. Bull. Var. Stars, No. 2016.
Okuda, T. 1983, Pub. Astr. Soc. Japan, 35, 235.
Oliversen, N.A. and Anderson, C.M. 1982, in The Nature of
    Symbiotic Stars, IAU Colloq. 70, eds. M. Friedjung and
    R. Viotti (Dordrecht: D. Reidel), p. 177.
Oliversen, N.G., Anderson, C.M., Stencel, R.E., and Slovak, M.H.
    1985, Ap. J., 295, 620.
Pringle, J.E. 1977, M.N.R.A.S., 178, 195.
Pringle, J.E. 1981, Ann. Rev. Astron. Ap., 19, 137.
Pringle, J.E. and Savonije, G.J. 1979, M.N.R.A.S., 187, 777.
Reimers, D. and Cassatella, A. 1985, Ap. J., 297, 275.
Shakura, N.I. and Sunyaev, R.A. 1973, Astron. Ap., 24, 337.
Shapiro, S.L. and Lightman, A.P. 1976, Ap. J., 204, 555.
Shima, E., Matsuda, T., Takeda, H., and Sawada, K. 1985,
    M.N.R.A.S., 217, 367.
Soker, N., Livio, M., de Kool, M., and Savonije, G.J. 1986,
    M.N.R.A.S., 221, 445.
Taranova, O.G. and Yudin, B.F. 1986, Astron. Zh., 63, 151.
Tylenda, R. 1981, Acta Astron., 31, 267.
Walker, R.L. 1985, Pub. U.S. Naval Obser., Vol. XXV-Part II,
    p. 16.
Wang, Y.-M. 1981, Astron. Ap., 102, 36.
Wannier, P.G., Redman, R.O., Phillips, T.G., Leighton, R.B.,
    Knapp, G.R., and Huggins, P.J. 1980, in IAU Symp. 87,
    Interstellar Molecules, ed. B.H. Andrews (Dordrecht: D.
    Reidel), p. 487.
Warner, B. 1972, M.N.R.A.S., 159, 95.
Weast, G.J., Durisen, R.H., Imanrura, J.N., Kylafis, N.D., and
    Lamb, D.Q. 1981, in Fundamental Problems in the Theory of
    Stellar Evolution, eds. D. Sugimoto, D.Q. Lamb, and D.N.
    Schramm (Dordrecht: D. Reidel), p. 234.
Whitelock, P.A. 1987, P.A.S.P., in press.

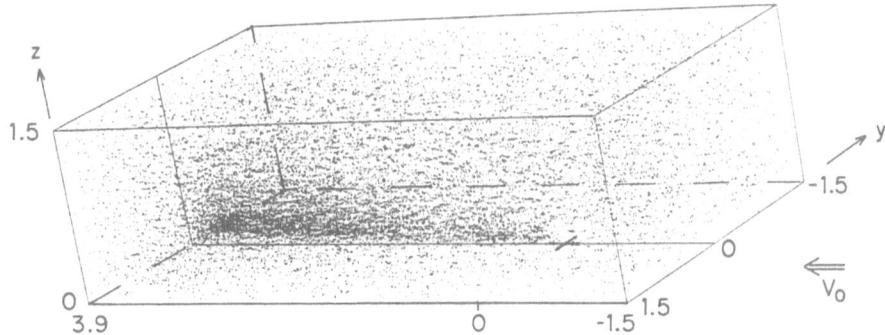

Fig. 1. The instantaneous location of the particles for accretion
from an inhomogeneous medium. The cross marks the accreting object.

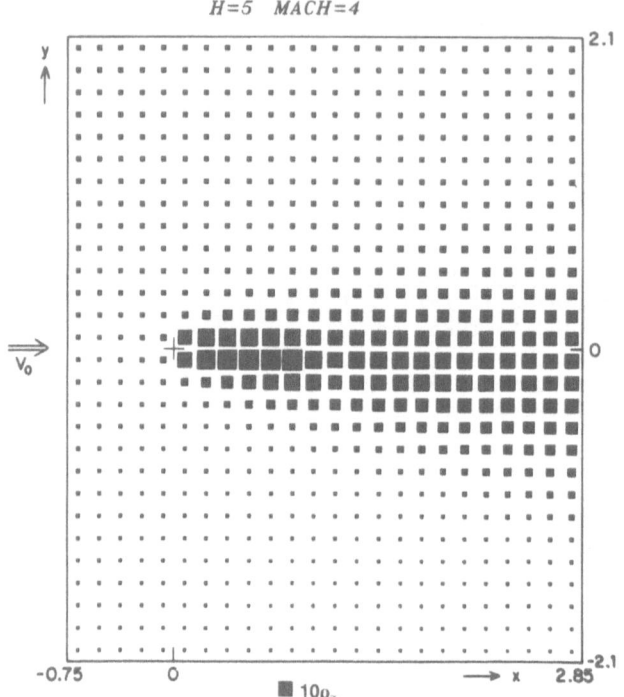

Fig. 2. The
density pro-
file for an
isothermal
flow. H is
the scale of
the density
gradient (in
units of the
accretion
radius).

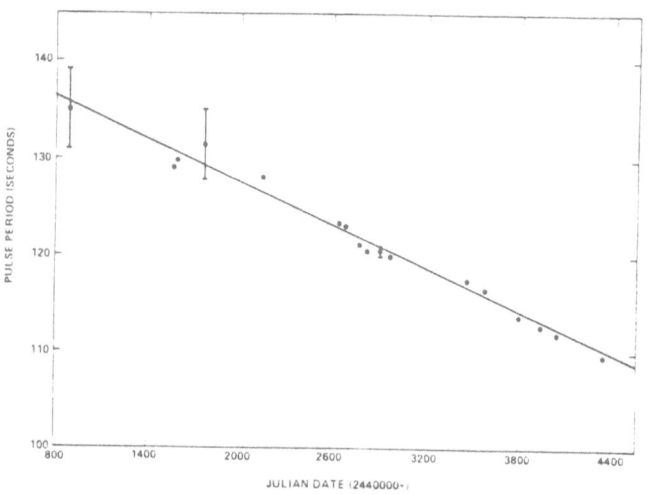

Fig. 3. The
x-ray pulse
period his-
tory of GX
1+4, taken
from Elsner
et al. (1985).

THERMONUCLEAR RUNAWAY MODELS FOR SYMBIOTIC NOVAE

Scott J. Kenyon
Smithsonian Astrophysical Observatory
Harvard-Smithsonian Center for Astrophysics
60 Garden Street
Cambridge, MA 02138 USA

ABSTRACT. This paper reviews the basic physics of thermonuclear runaways on the surfaces of accreting white dwarf stars, with a special emphasis on understanding the evolution of symbiotic novae.

1. Introduction

The eruptions of the small class of objects known as symbiotic novae are very different from those experienced by classical symbiotic binaries such as Z And and CI Cyg. As reviewed by Viotti in this volume (see also Kenyon 1986, Chapter 5), symbiotic nova eruptions are characterized by a slow rise to visual maximum (~ a few years) followed by a very tedious decline (~ many decades). Observations suggest the bolometric luminosity, L_{bol}, of a symbiotic nova remains roughly constant following visual maximum, although the visual luminosity, L_{vis}, decreases by a factor of ~ 100. This evolution of L_{bol} and L_{vis} with time is very similar to that observed in classical novae (see Gallagher and Starrfield 1978), which suggests a common eruption mechanism for these two types of novae.

It is well-established that eruptions of classical novae result from thermonuclear runaways on the surfaces of white dwarf stars. The basic physical model consists of a short period binary system (P_{orb} ~ hours), in which a lobe-filling red dwarf transfers material into an accretion disk surrounding a white dwarf. A hydrogen-rich atmosphere builds up on the white dwarf's surface, and eventually the pressure at the base of this envelope is sufficient to ignite the accreted material. Since the white dwarf atmosphere is degenerate, the ignition is explosive and material is accelerated outward. The white dwarf increases in luminosity by ~ 15 mag to M_V ~ -7 during the eruption, and the radius of the nova photosphere expands to envelop the entire binary system (Fujimoto 1982a,b; MacDonald 1983). Nuclear reactions and gravitational stirring of the nova photosphere by the red dwarf generate enough energy to expel a large fraction of the accreted envelope from the binary system at velocities of 300 to 2000 km s^{-1} (MacDonald 1980). Once the accreted material has been either burned or ejected, the nova fades to relative insignificance over a period of several months.

The major physical difference between a symbiotic nova and a classical nova is that the mass-losing star in a symbiotic binary is a red giant rather than a red dwarf. Observations indicate that symbiotic novae are detached binaries, so mass transfer occurs via a wind instead of tidal overflow. The large orbital periods of symbiotic novae, $P_{orb} \gtrsim$ 2-3 yr, also guarantee that the expanding white dwarf photosphere lies well within its inner Lagrangian surface throughout the eruption, and thus these systems do not become common envelope

161

J. Mikolajewska et al. (eds.), The Symbiotic Phenomenon, 161–168.

binaries at visual maximum. This fact may explain why symbiotic novae decline on a time scale which is more comparable to the nuclear time scale, as opposed to the classical novae, in which gravitational stirring of the nova photosphere by the red dwarf companion appears to cause a more rapid decline than can be accounted for by the combined actions of nuclear burning and mass ejection in a normal stellar wind. Thus, the symbiotic novae may provide a better laboratory in which to study thermonuclear runaways than classical novae.

II. Basic Physics

The outburst cycle of a symbiotic nova is conveniently divided into two phases, a quiescent or "off" phase and an eruptive or "on" phase. The white dwarf patiently accretes matter from its companion during quiescence, and its physical appearance to an outside observer depends on the mass and radius of the white dwarf, M_{wd} and R_{wd}, and the mass accretion rate, $\dot{M}$. Material settles onto the white dwarf fairly smoothly if $\dot{M}$ is fairly small (say, $\lesssim 10^{-10}$ $M_\odot$ yr^{-1}), while a hydrogen burning shell is created near the white dwarf's surface if the accretion rate is large. Paczyński and Żytkow (1978) demonstrated that there is a range in accretion rates such that material burns completely as it is accreted, and therefore does not accumulate on the white dwarf's surface. This *steady burning limit* was investigated in detail by Iben (1982), who found a minimum accretion rate of:

$$\dot{M}_{steady,min} \approx 1.32 \times 10^{-7} \, M_\odot \, yr^{-1} \left[\frac{M_{wd}}{1 \, M_\odot} \right]^{3.57} . \tag{1}$$

Since the material burns as it is accreted, this limit results in a white dwarf luminosity of

$$L_{steady,min} \approx 10^4 \, L_\odot \left[\frac{M_{wd}}{1 \, M_\odot} \right]^{3.57} . \tag{2}$$

The maximum luminosity that can be achieved by a steadily burning white dwarf is given by the Paczyński (1970) - Uus (1970) core mass-luminosity relationship

$$L_{plateau} \approx 59{,}250 \, L_\odot \, (M_{wd} - 0.52 \, M_\odot) , \tag{3}$$

which results in a maximum steady accretion rate of:

$$\dot{M}_{steady,max} \approx 6 \times 10^{-7} \, M_\odot \, yr^{-1} \, X^{-1} \, (M_{wd} - 0.52 \, M_\odot) , \tag{4}$$

where X is the mass fraction of hydrogen in the accreted envelope.

If the accretion rate exceeds the maximum steady rate ($\dot{M} > \dot{M}_{steady,max}$), then the additional (unburnt) material expands into an extended atmosphere similar to that of a red giant. On the other hand, if the accretion rate is less than the minimum steady rate ($\dot{M} < \dot{M}_{steady,min}$), then the radius of the stellar photosphere remains comparable to the white dwarf's radius (Iben 1982). Explosive shell flashes occur in both of these configurations. The helium burning shells of red giants are thermally unstable, resulting in recurrent *helium shell flashes* which will not be discussed here (Iben and Renzini 1983). When $\dot{M} < \dot{M}_{steady,min}$, both the hydrogen and helium shells are thermally unstable; the *hydrogen shell flashes* result in objects that are observed as classical novae and symbiotic novae, while the less frequent helium flashes may give rise to supernova eruptions (Iben and Tutukov 1983).

Various numerical and semi-analytical calculations have investigated the time evolution of accreting white dwarfs when $\dot{M} < \dot{M}_{steady,max}$ (Paczyński and Żytkow 1978; Sion, Acierno, and Tomczyk 1979; Kutter and Sparks 1980; Fujimoto 1982a,b; MacDonald 1980, 1983; Prialnik, *et al.* 1982; Nariai, Nomoto, and Sugimoto 1980; Starrfield, Sparks, and Truran

1984; Sion and Starrfield 1985; Prialnik 1986). Fujimoto and MacDonald have shown that hydrogen burning commences when the pressure at the base of the accreted envelope reaches a critical value. The material does not expand significantly during the early stages of shell burning, because the atmosphere is partially (in some cases, completely) degenerate. The temperature at the base of the accreted layer increases exponentially until the degeneracy is lifted and the atmosphere responds on a hydrodynamic time scale (~ minutes). Because the energy released in hydrogen burning (~ 10^{18} erg g^{-1}) exceeds the binding energy of the accreted envelope (~ 10^{17} erg g^{-1}), the shell flash can result in a rapid ejection of material if most of the hydrogen fuel can be processed via the CNO cycle before the atmosphere expands (and the resulting drop in the density and temperature of the burning shell quenches the runaway). Thus the most important parameter describing the evolution of the shell flash is the ratio of nuclear energy generated while the atmosphere is degenerate, E_{nuc}, to the envelope's binding energy, E_{bind}. Violent runaways occur if $E_{nuc}/E_{bind} \gtrsim 1$, and detailed numerical calculations yield expansion velocities of 1000-3000 km s^{-1}. Weak shell flashes are predicted if $E_{nuc}/E_{bind} \ll 1$, and such objects may not eject material dynamically.

The critical ratio, E_{nuc}/E_{bind}, depends on several physical parameters: the mass and luminosity of the accreting white dwarf (M_{wd} and L_{wd}), the mass accretion rate ($\dot{M}$), and the mass fraction of CNO nuclei in the accreted envelope (Z_{CNO}). Since E_{nuc}/E_{bind} increases if the accreted envelope is more degenerate, low white dwarf luminosities and low accretion luminosities favor violent eruptions. For a given L_{wd} and $\dot{M}$, stronger eruptions can be achieved on a more massive white dwarf, because (i) a lower mass envelope is required to initiate the runaway (thereby decreasing E_{bind}) and (ii) the pressure at the base of the accreted envelope is larger for a given M_{env} (thereby increasing E_{nuc}).

The relationship between E_{nuc}/E_{bind} and the CNO abundance depends on the nuclear physics involved with the main energy source for the runaway, the CNO cycle. For the temperatures that usually are reached in the hydrogen burning shell, $T_{shell} \sim 10^8$ K, stable C and N nuclei fuse into the β^+ unstable nuclei ^{14}O and ^{15}O. The runaway must then "wait" for these nuclei to decay before energy production can continue. However, the lifetimes of these nuclei (~ 180 sec) are longer than the local hydrodynamic time scale, so the atmosphere expands before the nuclei decay. Thus, E_{nuc}/E_{bind} is limited by the amount of energy which can be produced from the capture of 1-2 protons on every CNO nucleus:

$$\left[\frac{E_{nuc}}{E_{bind}}\right] \sim 0.01 \left[\frac{Z_{CNO}}{Z_{CNO,\odot}}\right], \tag{5}$$

where $Z_{CNO,\odot}$ is the solar CNO abundance. It is apparent that a hydrodynamic runaway requires $Z_{CNO} \gg Z_{CNO,\odot}$. Detailed numerical calculations have confirmed this basic point in detail: hydrodynamic or "fast" novae result if $Z_{CNO} \gg Z_{CNO,\odot}$ while hydrostatic or "slow" novae occur if $Z_{CNO} \sim Z_{CNO,\odot}$.

Important observational parameters characterizing shell flashes are the the time spent at quiescence, Δt_{off}, and the time spent in eruption, Δt_{on}. If the mass of the accreted envelope needed for the runaway is M_{env}, the quiescent time scale can be estimated as:

$$\Delta t_{off} \approx 10^4 \text{ yr} \left[\frac{M_{env}}{10^{-5} M_\odot}\right]\left[\frac{10^{-9} M_\odot \text{ yr}^{-1}}{\dot{M}}\right] \tag{6}$$

An upper limit to the time spent in the "on" state is the ratio of available nuclear energy, E_{nuc}, to the luminosity at maximum, $L_{plateau}$:

$$\Delta t_{on} \sim 40 \text{ yr} \left[\frac{M_{env}}{10^{-5} M_\odot}\right]\left[\frac{2 \times 10^4 L_\odot}{L_{plateau}}\right] \tag{7}$$

Table 1 - Time scales for Accreting 1 $M_\odot$ White Dwarfs (Iben 1982)

Accretion Rate ($M_\odot$ yr^{-1})	Δt_{off} (yr)	Δt_{on} (yr)
10^{-7}	38	18
10^{-8}	630	21
10^{-9}	15000	49
10^{-10}	330000	110

Characteristic values for Δt_{on} and Δt_{off} are given by Iben (1982) for a 1 $M_\odot$ white dwarf, and are summarized in Table 1. The ratio $\Delta t_{off}/\Delta t_{on} \gg 1$ as long as $\dot{M}$ does not greatly exceed a few x 10^{-8} $M_\odot$ yr^{-1}, implying that accreting white dwarfs undergoing thermonuclear runaways should spend the bulk of their time in a quiescent state. Rapid recurrence time scales are possible only at very high accretion rates; systems displaying fairly frequent outbursts are therefore expected to be very intense ultraviolet sources at quiescence.

III. Evolution of Symbiotic Novae

Applications of the basic theory of hydrogen shell flashes in the envelopes of accreting white dwarf stars to symbiotic novae have been described by Tutukov and Yungel'son (1976, 1982), Paczyński and Żytkow (1978), Paczyński and Rudak (1980), Iben (1982), Kenyon and Truran (1983), and Shustov and Tutukov (1985). Tutukov and Yungel'son (1976) noted that the high energy photons likely to be produced by a steady shell source could probably account for the blue continuum and high energy emission lines observed in most quiescent symbiotic systems, and suggested that slight increases in the mass accretion rate might result in an eruption (especially if $\dot{M}$ during the outburst is larger than $\dot{M}_{steady,max}$ as outlined above). They later investigated this model in more detail (Tutukov and Yungel'son 1982), and derived constraints on the red giant mass loss rate, $\dot{M}_{giant}$, and the binary separation, A, with the basic assumption of steady burning. Their results, 2 AU < A < 20 AU and 10^{-7} $M_\odot$ yr^{-1} < $\dot{M}_{giant}$ < 10^{-5} $M_\odot$ yr^{-1}, are in reasonable agreement with the orbital periods (Garcia, this volume) and mass loss rates (Taylor, this volume) found for most symbiotic stars.

Paczyński and Rudak proposed that outbursts of symbiotic stars naturally divide into two categories, which they called type I (Z And, CI Cyg, AG Dra) and type II (V1016 Cyg, V1329 Cyg, HM Sge). They suggested that type I symbiotics possess accreting white dwarfs with stable hydrogen burning shells; outbursts then result from variations in the accretion rate as described by Tutukov and Yungel'son. Type II symbiotic systems contain white dwarfs accreting at rates below $\dot{M}_{steady,min}$, and experience hydrogen shell flashes. Kenyon and Truran (1983) pointed out that this idea could be understood in terms of eruptions which occur in degenerate white dwarf atmospheres (some type II systems) and others which begin in non-degenerate envelopes (all type I systems and some type II objects). Thus, it is convenient to divide a discussion of the time behavior of symbiotic novae into the evolution of *degenerate* eruptions and the evolution of *non-degenerate* eruptions.

A. Degenerate Flashes

If the accretion rate onto a white dwarf is fairly low ($\dot{M} \leq 10^{-10}$ $M_\odot$ yr^{-1} or so), then the material can cool efficiently before it settles into the hydrogen-rich atmosphere (see Figure 1). In this situation, the temperature of the white dwarf atmosphere depends only on its luminosity, L_{wd}, which will be assumed to be fairly small ($L_{wd} \leq 0.1$ $L_\odot$). The white dwarf envelope is then *completely degenerate* when the runaway commences, which guarantees an explosive flash as noted above.

The rise to visual maximum during a degenerate shell flash is characterized by (i) a rapid increase in bolometric luminosity at nearly constant radius, followed by (ii) a slow expansion at constant bolometric luminosity into an A-F supergiant configuration (Figure 2). During phase (i) the visual brightness of the white dwarf remains fairly constant, although ionization of material expelled by the giant should produce optical emission lines. Once the eruptive star enters phase (ii), its visual brightness increases significantly and overwhelms that of its giant companion. The visual luminosity at maximum can be estimated from the core mass-luminosity relation of equation (3) to be ≈ 4000 $L_\odot$ for a typical (0.6 $M_\odot$) white dwarf, and can be compared to the 100-200 $L_\odot$ expected of a normal giant star.

The spectroscopic evolution of a degenerate shell flash can be outlined using spectrum synthesis techniques described by Kenyon and Webbink (1984). I will assume that the binary initially consists of an M2 III star in combination with a low luminosity, 0.8 $M_\odot$ white dwarf ($L_{wd} \sim 10^{-2}$ $L_\odot$) accreting material at a rate, $\dot{M} \sim 10^{-10}$ $M_\odot$ yr^{-1}. The optical/ultraviolet spectrum predicted for this configuration appears in the lower left panel of Figure 3.

The rise to visual maximum begins with an increase in effective temperature at nearly constant radius (phase (i)). The maximum effective temperature for a 1 $M_\odot$ object is $T_{hot} \sim 2 \times 10^5$ K, so the blackbody continuum of the rapidly evolving nova shifts from the UV into the X-ray spectral region. My models assume that all of the high energy photons from the hot object are locally absorbed by the red giant wind, so the nebular spectrum grows dramatically in strength as T_{hot} increases (Figure 3, upper left panel).

During the expansion phase (ii), the hot component evolves at constant luminosity from a very hot white dwarf into an A-F supergiant. The visual brightness of the system increases dramatically, and the emission lines become progressively weaker as T_{hot} decreases. The spectroscopic evolution during this constant luminosity phase can be visualized by comparing the upper left panel of Figure 3 ($T_{hot} \sim 2 \times 10^5$ K) with the upper right ($T_{hot} \sim 7 \times 10^4$ K) and lower right ($T_{hot} \sim 7 \times 10^3$ K) panels.

Following visual maximum, the hot component retraces its evolution in the H-R diagram. By analogy with classical novae, declining symbiotic novae should possess dense outflowing winds, and Wolf-Rayet features have been observed in several symbiotic novae (Kenyon 1986). Mass loss rates derived for these systems are typically $\sim 10^{-6}$ $M_\odot$ yr^{-1}, which is somewhat larger than the rate at which hydrogen is consumed in the burning shell ($\sim 5 \times 10^{-7}$ $M_\odot$ yr^{-1}). Thus, the length of time spent in the "on" phase, Δt_{on}, is apt to be somewhat shorter (perhaps a factor of ~ 3) than estimated by equation (7).

B. Non-Degenerate Flashes

If the accretion rate onto the white dwarf exceeds $\sim 10^{-9}$ $M_\odot$ yr^{-1}, then accreted material cannot cool before mixing with the white dwarf atmosphere. This heating of the white dwarf by newly arrived hydrogen-rich matter results in a non-degenerate envelope at the onset of the thermonuclear runaway, and causes a relatively weak shell flash.

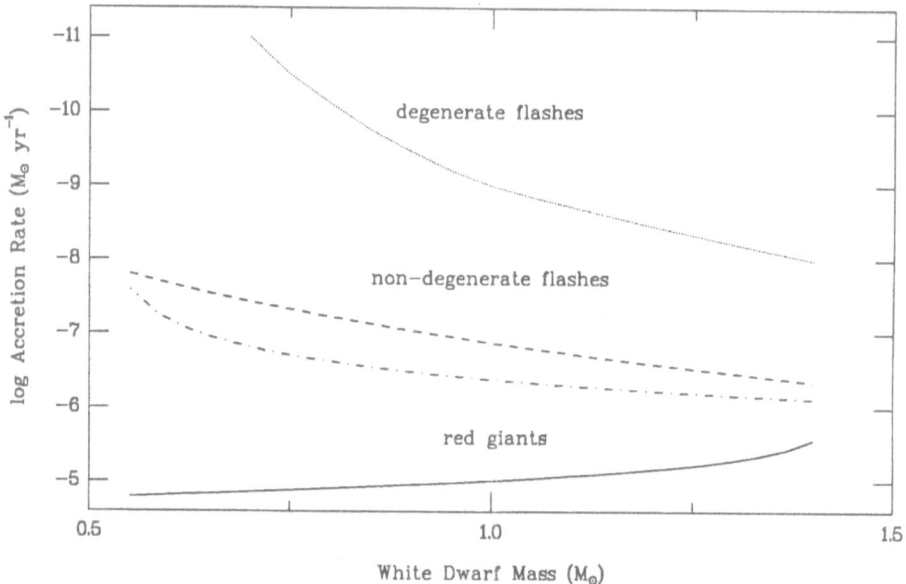

Figure 1 - Behavior of hydrogen shell flashes as a function of $\dot{M}$ and M_{wd} for solar CNO abundance and $L_{wd} \lesssim L_{accr} \lesssim 0.1 \, L_{\odot}$. Steady-burning configurations are allowed between the dashed and dot-dashed lines. The Eddington limit has been plotted as the solid line at the bottom of the Figure.

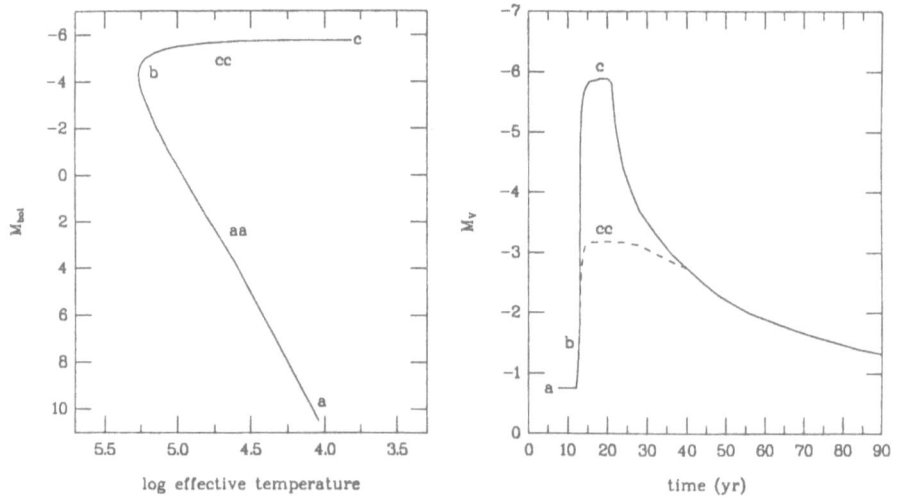

Figure 2 - Evolution of a thermonuclear runaway on a 0.8 $M_{\odot}$ white dwarf star. The left panel shows the behavior of the system in the H-R diagram, while the right panel displays theoretical light curves for degenerate (a-b-c) and non-degenerate (aa-b-cc) configurations.

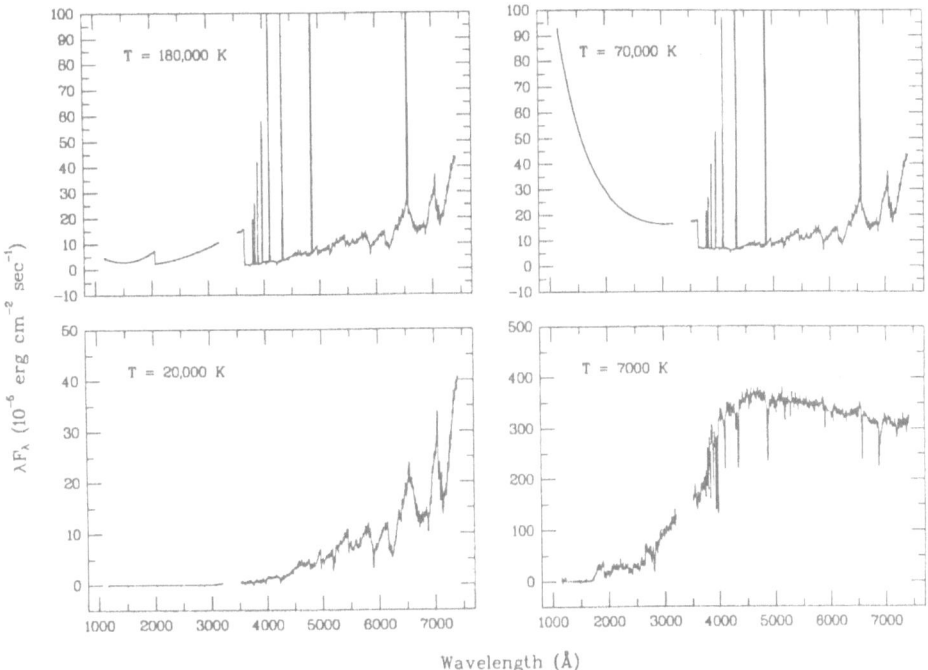

Figure 3 - Synthetic spectra of a symbiotic nova as a function of effective temperature during the eruption. The models combine spectra of an M5 giant (V = -1.5) with a hot stellar source at the indicated effective temperatures and an ionized nebula at a gas temperature of 10,000 K. The plotted spectra correspond to marked points in the H-R diagram of Figure 2.

Unlike the degenerate shell flashes described above, weak non-degenerate flashes do not evolve into A-F supergiants and remain very hot throughout the eruption. Although these systems reach the plateau luminosity defined in equation (6), they make smaller excursions in the H-R diagram (e.g., aa-b-cc in Figure 2) and have a correspondingly smaller increase in visual brightness (Iben 1982; Kenyon and Truran 1983).

The spectroscopic evolution of a non-degenerate shell flash during the expansion phase (i) is essentially identical to that of a degenerate flash, although objects with comparatively weak eruptions may require a longer period of time to reach bolometric (and visual) maximum. The change in the optical/ultraviolet spectrum during phase (i) can be visualized by comparing the left panels of Figure 3.

These systems spend a very short amount of time at temperatures, $T_{hot} \lesssim 10^5$ K in the constant luminosity phase (ii), and thus remain intense emission line sources throughout their eruptions. An example of a possible spectrum for such an object near visual maximum is shown in the upper right panel of Figure 3.

IV. The Symbiotic Novae

The theory of hydrogen shell flashes predicts that accreting white dwarf stars will display two types of eruptive behavior depending on the physical parameters of the white dwarf (M_{wd} and L_{wd}) and of the accreted material ($\dot{M}$ and Z_{CNO}). Degenerate flashes occur when the accretion luminosity ($L_{accr} \propto M_{wd} \dot{M}$) and the white dwarf luminosity are small ($L_{wd} \lesssim L_{accr} \lesssim 0.1\ L_\odot$) and the mass fraction of CNO nuclei is large ($Z_{CNO} > Z_{CNO,\odot}$). White dwarfs undergo non-degenerate flashes when L_{wd} or L_{accr} is large ($\gtrsim 1\ L_\odot$).

Detailed comparisons of the theory with observations suggest that both types of eruptions actually occur in symbiotic novae (Kenyon 1986; Viotti, this volume). It appears that AG Peg, RT Ser, RR Tel, and PU Vul have undergone degenerate flashes, and their evolution following visual maximum is consistent with the constant luminosity phase, $L_{bol} \sim L_{plateau}$, described above. Other symbiotic novae, V1016 Cyg, V1329 Cyg, and HM Sge, seem to have experienced non-degenerate flashes, as evidenced by their well-developed emission-line spectra at visual maximum.

Additional observations of these systems as they decline from visual maximum are important tests of the theory of hydrogen shell flashes. The newest symbiotic nova, PU Vul, is of special interest in this regard, because it has not yet evolved to high effective temperatures. Evolution of its spectrum, particularly in the X-ray and radio regions, should be most instructive.

References

Fujimoto, M.Y. 1982a. *Astrophys. J.*, **257**, 752.
Fujimoto, M.Y. 1982b. *Astrophys. J.*, **257**, 767.
Gallagher, J.S. and Starrfield, S.G. 1978. *Ann. Rev. Astr. Astrophys.*, **16**, 171.
Iben, I. Jr. 1982. *Astrophys. J.*, **259**, 244.
Iben, I. Jr. and Renzini, A. 1983. *Ann. Rev. Astr. Astrophys.*, **21**, 271.
Kenyon, S.J. and Truran, J.W. 1983. *Astrophys. J.*, **273**, 280.
Kenyon, S.J. and Webbink, R.F. 1984. *Astrophys. J.*, **279**, 252.
Kutter, G.S. and Sparks, W.M. 1980. *Astrophys. J.*, **239**, 988.
MacDonald, J. 1980. *Mon. Not. Roy. Astr. Soc.*, **191**, 933.
MacDonald, J. 1983. *Astrophys. J.*, **267**, 732.
Nariai, K., Nomoto, K., and Sugimoto, D. 1980. *Pub. Astr. Soc. Japan*, **32**, 472.
Paczyński, B. 1970. *Acta Astr.*, **21**, 417.
Paczyński, B. and Rudak, B. 1980. *Astr. Astrophys.*, **82**, 349.
Paczyński, B. and Żytkow, A. 1978. *Astrophys. J.*, **222**, 604.
Prialnik, D. 1986. *Astrophys. J.*, **310**, 222.
Prialnik, D., Livio, M., Shaviv, G., and Kovetz, A. 1982. *Astrophys. J.*, **257**, 312.
Shustov, B.M., and Tutukov, A.V. 1985. *Recent Results on Cataclysmic Variables* (ESA SP-236), p. 113.
Sion, E.M., Acierno, M.J., and Tomczyk, S. 1979. *Astrophys. J.*, **232**, 832.
Sion, E.M., and Starrfield, S.G. 1985. in *Recent Results on Cataclysmic Variables* (ESA SP-236), p. 17.
Starrfield, S., Sparks, W.M., and Truran, J.W. 1984, *Astrophys. J.*, *291*, 136.
Tutukov, A.V. and Yungel'son, L.R. 1976. *Astrofizika*, **12**, n21.
Tutukov, A.V. and Yungel'son, L.R. 1982. in *IAU Colloquium No. 70, The Nature of Symbiotic Stars*, ed. M. Friedjung and R. Viotti (Dordrecht: Reidel), p. 283.
Uus, U. 1970. *Nauk Informatsii*, **17**, 32.

On the outburst of symbiotic stars

H. Nussbaumer and M. Vogel
Institute of Astronomy
ETH Zentrum
8092 Zürich (Switzerland)

SUMMARY. The most commonly advocated models for explaining the outburst of symbiotic systems are thermonuclear runaways, modulated mass accretion through a disc, or
accretion disc instabilities. We show that the onset of a strong stellar wind on the cool
component can also produce the observed increase in luminosity.

1. The proposed mechanism

In 1983 Nussbaumer and Schmutz published a study on HBV 475 (= V 1329 Cyg) in which
they suggested, as one of several possibilities, that the nebula observed in symbiotic systems
may actually be the wind of the cool star, ionized by the radiation of the hot component.
Nussbaumer and Vogel (1987) have carried that idea through a properly calculated model.
As a qualitative new idea they propose that the outburst of symbiotic systems can also be
explained by the onset of a strong stellar wind on the cool component. The growing wind
presents an increasing target for the ionizing photons. A wind of 20 km/s creates within
one year a target with a radius of $\approx 6 \cdot 10^{13}$cm, which is probably a typical distance for the
binary separation in symbiotic systems. The outburst of symbiotic stars occurs typically
over a period of several months, up to well over one year. There are two critical questions:

(1) How likely is the simultaneous presence of an M–giant in a state of heavy mass–loss,
    and a hot component with a radiation temperature of $T^* \gtrsim 10^5$K.
(2) What are the outburst magnitudes, ΔM, that can be reached.

White dwarfs are candidates for hot components. Their lifetimes at $T^* > 10^5$K, however, is
only about 100 000 years (Koester and Schönberner, 1986). The simultaneous presence of a
hot white dwarf and a red giant with massive mass–loss would therefore be best realised in
double star systems where both components are of about equal mass. Their evolutionary
times would therefore be about equal.

J. Mikolajewska et al. (eds.), The Symbiotic Phenomenon, 169–170.

Figure 1.
Luminosity variation in the B–band as a function of mass–loss rate. The dashed line shows the contribution of the nebular continuum to the total luminosity. The Figure does not include the contribution of the cool star. For mass loss rates $\gtrsim 10^{-6}$ $M_\odot$ yr^{-1} the luminosity in the emission lines is reduced due to collisional de–excitation.

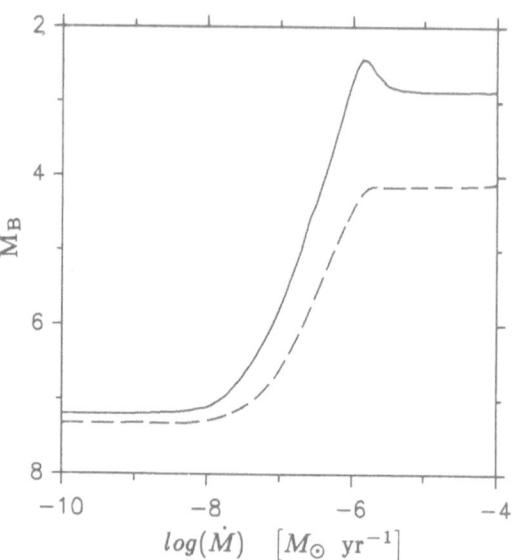

In Fig.1 we show the luminosity increase in the B–band due to the onset of a strong stellar wind. The model assumes the following physical conditions : A hot star with radius $R^* = 9 \cdot 10^8$ cm radiates as a blackbody with radiation temperature $T^* = 1.5 \cdot 10^5$K. At a distance $p = 5 \cdot 10^{13}$cm a cool star is losing mass through a wind of the form

$$\dot{M} = 4\pi r^2 \mu m_H N(r) v_\infty \left(1 - \frac{R}{r}\right) \quad .$$

The mean molecular weight is taken to be $\mu = 1.4$, the terminal velocity is $v_\infty = 80$ km/s and the origin of the stellar wind lies at $R = 50 R_\odot$. $N(r)$ is the hydrogen density by number, r is the distance from the center of the cool star and m_H is the mass of the hydrogen atom.

2. Conclusions

The outburst–model presented here is not meant to be exclusive. Thermonuclear runaways and unstable disc accretion are theoretically viable candidates for producing the observed outbursts. However, the symbiotic phenomenon may not be due to a unique event, valid in every system. Our wind outburst model has the right properties to match observations. We are not aware of any observational or theoretical objections excluding the possibility of a rapid onset of massive mass–loss through a wind in red giants.

Acknowledgments: M. Vogel thanks the Swiss Science Foundation and the Swiss Society of Astronomy and Astrophysics for financial support.

References:

Koester, D., Schönberner, D.: 1986, *Astron.Astrophys.* **154**, 125
Nussbaumer, H., Schmutz, W.: 1983, *Astron.Astrophys.* **126**, 59
Nussbaumer, H., Vogel, M.: 1987, *Astron.Astrophys.* **182**, 51

THERMONUCLEAR RUNAWAYS ON ACCRETING HOT WHITE DWARFS

Dina Prialnik
Racah Institute of Physics, Hebrew University
Jerusalem, Israel

Oded Regev
Department of Astronomy, Columbia University
New York, NY 10027, U.S.A.

ABSTRACT. Preliminary results of 1–D fully hydrodynamic numerical calculations including accretion and mass loss on a $1 M_\odot$ hot white dwarf are presented. For the case of $\dot{M} = 10^{-8} M_\odot/\text{yr}$ and $T_{eff} = 125000°\text{K}$ we get flashes of duration of $\sim 30 - 40$ yrs which reccur every ~ 1500 years. During each flash $\sim 6 \times 10^{-6} M_\odot$ is lost ($\sim 40\%$ of the accreted envelope.

Previous work on thermonuclear runaways on hot white dwarfs suggested their applicability to symbiotic stars. See *e.g.* Paczyński & Rudak (1980), Iben (1982), Kenyon and Truran (1983). None of these works are both hydrodynamic and include accretion.

We use the method and computer code described in Prialnik (1986) to follow hydrodynamically the accretion and subsequent evolution of the outburst, including the fate of the envelope. In the project we intend to survey the $\dot{M} - T_{eff}$ parameter space systematically to find the regions of steady burning, nondegenerate flashes and degenerate flashes. Of interest is also the amount of mass lost in these events and their temporal characteristics.

Here we report the results of one case ($\dot{M} = 10^{-8} M_\odot/\text{yr}, T_{eff} = 125000°\text{K}$). The results are presented in the figures. In figures 1 and 2 the luminosity and the nuclear luminosity respectively is shown as a function of time. Two flashes with an interval of ~ 1500 years between them are displayed. The structure of a flash is shown in figures 3 and 4. Figure 3 shows the typical behavior of the luminosity (full curve) and the nuclear luminosity (dashed curve) as a function of time during a flash and in figure 4 the same is shown with regard to the effective temeprature $-T_{eff}$.

These results are similar to previous work (*e.g.* Iben 1982). During each flash we have computed, a mass of $\sim 6 \times 10^{-6} M_\odot$ was lost from the system (achieved escape velocity and was removed from the calculation).

We intend to continue these calculation also for other values of M and T_{eff}.

REFERENCES: 1. Paczyński, B. and Rudak; 1980, *Astr. Astrophys.*, **82**, 349.
2. Iben, I., 1982, *Astrophys. J.*, **259**, 244.
3. Kenyon, S.J. and J.W. Truran, 1983, *Astrophys. J.*, **273**, 280.
4. Prialnik, D., 1986, *Astrophys. J.*, **310**, 1986.

J. Mikolajewska et al. (eds.), The Symbiotic Phenomenon, 171–172.
© 1988 by Kluwer Academic Publishers.

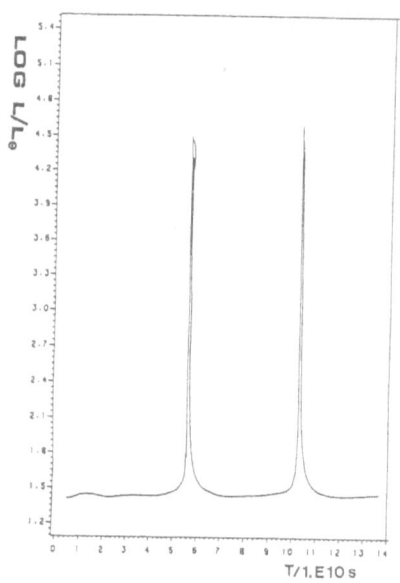

Figure 1
Luminosity as function of time

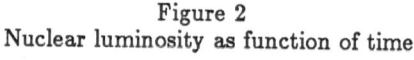

Figure 2
Nuclear luminosity as function of time

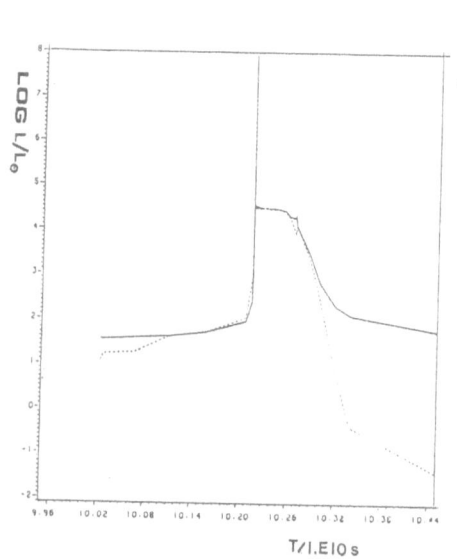

Figure 3
Luminosity as function of time

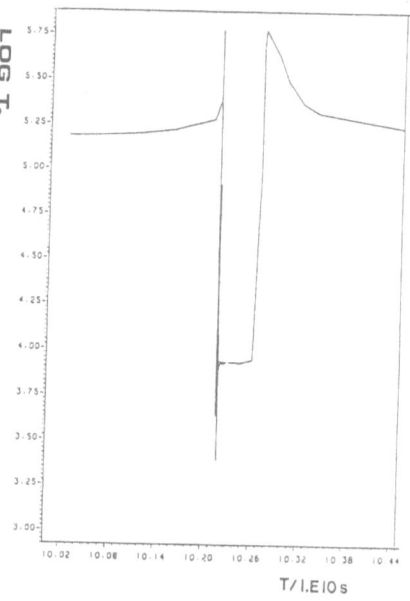

Figure 4
T_{eff} as function of time

THE DECAY TIME AFTER A THERMONUCLEAR FLASH

Marina Orio
Dept. of Physics
Technion
32000 Haifa
Israel

ABSTRACT. The drag luminosity produced by a secondary embedded in a nova envelope is calculated and it is assumed to liberate gravitational energy. The results of a preliminary calculation show that the time for envelope ejection and therefore for the light decay varies sharply with the binary distance. It would be considerably reduced if the binary distance is a < 2 R_0. It is suggested that this is the main difference between the thermonuclear runaway of symbiotic stars and nova systems, instead of other physical parameters.

Kenyon & Truran (1983) pointed out at the importance of a known difference between symbiotic systems and nova systems: the dimensions of the binary systems. The orbital periods of symbiotic stars are of years, those of novae of hours. The different order of magnitude of the energetics of novae can be understood considering that the binary separation in novae (10^{11} cm vs. 10^{13} cm of symbiotics) implies interaction of the secondary with the expanding envelope, since the time of the optical miximum. McDonald (1980) and McDonald et al. (1985) estimated that this additional energy contributes to blow off the nova envelope in a much shorter time than it would be otherwise, switching off the nuclear reactions and causing the rapid decay of many novae.
We considered the structure of a slow nova shell (v = 300 km/sec) obtained in a full scale hydrodynamical calculation of Prialnik (1985), with luminosity at maximum L = 3.41 10^4 $L_\odot$ and effective temperature T = 7128 K. We assumed mass outflow at a constant rate $\dot{m}_0$ and variations of the density profile with a law that matches the subsequent evolution of the model. The white dwarf, of 1.25 $M_\odot$, had accreted 10^{-7} $M_\odot$ at a rate 10^{-11} $M_\odot/yr$ before the outburst. Envelope ejection was completed in about 345 days.
Supposing that a secondary of 0.5 $M_\odot$ orbits in the ejected shell, the drag luminosity is:

$$L_D = \pi R_2 \rho V_{orb}^3 \qquad (1)$$

where we assumed for the density ρ the average value in the layers of

173

nova shell around the secondary, that rotates with a period almost equal to the orbital one (a few hours). If all the drag energy contributes to liberate gravitational energy, additional mass flows from the radius $R = a$ (where a is the binary separation) at a rate:

$$\dot{m}_D = \frac{L_D a}{GM} \qquad (2)$$

The total mass outflow from a fixed radius (the photospheric boundary at maximum) is in this case:

$$\dot{m} = \dot{m}_o + \dot{m}_D \qquad (3)$$

The density profile also varies more sharply beyond the secondary. When all the mass beyond the radius $r = a$ has been lost, it is just:

$$\dot{m} = \dot{m}_o \qquad (4)$$

If not all the drag energy liberates gravitational energy, it also contributes to increase the temperature, effecting the radiation pressure. It was not calculated yet how the mass outflow would vary in this case. We also neglected accretion of material by the secondary, whose accretion radius is of the same order of the stellar radius (the Roche lobe radius). It was verified that accretion with the Bondi Law would mean that most of the envelope (95%) is accreted by the secondary and not ejected from the system: this is very unrealistic. It is not clear if and how much material is accreted and only a full scale hydrodynamical calculation could give an answer. We solved the angular momentum equation at each time step and checked that the binary distance varied of an irrelevant factor(10^{-6}%). Shara et al. have shown that accretion of a small fraction of the envelope would cause an increase of the separation, but we checked that this uncertainty does not significantly change the result that we show in Fig. 1.

The mass outflow rate due to the drag luminosity is, on the contrary, comparable to $\dot{m}_o$ and very significant. If the time required for ejection of the envelope is the typical decay time of the light curve, that for return at the pre-outburst magnitude, the drag luminosity can transform a slow nova in a very fast nova, if the secondary is at distance $a < 2\ R_o$. We show the variation of the decay time with the binary distance in Fig. 1. In a successive calculation we intend to check the effect of the drag energy on a symbiotic-type shell produced by a weak thermonuclear flash like in the model of Prialnik & Regev (1987). The significant difference between the slow flash of a symbiotic and the sudden burst of a nova could not only (or not at all?) be due to the higher luminosity of the white dwarf and to the mass transfer rate higher than $10^{-8}\ M_\odot/yr$ (see Paczynski & Rudak, 1983, Iben, 1977, Prialnik & Regev, 1987), but mainly to the greater binary distance of symbiotic star.

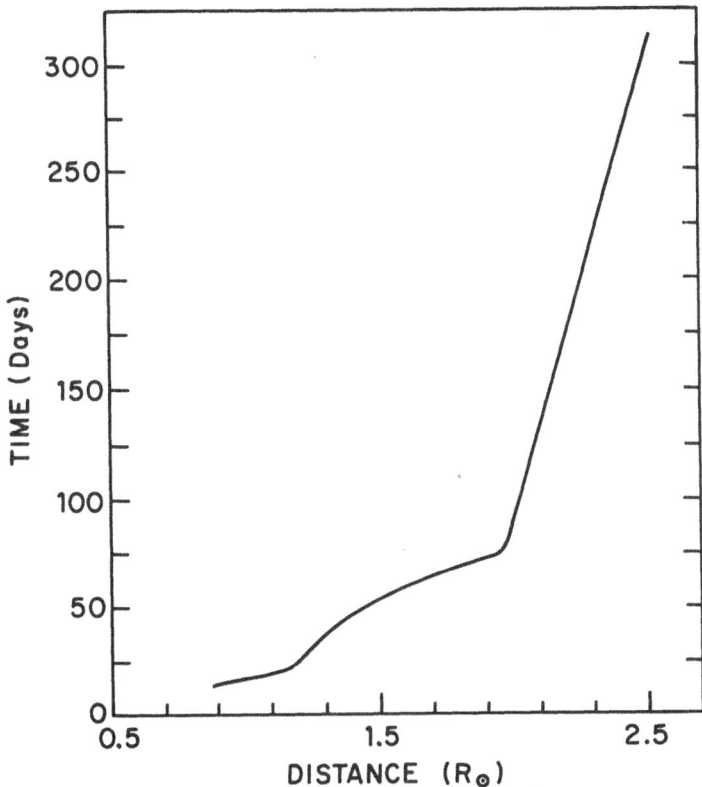

Figure 1. Variation of the time for envelope ejection with the binary
distance.

References

Kenyon S.J., Truran J.W., 1983, Ap. J., 273, 280.
Iben I., 1977, Ap. J., 217, 788.
Mc Donald J., 1980, M.N.R.A.S., 182, 35.
Mc Donald J., Fujimoto M.Y., Truran J.W., 1985, Ap. J., 294, 263.
Paczynski B., Rudak B., 1980, Astr. Ap., 82, 349.
Prialnik D., Regev O., 1987, this conference.
Shara M.M., Livio M., Moffat A.F.J., Orio M., 1986, Ap. J., 311, 163.

GENERAL DISCUSSION ON THE PHYSICS OF SYMBIOTIC STARS

Scott J. Kenyon
Smithsonian Astrophysical Observatory
Harvard-Smithsonian Center for Astrophysics
60 Garden Street
Cambridge, MA 02138 USA

ABSTRACT. This paper reviews the general discussion on the physics of symbiotic stars.

R. Webbink led this discussion, which opened when H. Nussbaumer wondered if disc accretion leads naturally to bipolar outflows and coronae (among other high energy phenomena). Webbink and M. Livio replied that calculations of discs in cataclysmic variable systems (CVs) suggest that coronae do form in discs. S. Kenyon and P.-L. Selvelli said that winds are observed in other systems known to possess discs, such as CVs, T Tauri stars, and the FU Orionis variables. A disc wind is required to understand line profiles in F Ori objects, but not for CV's or T Tauri stars. Disc winds may still be important for CVs, but have not been demonstrated observationally.

The discussion then moved on to a consideration of interacting winds. B. Yudin commented that we generally think of mass transfer from the cool star to the hot star, rather than the reverse. Webbink remarked that in the case of AG Dra, the observations suggest that the hot component is always quite luminous and wondered if a hot star wind might prevent accretion of needed material from the cool companion. R. Stencel suggested that instabilities in the interaction region might lead to some accretion.

As the next topic, Webbink noted that the observations regarding mass loss rates and infrared excesses indicated that many D-type symbiotics contain extreme asymptotic giant branch (AGB stars) near the end of their red giant evolutionary phase, and asked if these stars (i) were normal and (ii) if they then could give us any information concerning the late stages of single star evolution. These questions provoked a lively discussion. D. Allen commented that the single star counterparts to the Mira symbiotics are obscured *IRAS* sources and questioned whether there are enough D-types for a statistically significant sampling of single star evolution. P. Whitelock commented that the obscuration events observed in Mira symbiotics complicate the interpretation, and said (in response to a question by O. Regev) that single Miras have much smaller events (if they have them at all). A. Magalhaes remarked that the semi-regular variable L_2 Puppis displays activity similar to that observed in the Mira symbiotics described by Whitelock, but Whitelock emphasized that normal Miras do not display such behavior. Nevertheless, I think it is important to understand the events in L_2 Puppis and to determine if they have any relation to the obscuration events observed in Mira symbiotics.

Livio continued the debate by asking if red giants can lose mass more rapidly than would be estimated by a simple Reimers law (for example). R. Tylenda reminded the audience that the transition from a normal AGB star into a planetary nebula is believed to be achieved as a result of a "superwind" which transforms a Mira variable into an OH/IR star

J. Mikolajewska et al. (eds.), The Symbiotic Phenomenon, 177–178.

and then into a planetary nebula nucleus. Allen noted that OH/IR stars possess mass loss rates of roughly 10^{-4} $M_\odot$ yr^{-1} and suggested H1-36 as an example of an OH/IR star which just happens to have a hot white dwarf companion star.

Changing topics, Webbink pointed out that the radial velocity data collected by Garcia and Kenyon demonstrate that reliable orbits can now be obtained for the cool components and cautioned that published analyses of orbital solutions based on emission line radial velocities are far from convincing. Given the complicated emission line profiles, Webbink asked, how is it possible to determine accurate mass ratios for symbiotic stars? M. Slovak said that reasonable radial velocity solutions can be obtained using the wings of strong emission lines in CVs, but Webbink noted that lines in CVs are certainly formed in or near the disc, while the location of the line-emitting region in symbiotics is not as obvious. Stencel agreed, and suggested that emission lines might move with the cool component in some objects. Chochol followed by commenting that the centroid of a broad, variable emission line is rather difficult to determine whether it follows the motion of the cool component or the hot component. J. Mikolajewska proposed that since the eclipse behavior of He II in CI Cyg demonstrates an association with the hot component, one might begin with a detailed study of emission line velocities in this object. A. Michalitsianos replied that narrow, rather easily measured emission lines made SY Mus a good candidate for radial velocity observations. Several cautionary (pessimistic!) voices reminded the audience that it is (i) necessary to measure relative motions of various emission lines and red giant absorption lines very carefully (Nussbaumer), (ii) important to remember that the relative phases of radial velocity maxima of emission lines such as H I and He I are known to change with time (e.g., AG Peg - Kenyon), and (iii) essential to keep an open mind concerning the location and geometry of the ionized region and the importance of variability (Allen).

This fast-paced discussion closed with a plea from R. Viotti to compare high quality observational data with equally high quality models.

SESSION 3. PHYSICS OF INDIVIDUAL OBJECTS

"Nature is malicious"

Pier Luigi Selvelli

The famous H. Draper astrograph, since 1947 at the Torun Observatory.
The first symbiotics were discovered by A. Cannon with this telescope
during her work on the HD catalogue.

Z ANDROMEDAE: QUIESCENCE AND ACTIVITY

A. Cassatella
IUE Observatory, ESA,
P.O. Box 54065, 28080 Madrid, Spain

T. Fernandez-Castro
Planetario de Madrid, Parque Tierno Galvan
28045 Madrid, Spain

N. Oliversen
Astronomy Programs, Computer Sciences Corporation
NASA Goddard Space Flight Center, Greenbelt,
Maryland 20771, U.S.A.

1. INTRODUCTION

Z Andromedae, often considered as the prototype of symbiotic stars, experimented, after several years of quiescence, a small outburst in March-April 1984, followed by a larger one in September-October 1985. The imminence of a new activity phase was predicted by Viotti et al. (1982). Z And is, together with AG Draconis, the only symbiotic star observed both during quiescence and activity with the IUE satellite. The early photometric and spectroscopic history of Z And has been recently reviewed by Kenyon (1986).

2. Z And during quiescence

The behaviour of Z And in the ultraviolet during quiescence has been studied by Fernández-Castro et al. (1988), on the basis of data obtained by the International Ultraviolet Explorer from 1978 to 1982. In that period of time, the UV continuum and the emission line fluxes varied quasiperidically with a period of about two years, in phase with the Hα variability found by Altamore et al. (1979), with the UBV photometry by Belyakina (1985) and also in

1) Based on observations by the International Ultraviolet Explorer; (2) Affiliated with the Astrophysics Division, Space Sciences Department

J. Mikolajewska et al. (eds.), The Symbiotic Phenomenon, 181–186.
© *1988 by Kluwer Academic Publishers.*

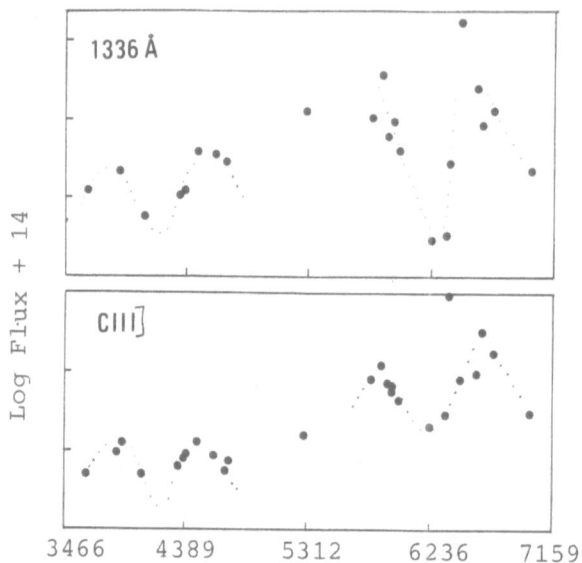

Figure 1: Time evolution (JD-2440000) of the continuum flux at 1336 A and of the emission line flux in C III] 1909 A.

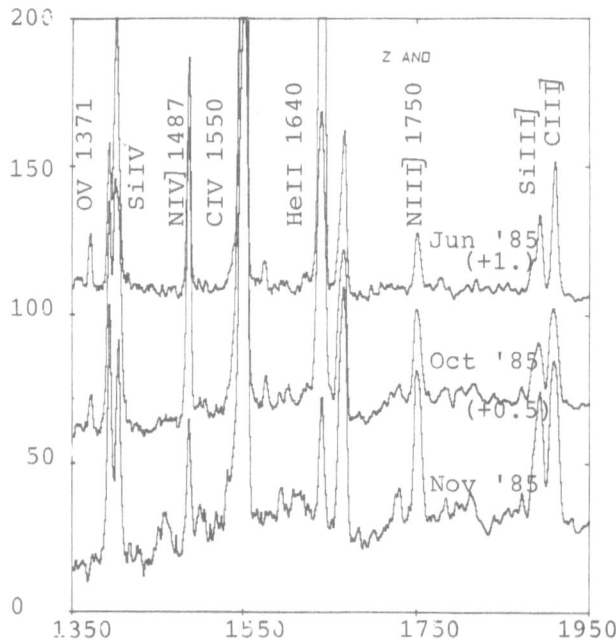

Figure 2: Spectral variability of Z And in the far UV shortly before (June 1985) and during the outburst (October and November 1985). Ordinates are observed fluxes in units of 10^{-14} erg cm^{-2} s^{-1} A^{-1}.

agreement with the ephemeris given by Kenyon and Webbink
(1984). The electron density derived from Si III]/C III]
flux ratio, also varied in phase with the UV flux. In
particular, a correlation was found between the electron
density variations, and the UV continuum flux at 2900 A.
This fact, together with the presence of a Balmer jump in
emission, indicates that the principal contributor at those
wavelengths is nebular emission, mainly free-bound
transitions.

The temperature of the hot source
($T_h \geq 10^5$ K), estimated with the Zanstra method
using the He II 1640 A emission line and the continuum flux
at ~ 1400 A, remained roughly constant, suggesting that no
significant changes of the ionization temperature occurred
during quiescence.

In the near-infrared, Z And remained practically
constant, even during activity phases. From IRAS data,
there is no significant indication of infrared excess due
to dust emission in the 12 μm and 25 μm bands (Kenyon,
Fernández-Castro and Stencel 1987).

The IR reddening corrected (E(B-V) = 0.35; Viotti et
al. 1982) energy distribution of Z And in quiescence can be
fitted by a black body of 3200 K which corresponds to a
spectral type M4 on the Ridgway et al. (1980) scale, in
good agreement with the recent determination by Kenyon and
Fernández-Castro (1987a), who derived a spectral type M3.5
from the analysis of molecular absorption bands in red
spectra.

Although the assumption of a luminosity class is very
crucial for the model, we can adopt with reasonable
confidence that the cool component of Z And is a giant, as
determined by Kenyon and Fernández-Castro (1987a).

The radio spectral index of Z And in quiescence, as
observed by Seaquist, Taylor and Button (1984), is α =
0.62, close to the expected value for a completely
photoionized wind. The corresponding mass loss rate,
calculated from Wright and Barlow (1975) is 2.0 x
10^{-7} M_0/yr, while the accretion rate onto the
compact object represents a small fraction of the total mass
lost (2 %). Therefore, the accretion luminosity is at least
~ 40 times smaller than that provided by the recombination
continuum, for any reasonable range of masses assumed,
implying that accretion effects from the giant play a
negligible role in the energetics of the system, as in the
case of RW Hya discussed by Kenyon and Fernández-Castro
(1987b).

The case of Mira Ceti is substantially different from
that of Z And. In Mira Ceti the accretion luminosity is of
the same order as that of the recombination continuum.
Reimers and Cassatella (1985) and Cassatella et al. (1985)
suggested that in Mira Ceti accretion from the wind of the

184

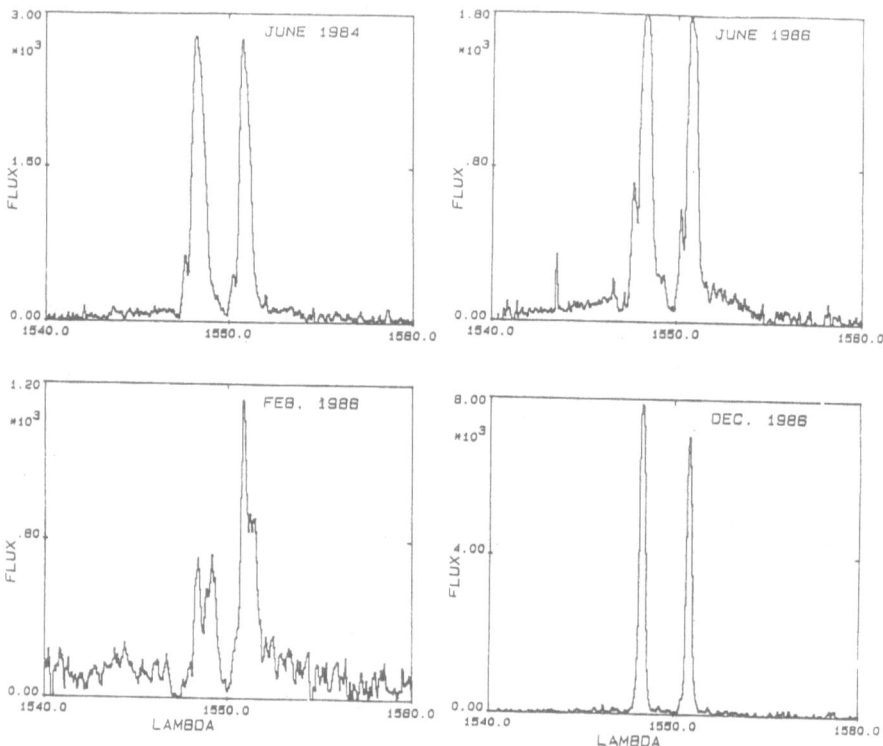

Figure 3: Evolution of the line profiles of the C IV 1550 A
doublet at four different dates during the outburst of Z
And. Note the complex structure of the profiles in February
1986.

red giant forms a optically thin accretion disk around the
white dwarf which gives raise to the broad emission lines
and the Balmer continuum observed in the UV. Z And does not
present in the UV, at least during quiescence, any feature
that could be ascribed to line formation in an accretion
disk.

3. THE OUTBURST OF Z AND

Z And, after a long period of quiescence, entered a new
active phase undergoing a small outburst in March-April
1984, lasting less than 200 days, followed by a larger one
in September-October 1985 with a duration of about 250 days.
    The behaviour of Z And in the ultraviolet during the
recent activity phases can be summarized as follows:
    a) The regular variations of the UV continuum and of
the emission line fluxes observed during quiescence continue
during activity, but are superimposed to variations having
the time scale of the V light curve (Mattei 1987).Examples
of the time evolution of the UV continuum and of the
emission lines in the period 1978-1987 are shown in Fig. 1.
    b) The UV energy distribution of Z And in outburst
differs strongly from that in quiescence: the hot stellar
continuum almost disappeared during activity. The fainting
of the hot continuum is accompanied by a decrease of the
degree of ionization of the emission lines. An example of
these changes for the period June-November 1985 is shown in
Fig. 2.
    c) A general broadening of the emission lines is seen in
the high resolution IUE spectra obtained during the active
phase in 1985-1986. The same behaviour was observed by
Viotti et al. (1984) during the outburst of AG Draconis in
1981-1982. In February 1986, the profiles of the emission
lines were broad and had a complex structure, as it can be
realized from Fig. 3, showing the C IV 1548-1550 A
doublet. Another interesting aspect is that the flux ratio
CIV 1548/ CIV 1550 A changed dramatically during activity
and, in February 1986, became lower than what expected in
the optically thick case. This fact might be indicative of
the development of a high velocity wind from the hot
component during outburst.

4. THE MODEL OF Z ANDROMEDAE

The UV to near infrared energy distribution of Z And can be
modeled, according to Fernandez-Castro et al. (1988), in
terms of three components: a nonvariable cool giant (M3.5),
a hot component with $T_h \geq 10^5$ K and
$R_h = 0.07$ R_0 which accretes only a small fraction

of the mass lost by the red giant, and a nebula, photoionized by the hot source. This model is analogous to that proposed by Boyarchuk (1968). The nebula is bow-shaped, as also supported by radio data (Seaquist, Taylor and Button 1984), and is responsible for the observed gas continuum and for the emission lines. The regular variability during quiescence could be ascribed to partial occultation by the cool star or to eccentricity effects. A discussion on the possible models of Z And is given by Yudin (1986).

As for the active phases of Z And, the data recently obtained are being analyzed in order to gain insight on the outburst mechanism.

REFERENCES

Altamore, A., Baratta, G.B. and Viotti, R.: 1979, Inf. Bull. Var. Stars, No. 1636.
Belyakina, T.S.: 1985, Inf. Bull. Var. Stars, No. 1611.
Boyarchuk, A.A.: 1968, Soviet Astron. 11, 818
Cassatella, A., Holm, a., Reimers, D., Ake, T. and Stickland, D.J.: 1985, M.N.R.A.S. 217, 589.
Fernández-Castro, T., Cassatella, A., Giménez, A. and Viotti, R.: 1988, Astrophys. J., 324, in press
Kenyon, S.J.: 1986, The Symbiotic Stars, Cambridge Univ. Press
Kenyon, S.J. and Fernández-Castro, T.: 1987a, Astron. J. 93, 938.
Kenyon, S.J. and Fernández-Castro, T.: 1987b, Astrophys. J. 316, 427.
Kenyon, S.J., Fernández-Castro, T. and Stencel, R.E.: 1987, Astron. J., submitted.
Kenyon, S.J. and Webbink, R.F.: 1984, Astrophys. J., 279, 252.
Mattei, J.: 1987, these Proceedings
Reimers, D. and Cassatella, A.: 1985, Astrophys. J. 297, 275.
Ridgway, S.T., Joyce, R.R., White, N.M., Wing, R.F.: 1980, Ap. J. 235, 126
Seaquist, E.R., Taylor, A.R. and Button, S.: 1984, Astrophys. J. 284, 202.
Viotti, R., Altamore, A., Baratta, G.B., Cassatella, A. and Friedjung, M. 1984, Astrophys. J. 283, 226.
Viotti, R., Giangrande, A., Ricciardi, O. and Cassatella, A.: 1982, in IAU Coll. 70, The Nature of Symbiotic Stars, ed. M. Friedjung and R. Viotti (Dordrecht: Reidel), p. 125.
Yudin, B.F.: 1986, Sov. Astron. 30, 84
Wright, A.E. and Barlow, M.J.: 1975, M.N.R.A.S. 170, 41.

CI CYGNI - THE WELL UNDERSTOOD SYMBIOTIC BINARY ?

J. Mikolajewska & M. Mikolajewski
Institute of Astronomy, Nicolaus Copernicus University
Chopina 12/18, PL-87100 Torun, Poland

ABSTRACT. The observed properties of CI Cyg are reviewed. Spectroscopic and photometric changes due to orbital effects as well as intrinsic variability of the components are discussed.

1. GENERAL CHARACTERISTICS

CI Cyg is among the best studied "prototype" symbiotic stars. It is also important as one of the few eclipsing symbiotics. Its physical nature seems to be well established and beyond all question (Roche lobe overflow, disk accreting main sequence star).

Since its discovery on a Harvard objective prism plate by A. Cannon (Shapley 1922), CI Cyg has been studied many times, both photometrically and spectroscopically. Its main optical characteristics are: a composite spectrum with Balmer series and high excitation emission lines (e.g. HeII, HeI, [OIII], [FeVII], [NeV]) superimposed on a late type continuum with low excitation absorptions (TiO bands, FeI), and quasi-periodic activity with outbursts observed in 1911, 1937, 1971-73 and 1975.

The optical spectrum of CI Cyg in outburst is radically different from that observed in quiescence. The rise in brightness is accompanied by an increase in the intensity of the blue continuum and a weakening of the high ionization emission line intensity as well as of the relative intensity of the TiO absorption bands. At maximum light, the spectrum resembles that of an FO-5 supergiant with superimposed HI and HeI emission lines. As the star declines from maximum and the blue continuum diminishes in intensity, the excitation of the emission line spectrum increases and the TiO absorption bands return.

The *IUE* spectra of CI Cyg show a bright flat continuum, up to about 2500 Å dominated by the Balmer continuum ($T_e \sim 20000$ K), with numerous strong emission lines. Kenyon & Webbink (1984) found that it is understood best as an accretion disk surrounding a main seaquence star. Their derived accretion rate, $\sim 10^{-5}$ $M_\odot yr^{-1}$, requires a Roche lobe filling giant companion. The available IR and optical data suggest the cool star may have supergiant dimensions (luminosity class II) and fill its tidal lobe (Kenyon & Gallagher 1983; Kenyon & Fernandez-Castro 1987). Additional evidence for a very evolved red giant is provided by the 10 μm excess.

J. Mikolajewska et al. (eds.), The Symbiotic Phenomenon, 187-192.

The summary of CI Cyg and bibliography has been recently presented by Kenyon (1986) and Mikolajewska (1985).

2. ORBITAL EFFECTS

Whitney (Aller 1954) first noticed periodic minima in the light curve of CI Cyg, following the ephemeris: $JD\ MIN=2411902+855.25\ E$. Extensive and systematic UBV photometric studies carried out since 1970 in Crimea have allowed Belyakina (1979) to confirm the eclipse character of the minima: the hot component is occulted by a cool giant (Fig.1). The mid-eclipse brightness (V=11.1; B=12.6-12.8) is practically independent on the phase of activity. The eclipse observed during the 1975 outburst also permitted to ascertain that the hot component is responsible for the outbursts. The duration of the eclipse allows for estimation of relative sizes of the components: $R_{cool}/A \gtrsim 0.37$, $R_{ecl}/A \lesssim 0.11$ (Mikolajewska & Mikolajewski 1983). Reasonable mass ratios require the occulting star to be near to filling its Roche lobe.

IR photometry (e.g. Taranova & Yudin 1981, 1986) revealed low amplitude, irregular fluctuations of brightness (ΔJ<0.3). These variations do not depend on the phase of activity and are probably due to intrinsic variability of the cool giant. On the other hand Kenyon & Fernandez-Castro (1987) detected periodic changes of relative depths of TiO absorption bands, consistent with the larger absorption when the cool giant lies in front of the hot component. Their result suggests that (1) radiation from the hot component raises the observed continuum level (thus changing the relative depth) and/or (2) the hot companion illuminates the facing hemisphere of the giant. To produce the observed effect in TiO bands the continuum level ($\lambda\lambda$~6000-7000 Å) should vary by about 10% which is consistent with the eclipse effects observed in the R band (Mikolajewska & Mikolajewski 1983). Simultaneously, if the hot component is embedded in an accretion disk it cannot directly illuminate the cool companion.

IUE and optical spectroscopic observations (Stencel et al. 1982; Barratta et al. 1982; Oliversen, Kenyon & Stencel 1987, in preparation; Mikolajewska & Mikolajewski 1983; Oliversen & Anderson 1983; Mikolajewska 1985) show that the shortwavelength UV continuum disappears during the eclipses while the Balmer continuum as well as many emission lines (e.g. HI, HeI, HeII, UV intercombination lines) are partially eclipsed.

Fig.2 presents secular and periodic variations of integrated line fluxes (optical and IUE) and the Balmer continuum brightness (λ3000 Å) in 1979-84, and suggests that the nebula in CI Cyg is very complex with a few distinct line emission

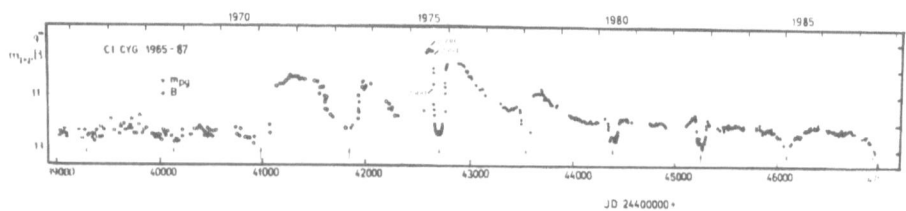

Figure 1. The B light curve of CI Cyg in 1965-87 (Belyakina 1987, presented at IAU Coll. No. 103). Arrows indicate times of minima from the Whitney's ephemeris.

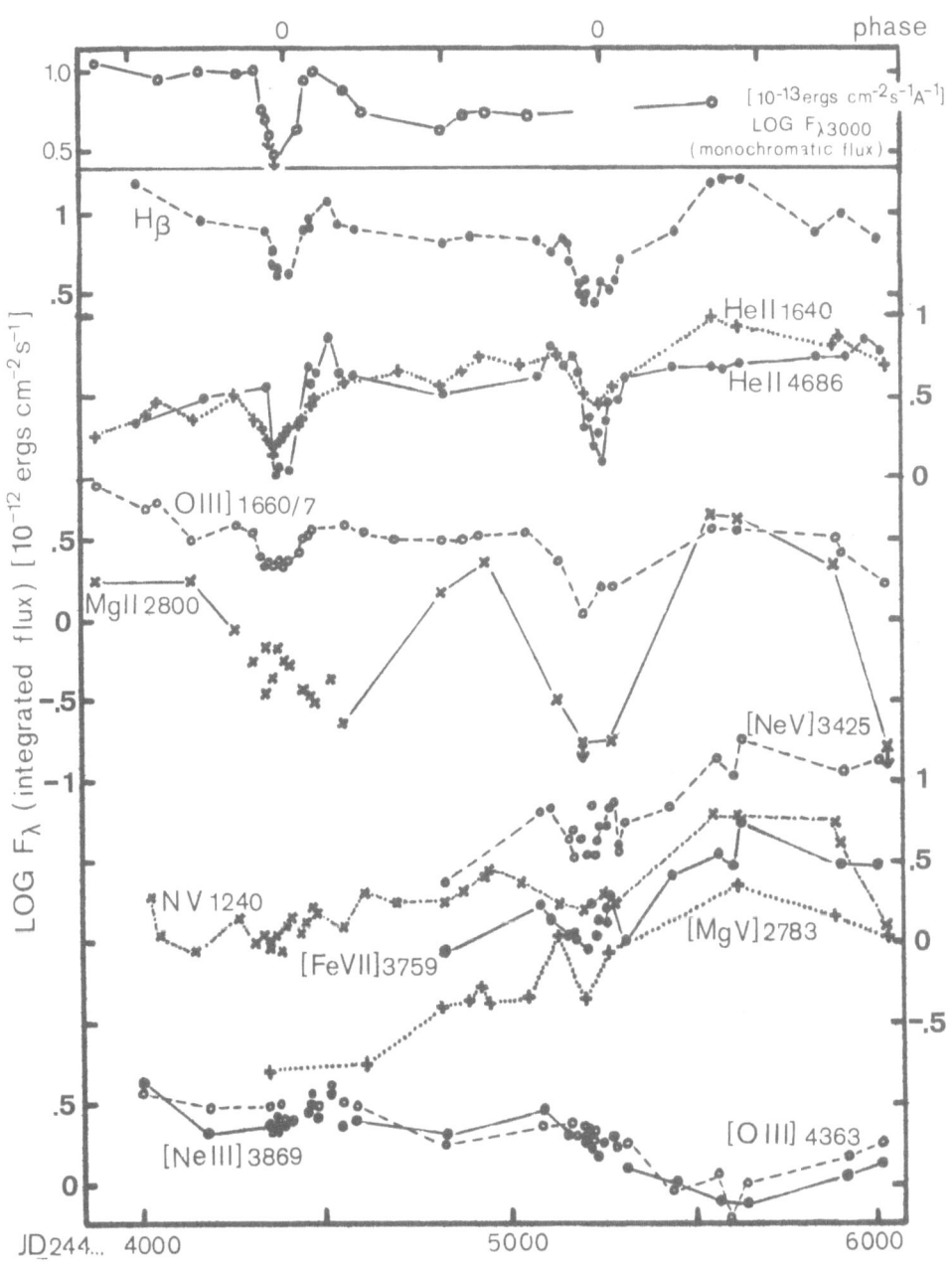

Figure 2. Secular and periodic variations of continuum brightness (o) and integrated emission line fluxes of H_β (●), HeII 4686 (●), HeII 1640 (+), OIII] 1660/7 (o), MgII 2800 (x), [NeV] 3425 (o), NV 1240 (x), [FeVII] 3759 (●), [MgV] 2793 (+), [NeIII] 3869 (●) and [OIII] 4363 (o). The unpublished *IUE* line fluxes were provided by Oliversen, Kenyon and Stencel.

regions. The eclipses in optical emission lines and the Balmer continuum are deeper and narrower than in the UV intercombination lines and HeII 1640. This probably reflects difference in sizes of the optical and UV line emission regions and/or different physical properties of the cool giant atmosphere in the optical and UV range (see Mikolajewska *et al.* 1983). The bulk of the HI, HeII and UV intercombination line emission is produced close to the hot component, in a region with $T_e \simeq 20000$ K, $n_e \simeq 10^{10}$ cm^3 and r $\simeq$ 100-300 $R_\odot$, while the nebular [OIII] and [NeIII] lines are preferentially emitted in a much more extended region (lack of any orbital effects) with $n_e \simeq 10^7$ cm^{-3} and r > 1000 $R_\odot$ (Mikolajewska 1985). The fluxes of the high ionization UV resonance lines (NV, CIV, SiIV) as well as those of the high ionization forbidden lines ([MgV], [NeV] and [FeVII]) are strongly correlated (very weak, shifted eclipse effects?) and show increasing secular trend, contrary to the optical nebular lines ([OIII] and [NeIII]). These high ionization lines may be formed either in an extended thin shock region between the components, or in a corona above the accretion disk (see Mikolajewska 1985). The strong variation of the MgII lines with the orbital period (not due to the eclipses!) is very interesting and suggests that they are produced in vicinity of the L_1 point in the extended atmosphere of the M giant.

   Garcia & Kenyon (1987, this volume) have recently presented very reliable orbital radial velocity curve of the M giant component of CI Cyg, and confirm its eclipsing nature. Although the orbital radial velocity curve based on the emission lines can be uncertain, the eclipse behaviour of the HeII and HI lines suggests that these lines may reflect the orbital motion of the hot component. The radial velocities of HI and HeII 4686 show phasing correct for eclipse (Mikolajewska 1987) and systemic velocity ($\gamma \approx$ 10-20 km/s) consistent with that found by Garcia & Kenyon. The amplitude of the M giant radial velocity curve (K_{cool}=6.5 km/s) combined with that of the HeII and HI radial velocities ($K_{hot} \approx$ 20 km/s) leads to relatively low masses: $M_{cool} \sin^3 i \simeq 1.2 M_\odot$, $M_{hot} \sin^3 i \simeq 0.4 M_\odot$, and a separation $A \sin i \simeq 440 R_\odot$. Simultaneously, any increase of the component masses leads to very high mass ratios ($q = M_{cool}/M_{hot} \gtrsim 5.5$; $M_{hot} \gtrsim 1 M_\odot$).

3. THE 1975 EPISODE

The eclipse observed in 1975 was particularly spectacular since it occured in the middle of the largest outburst in CI Cyg. A bright F-type continuum as well as most of the absorption lines observed at the maximum activity practically disappeared during the eclipse and the spectrum was dominated by the absorption features of the M giant. The star brightness was the same as during eclipses observed in quiescence (Belyakina 1979).

   Variations of the H_β and H_γ emission profiles during the eclipse ingress were analysed by Mikolajewska & Mikolajewski (1985). The profiles were double-peaked with a central absorption. During the eclipse ingress the violet peak rapidly declined while the red peak remained unaffected (Fig.3). Such behaviour is consistent with the occultation of a rotating disk-like envelope.

   Audouze *et al.* (1981) noted the peculiar systematic trend in radial velocities as a function of excitation potential in the spectra taken in July-August 1975. Kenyon *et al.* (1982) interpreted this phenomenon as due to the eclipse of an optically thick, differentially rotating accretion disk. Fig.3 presents fragments of the same spectra as those analysed by Audouze *et al.* The spectra were taken before the first contact, close to the first contact and in the mid-ingress, respectively (see Fig.1). The continuum and Balmer lines in the first spectrum (GA 2508) are

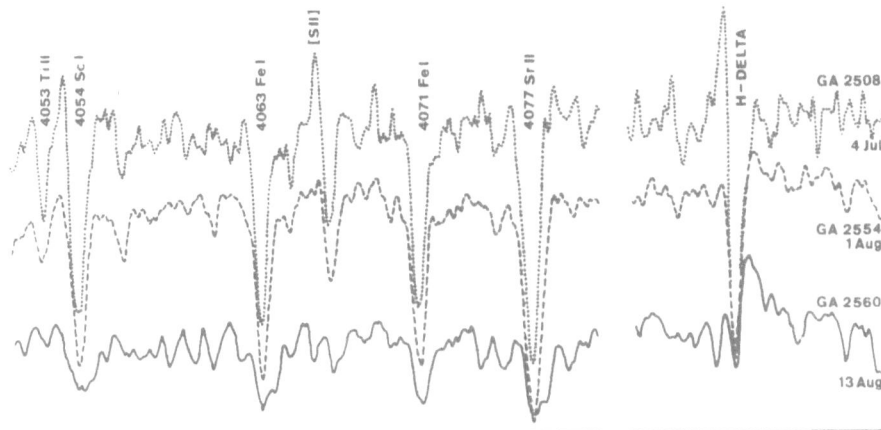

Figure 3. Changes in the blue spectral region (4100-4050 A) in 1975. The spectra are presented in the same intensity scale, so decrease of the continuum level reflects the real changes of brightness during eclipse ingress.

practically unaffected by the eclipse but the peculiar trend in radial velocities is present. The next spectrum (GA 2554) shows the radial velocities redshifted by about 5 km/s with respect to GA 2508 and similar dependence on the excitation potentials. So, the peculiar radial velocity structure cannot be explained as the partial occultation of the accretion disk (Kenyon *et al.* 1982). The last spectrum (GA 2560) exhibits the steepest dependence of the radial velocities on the excitation potentials. According to the interpretation of Kenyon *et al.* the profiles should be asymmetric during eclipse ingress, as the blue-shifted portion of the disk is the first to be occulted. Such asymmetry is not observed, instead of that additional redshifted components appear in most absorption lines (Fig.3). In addition, the radial velocities in all spectra are redshifted by about 25-40 km/s with respect to the systemic velocity ($\gamma \simeq$ 18 km/s; Garcia & Kenyon 1987, this volume). The discussed behaviour of the profiles suggests that the peculiar velocity structure of the aborption lines is due to velocity stratification effects in a matter streaming seen against a background of the accretion disk. The analysis of the spectra is complicated by a possible atmospheric eclipse (GA 2508 & GA 2554) and considerable contribution of the M giant absorption lines (GA 2560).

4. CONCLUDING REMARKS

The new radial velocity curve of the M giant presented by Garcia & Kenyon (this volume) allows to revise the masses of the components and suggests that the M giant is near to filling its Roche lobe ($R_{lobe} \simeq$ 200 $R_\odot$). The assumption that the accretor is a low mass ($\sim$0.5 $M_\odot$) main seaquence star leads to almost the same accretion rates as derived by Kenyon & Webbink (1984), 10^{-5} $M_\odot yr^{-1}$ in quiescence and 10^{-4} $M_\odot yr^{-1}$ (close to $\dot{M}_{cr} \simeq 6 \times 10^{-4}$ $M_\odot yr^{-1}$) in outburst.

A resumption of old observational data might be very interesting, especially:
(1) a thorough reanalysis of changes of the UV and optical continuum and emission lines due to the secular and orbital effects (including eclipses);

(2) a detailed study of the spectra taken in the 1971–75 active phase particularly emphasizing eclipses.
Preliminary analysis of these observations suggest much more complicated picture than that proposed by Kenyon *et al.* (1982).

We thank N.A. Oliversen, S.J. Kenyon and R.E. Stencel for providing their unpublished *IUE* measurements of line fluxes and T.S. Belyakina for poviding the light curve. We also acknowledge A. Woszczyk who made the 1975 spectra accessible.

REFERENCES:

Aller, L.H., 1954, *Publ. DAO Victoria*, **9**, 321.
Audouze, J., Bouchet, P., Fehrenbach, C., Woszczyk, A., 1981, *Astron. Astrophys.*, **93**, 1.
Barratta, G.P., Altamore, A., Cassatella, A., Friedjung, M., Ponz,D., Ricciardi, O., 1982, in *The Nature of Symbiotic Stars*, eds. M. Friedjung and R. Viotti, D. Reidel, 145.
Belyakina, T.S., 1979, *Izv. Krym. Astrofiz. Obs.*, **59**, 133.
Kenyon, S.J., 1986, *The symbiotic stars*, Cambridge Univ. Press.
Kenyon, S.J., Fernandez-Castro, T., 1987, *Astron. J.*, **93**, 938.
Kenyon, S.J., Gallagher, J.S., 1983, *Astron. J.*, **88**, 666.
Kenyon, S.J., Webbink, R.F., 1984, *Astrophys. J.*, **279**, 252.
Kenyon,S.J., Webbink,R.F., Gallagher, J.S., Truran, J.W., 1982, *Astron. Astrophys.*, **106**, 109.
Mikolajewska, J., 1985, *Acta Astr.*, **35**, 65.
Mikolajewska, J., 1987, *Astrophys. Space Sci.*, **131**, 713 (IAU Coll. No.93).
Mikolajewska, J., Mikolajewski, M., 1983, *Acta Astr.*, **33**, 403.
Mikolajewska, J., Mikolajewski, M., 1985, in *Recent Results on Cataclysmic Variables* ESA SP-236, 201.
Mikolajewska, J., Mikolajewski, M., Krelowski, J., 1983, *I.B.V.S.*, No. 2356.
Oliversen, N.A., Anderson, C.M., 1983, *Astrophys. J.*, **268**, 250.
Shapley, H., 1922, *Bull. Harv. Coll. Obs.*, Nos. 762 and 778.
Stencel, R. E., Michalitsianos, A. G., Kafatos, M., Boyarchuk, A. A., 1982, *Astrophys. J. Letters*, **253**, L77.
Taranova, O.G., Yudin, B.F., 1981, *Astr. Zh. Akad. Nauk SSSR*, **58**, 1051.
Taranova, O.G., Yudin, B.F., 1986, *Astron. Tsirk.*, **1454**.

A P Cygni Profile for the He 10830 A line of CI Cyg in Eclipse

S. Bensammar[1], M. Friedjung[2], N. Letourneur[1], J.P. Maillard[2]
[1] Observatoire de Paris, section de Meudon
UA 337 du CNRS, 92195 Meudon Principal Cedex, France
[2] Institut d'Astrophysique de Paris, CNRS
98 bis, bvd Arago 75014 Paris, France

ABSTRACT. Infrared spectroscopic observations of CI Cyg in eclipse show a P Cygni profile for He I 10830 Å, with a wind velocity of the order of 150 km/s, not seen in hydrogen Brackett and Paschen emission lines, which are single peaked.

1. OBSERVATIONS

We report here very early results of high-resolution observations of the eclipsing symbiotic star CI Cyg in the infrared (1 to 2.5 μm) made in July 1987, just at the time of an eclipse of the hot component by the cool component (phase 0). The observations were performed with the Cassegrain Fourier Transform Spectrometer of the Canada - France - Hawaii Telescope (Maillard and Michel 1982), through the standard filters J (1.25 μm), H (1.65 μm) and K (2.25 μm), at a resolution ranging roughly from 20,000 to 40,000. These observations follow similar observations made during the same cycle at phase 0.5 (Bensammar et al 1987), allowing a comparison of the observed features, and in particular the hydrogen and helium lines, in a different configuration of the binary system. The most striking fact in the phase 0.5 observation was the detection of Brγ(2.15μm) seen in absorption, which was interpreted as the absorption of the infrared thermal radiation of the cool component by an accretion disk surrounding the hot component.

In the new spectra, the most surprising feature is the profile of the He 10830 Å line, which exhibits a very strong P Cygni shape (fig. 1), which was absent at phase 0.5. Such a profile to our knowledge, is reported for the first time for this star. The terminal velocity of the blue absorption is as high as 150 km/s.

Other prominent lines in these spectra, which all belong to hydrogen, Brγ, Paβ and Paγ (fig 2,3 and 4), are detected in emission. These lines are not totally eclipsed, which is consistent with previous observations in the the visible of the Balmer lines (Oliversen et al 1982, Micholajewska 1985) at phases in and near eclipse, while the Balmer continuum at 3000 A is totally eclipsed (Stencel et al 1982). The Brγ line has a single peak, while asymmetrical double peak structure, with a deep blueshifted reversal near the time of the total eclipse was observed in Hα (Oliversen et al 1982). No high-resolution profiles of higher Balmer lines are reported to allow a comparison. The strong absorptions seen, one in the red wing of Paβ (fig.3), the other one in the blue wing of Paγ (fig.4), are not part of the hydrogen line profiles. The profiles presented result from the division of the observed spectrum by that of a reference star, HR 6702, chosen to have a spectral type as close as possible to the type reported for the cool giant of CI

J. Mikolajewska et al. (eds.), The Symbiotic Phenomenon, 193–195.
© *1988 by Kluwer Academic Publishers.*

Cyg. From Kenyon and Fernandez-Castro (1987) the giant is quoted as of type M4.9±0.7 and class II whereas HR 6702 is identified in catalogues as M5 II. Taking the ratio of the CI Cyg spectrum to the giant star's spectrum should allow one to obtain a pure spectrum of the ionized component, providing the cool component is identical with the reference giant. In addition, the respective Doppler shifts have to be perfectly compensated. A better compensation of Doppler shift should reduce the residual absorptions, on the figures shown. However, comparing the spectra before division shows that differences appear between the two spectra. Some lines are deeper or present only in the CI Cyg spectrum, which makes a perfect compensation impossible. This is particularly true in the Paγ region. At the present stage of analysis, no conlusion can be drawn as to whether the Paschen lines are symmetrical or not. It can only be stated that they do not exhibit a P Cygni profile and most probably are also single peaked.

2. DISCUSSION

The stationary absorption detected at phase 0.5 for Brγ is no longer present at phase 0, in agreement with our prediction that it should not be seen at phases other than 0.5, if it is produced by material associated with an accretion disk. The fact that Paβ and Paγ are strong in these last spectra, unlike the phase 0.5 observation, can be interpreted as due to the absence of the previous absorption which no longer hides the emission.

The clear difference of profile between hydrogen and helium raises the questions of the excitation mechanism of these lines by the radiation of the hot component, the origin of the P Cygni profile seen in the helium line, and of the respective sizes of the HI and HeI emission regions. A P Cygni profile is traditionally interpreted as the sign of a line formed in an expanding envelope. If the emission lines are formed as a result of the ionization of the cool star's envelope, the ionization front must reach uneclipsed regions of this envelope. The expansion velocity ($\sim$ 150 km/s) measured in helium between the observer and the cool star, cannot be attributed to the normal expansion velocity of the cool star's wind, which should be rather of the order of 15 km/s (Bernat et al 1979). Another mechanism to accelerate the gas has to be advocated, such as a shock wave. The presence of a P Cygni profile for the He 10830 Å could be connected with the fact that the lower level is metastable, which may permit formation of the absorption component at large distances from the binary.

The difference between the cool component spectrum and a normal M giant star spectrum, only detectable at high resolution, suggests a different evolutionary status for the giant component of the symbiotic star. A complete study of the stellar absorbers has to be performed in order to deduce relevant abundance ratios like C/O, $^{12}C/^{13}C$ which should lead to decisive clues on this important question.

REFERENCES

Bensammar, S., Friedjung, M., Letourneur, N., Maillard, J.P., 1987, Astron. Astrop., in press

Bernat, A.P., Hall, D.N.B., Hinkle, K.H., Ridgway, S.T., 1979, Ap. J. (Letters), 233, L135.

Kenyon, S.J., Fernandez-Castro, T., 1987, Astron. J., 93, 938.

Maillard, J.P. and Michel, G., 1982, in IAU Colloquium No 92, Instrumentation for Astronomy with large Telescopes, ed. C.M. Humphries, (D. Reidel Pub.), p. 213.

Mikolejewska, J., Acta Astron. 35, 65.

Oliversen, N.A., Anderson, C.M., Nordsieck, K.H., 1982, in IAU Colloquium No 70 The Nature of Symbiotic Stars, ed By M. Friedjung and R. Viotti (D. Reidel Pub), p.153.

Stencel, E.S., Michalitsianos, A.G., Kafatos, M., Boyarchuk A.A., 1982, Ap. J. Letters,253, L77.

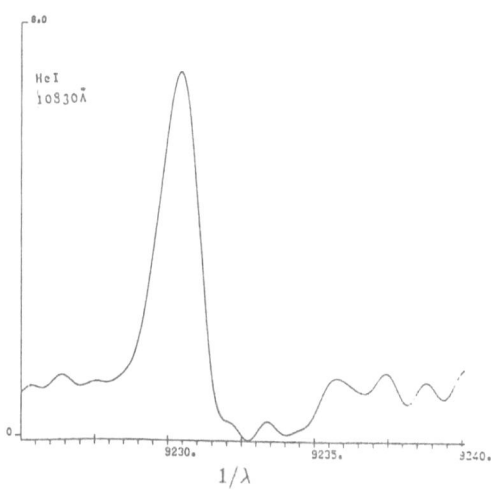

Fig. 1

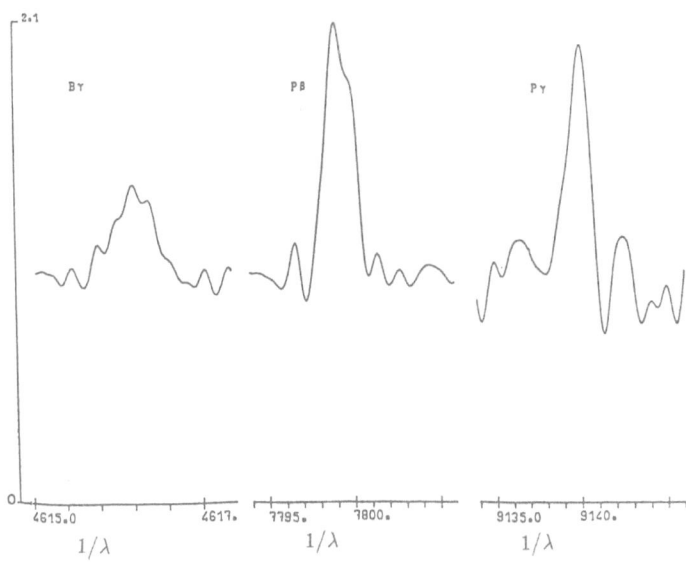

Fig. 2   Fig. 3   Fig. 4

SYMBIOTIC ECLIPSING BINARY STAR CI CYG. THE COLD COMPONENT
VARIABILITY.

T.S.Belyakina
Crimean Astrophysical Observatory
334413,Nauchny,Crimea,USSR

ABSTRACT.It has been shown that the cold component of CI
Cyg eclipsing binary system is a variable star. The amplitude of its light variations is close to $0\overset{m}{.}4$ with the time
interval of 40-60 days.This star being the red giant M4 can
be attributed as SR type variability. The light variations
is caused by its temperature variations.

J. Mikolajewska et al. (eds.), The Symbiotic Phenomenon, 197.
© *1988 by Kluwer Academic Publishers.*

AG DRA A SYMBIOTIC STAR WITH AN UNCOMMON COOL COMPONENT

M. Friedjung
Institut d'Astrophysique
98 bis, Bd Arago
F-75014 PARIS

SUMMARY.   AG Dra is probably a metal poor symbiotic binary of the galactic halo. The luminosity of the cool component is uncertain, which contributes to doubts about the correct model. Behaviour in activity is non classical with high ionization emission lines still having been seen then in the spectrum. This may perhaps most easily be explained by a weak thermonuclear event.

1. INTRODUCTION

In order to find out whether one "symbiotic phenomenon" exists, we need to examine symbiotic stars of very different types. I shall talk about a star whose cool stellar component appears to be less cool than that of most other symbiotic stars. It also seems to be metal deficient and the radial velocity of AG Dra indicates that it is not of population I. Such differences compared with more "classical" symbiotic stars can pose problems if they are not taken into account in interpretation and in the comparison of the "symbiotic phenomenon" with that of other stars.

In this short review I shall not describe all the work done on AG Dra (for detailed summaries of previous work see Kenyon 1983,1986), but I shall try to emphasize what is distinctive and what can teach us about symbiotic stars in general. The properties in quiescence will first be considered and then those of active phases. Finally possible interpretations and problems will be discussed.

2. PROPERTIES IN QUIESCENCE

What are presumably orbital radial velocity variations of the cool component were measured by Garcia (1986). They were consistent with the photometric period of 554 days (Meinunger 1979). According to Garcia the radial velocity curve is sinusoïdal indicating an orbital eccentricity near 0.0. The semi amplitude K equals 7.5 km s^{-1} assuming a mass of 1.5 M_O for the cool component and 1 M_O for its companion, Garcia found an orbital separation of 400 R_O. The periodic photometric variations (Meinunger 1979, Oliverson and Anderson 1982) had when observed by the last authors an amplitude of $\simeq$ 0.8 mag in U, with weaker variations in B and V. The variations are continuous, and do not indicate classical

J. Mikolajewska et al. (eds.), The Symbiotic Phenomenon, 199–204.
© *1988 by Kluwer Academic Publishers.*

"eclipses". Recently Kaler (1987) measured the variations in 11 intermediate and narrow wavelength bands from the near ultraviolet to the far
red. He found periodic variations largest at 3470 Å, decreasing towards
the red but also present to some extent for emission lines while indications of "secondary eclipse" at some wavelengths were also noted.

The systemic radial velocity is according to Garcia (1986) - 146 km
s^{-1}.Such a high radial velocity already given by measurements of Roman
(1953) and Eggen (1964) suggests that AG Dra is a halo star. Stars with
such velocities may be expected to be metal deficient, and this according to Iijima et al.(1987) can explain the contradictions in spectral
classifications of the cool component. These include G7 III (Smith and
Bopp 1981), K1 (Allen 1982), K3 III (Boyarchuk 1966) and K5 III
(Belyakina 1969) among others. Iijima et al. point out that the absorption spectrum is that of a late G dwarf, while the in particular
infrared colours correspond to those of a K3-S III star (Taranova and
Yudin 1982, Viotti et al 1983). The colour but not the absorption
spectrum then gives the effective temperature.

It is naturally perhaps even more difficult to avoid uncertainty in
the determination of the luminosity class. Huang (1982) gives a class of
KO Ib, i.e. the cool component would then be a supergiant. This high
luminosity was suggested by the strength of Ti II, Sr II and Ba II
absorption lines. Other classifications mentioned above suggest a giant
III luminosity class. The recent accurate classification of Kenyon and
Fernandez-Castra is one of < K4 III, the luminosity classification
unlike that of temperature being however not quantitative. Nevertheless
the luminosity classification is probably also affected by a metal
underabundance. The last authors used features due to CN between 8000
and 8100 Å and blends of Ti O and Fe at 8308 and 8330 Å. The sensitivity
of CN to what appeared to be differences in metal abundance was shown by
O'Connel (1973), so one cannot be sure of the lumiosity class of AG Dra
before a detailed high dispersion abundance study is undertaken. Other
anomalies may be present; Lutz et al. (1987) suggest that the cool
component of AG Dra might be a barium star.

Knowing the luminosity class is necessary, in order to know whether
the cool component fills its Roche lobe, Garcia (1986).Taking a radius
for it of 20-40 R_O (the former value is that of a K3 III star according to the Landölt-Borstein tables), Garcia found the Roche lobe to be
underfilled by factors of more than 3.5 for different assumptions about
the mass ratio and the mass of the compact companion star. In principle
one can also obtain information about the radius of the cool star if one
can give limits to its distance. A K3 III star having a V of 9.8 (the
magnitude in quiescence when the hot continuum can be expected to be
least important) would for an E (B-V) of 0.06 (Viotti et al., 1983) be
at a distance of 0.73 kpc and have a distance of 0.5 kpc from the
galactic plane. If the cool component fills its Roche lobe its distance
would be of the order of 3 kpc. The weakness of the interstellar C IV
lines in the ultraviolet spectrum (Viotti et al. 1983) may be an argument against such a distance.

Turning now our attention to other components of the AG Dra system
emission lines of H, He I and He II in particular are observed in the
optical spectrum, while excess continuum emission can be explained by
emitting gas (Boyarchuk, 1966). Emission lines of other ions including

[O I] O III, Fe I, Fe II, [Fe V] and [Fe VII] were reported by Smith and Bopp (1981). These authors studied profile variations of Hα. A blue asymmetry was found to be strongest near photometric minimum. The main red component is what varied, it being weakest when the asymmetry was strongest. Other Balmer lines also seem to show the blue asymmetry.

Postponing discussion of IUE ultraviolet observations to the next section on active phases of AG Dra, the detection of X ray emission before the peak of the 1980 outburst, still needs to be emphasized. This detection by the Einstein observatory was reported by Anderson et al. (1981, 1982). The source was both very strong and very soft, exceeding the X ray luminosity of normal late type giants by roughly two orders of magnitude.

Taking into account various constraints, the authors explained the X ray emission as either due to hot plasma with a temperature between 2×10^5 and 1.6×10^6K, or due to a spherical black body with a temperature of 2.5×10^5K and a radius of 1000 km (the last best fit assumes an interstellar hydrogen column density in front of the source of 3×10^{20}cm^{-2}). The black body fit could be stretched to one of radius 1.4×10^4km at a temperature of 1.5×10^5K.

3. PROPERTIES IN ACTIVE PHASES

The photographic light curve over many years was given by Robinson (1969) (see Fig.1). This curve appears rather to be one of the Z And type than one of a "symbiotic nova". It is possible as will be seen that this appearance is misleading, when one attemps to understand the nature of the activity of AG Dra. It may also be noted that Iijima (1987a) found a periodicity of about 15 years for times of strong activity since 1930.

The spectrum during the activity following the 1980 outburst has been studied in several papers. It must be emphasized that high ionization emission lines were still present during activity of this star; this behaviour is not that of a classical symbiotic star. He II was still seen in the optical, while the same effect is clearly shown by the ultraviolet spectrum, especially observed after outburst. Viotti et al. (1983) report ultraviolet permitted and intercombination emission lines including N V as well as He II, C III], C IV, N III], N IV], O I, O III, O III], O IV], O V, Mg II, Al III, Si II, Si III], Si IV, Si IV] and S V. The relative intensities of the O IV] lines suggested formation in a region with an electron density of 1×10^{10}cm^{-3}, where radiation was very dilute (at a radius of perhaps at least 10^2 of the hot ionizing source). The density, combined with the He II 1640 Å emission measure, suggested a radius for the formation zone of 3×10^{12}cm, if the He II and O IV] were formed in the same region. The 1981 ultraviolet continuum had two components, a steep hot one shortwards of 1600-1700 Å and a flatter one at longer wavelengths. Lutz et al. (1987) also studied the ultraviolet spectrum, finding similar electron density values from the OIV] lines, with a possible increase from 2.5×10^9cm^{-3} before outburst to one near 10^{11}cm^{-3}. Analysis of Si III] lines however gave much lower values near 10^5cm^{-3}, suggesting stratification effects or something wrong in the interpretation of the line fluxes.

The time variation of the ultraviolet spectrum following the 1980

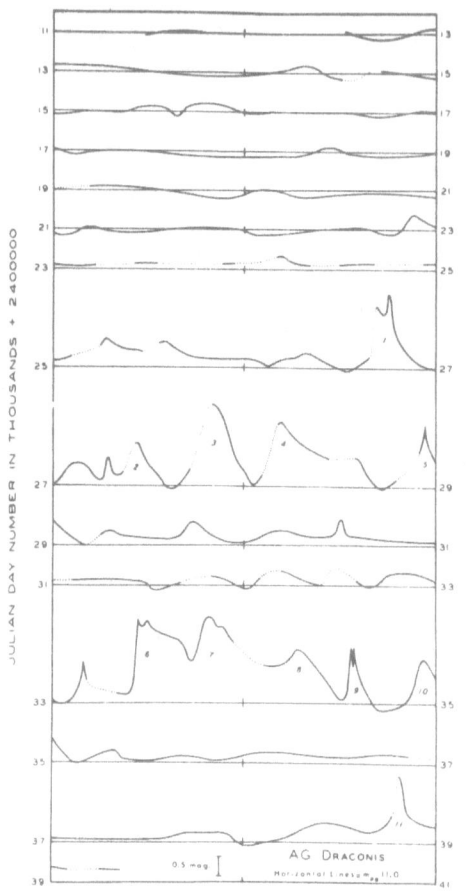

Figure 1. Photographic light curve of AG Dra
from Robinson (1969).

outburst was discussed by Viotti et al. (1984). A minimum in the ultraviolet continuum was slightly later than the predicted photometric
minimum from the pre-outburst light curve; it is not clear whether this
variation is or is not related to that in quiescence. If the ionization
is measured by the NV/CIV ratio it did not change, if measured by the
HeII/continuum ratio some variation occurred.

Another feature of the ultraviolet spectrum must also be pointed
out. The lines of the NV doublet have been observed to have P Cygni
profiles. This could be caused by a wind with a terminal velocity of 170
km s^{-1}. No other lines were observed to have P Cygni profiles, but HeII
1640 Å was seen to have broad emission wings. The NV and HeII emission
lines were observed to be broader when the luminosity was higher.

X ray observations with EXOSAT (Viotti, 1987) show an anticorrela-
tion of observed flux with activity. Indeed no X ray flux was detected

in February 1986, following a new increase in luminosity. The question is whether the X ray flux really decreased, or just become softer.

Radio emission was detected by Torbett and Campbell (1986). The image appeared to be extended and asymmetric, which could suggest bipolar flow or a jet.

4. INTERPRETATIONS

Kenyon and Webbink (1984) interpreted the energy distributions of symbiotic stars in terms of a hot stellar companion, or assuming an accretion disk around the companion. In the case of AG Dra they found that they could only fit the energy distribution, by supposing the presence of a hot star. This apparent absence of an optically thick accretion disk might then suggest that the cool component of AG Dra does not fill its Roche lobe, and produces no overflow. The temperature according to Kenyon and Webbink's calculations of the order of 1.1×10^5 K increased slightly during the outburst, the radius 0.014 R_0 typical of a white dwarf before outburst (assuming a cool component of a normal K3 III type) then increasing by a factor of 2-3. They considered that the 1980 outburst of AG Dra was thermonuclear, but substancially less developed than the large scale events of other stars. The white dwarf would not then develop a very extended envelope, and its effective temperature would remain high. It may be noted that Iijima et al. (1987) did not find any published thermo-nuclear runaway model which fitted AG Dra; more model calculations may be needed. In any case the uncertainty in the luminosity of the cool component will affect the determination of that of the hot component; it is better to be sure of the correct luminosity before attempting a fit !

Garcia (1986) calculated the cool component mass loss rate required, if the X ray luminosity was converted gravitational energy due to capture of part of the wind from this component. He obtained the exceptionally high mass loss rate (for a K giant) of 10^{-7} M_0 yr^{-1}. This shows that the X rays are most easily explained by another mechanism such as radiation from an intrinsically hot white dwarf. However some uncertainty remains because of doubts concerning the nature of the cool component. The weakening of X ray emission during active phases does not have an obvious explanation, though one might suppose a drop of the effective temperature of the hot component. It remains to be seen whether this might be reconciled with the interpretation of the emission line fluxes.

The causes of the photometric and line profile variations in quiescence are not clear. Iijima (1987b) proposed that the orbit was elliptical, and that the photometric variations were due to a variation in the mass transfer rate during the orbital cycle. The low eccentricity found by Garcia (1986), if correct, seems to contradict this. Viotti et al. (1983) suggested an explanation by a "reflection" effect; in fact one could conceive of variations in the visibility of ionized regions of a "chromosphere" of the cool component (it is unlikely that the wind could give enough emission) present in quiescence. According to Garcia (1986) photometric maximum occurred at a phase of the radial velocity consistent with this interpretation.

The P Cygni profiles only seen for the NV doublet may suggest the presence of a rather low velocity wind of uncertain origin near the hot star.This wind could indeed be that from the cool component, accelerated by the pressure of radiation from the hot one. Such a wind might also play a role in producing the blue component of the Hα profile observed by Smith and Bopp (1981). A model with a "warm wind" from the cool component (temperature $> 10^5$K) proposed by Viotti et al. (1983) also to explain X ray emission, now seems less likely.

To conclude many interpretations of AG Dra are unsure. In order to test them we must be more certain about "elementary things" such as the luminosity of the cool component, and the elements of the binary orbit.

REFERENCES

Allen, D.A., 1982, in "The Nature od Symbiotic Stars, Eds. M. Friedjung, R. Viotti; Reidel, Dordrecht, p.27
Anderson, C.M., Cassinelli, J.P., Sanders, W.T., 1981, Astrophys.J. 247, L127
Anderson, C.M., Cassinelli, J.P., Oliversen, N.A., Myers, R.V., Sanders, W.T., 1982 in "The Nature of Symbiotic stars", Eds. M. Friedjung, R. Viotti; Reidel, Dordrecht, p.117
Belyakina, T.S., 1969, Izu Krymskoj Astrofiz. Obs. 40, 39
Boyarchuk, A.A., 1969, Astrofizika 2, 101
Eggen, O.J., 1964, Bull. Roy. Green. Obs. n° 84
Garcia, M.R., 1986, Astron. J. 91, 1400
Huang, C.C., 1982 in "The Nature of Symbiotic Stars", Eds. M. Friedjung, R. Viotti; Reidel, Dordrecht, p.185
Iijima, T., 1987a, Astrophys. Space Science, 131, 759
Iijima, T., 1987b, in "Circumstellar Matter", Eds. I. Appenzeller, C. Jordan; Reidel, Dordrecht, p.481
Iijima, T., Vittone, A. Chochol, D., 1987, Astron. Astrophys. 178, 203
Kaler, J.B., 1987, Astron. J. 94, 437
Kenyon, S.J., 1983, "A Collected History of Symbiotic Stars".
Kenyon, S.J., 1986, "The Symbiotic Stars", Cambridge University Press
Kenyon, S.J., Fernandez-Castro, T., 1987, Astrophys. J. in press
Kenyon, S.J., Webbink, R.F., 1984, Astrophys. J. 279, 252
Lutz, J.H., Lutz, T.E., Dull, J.D., Kolb, D.D., 1987, Astron. J. 94, 463
Meinunger, L., 1979, Inf. Bul. Var. Stars, n° 1611.
O'Connell, R.W., 1973, Astron. J., 78, 1074
Oliversen, N.A., Anderson, C.M., 1982, in "The Nature of Symbiotic Stars", Eds. L. Friedjung, R. Viotti; Reidel, Dordrecht, p.177
Robinson, L., 1969, Perem. Zvezdzy, 16, 507
Roman, N.G., 1953, Astrophys. J., 117, 467
Smith, S.E., Bopp, B.W., 1981, Mon. Not. R. Astron. Soc. 195, 733
Taranova, O.G., Yudin, B.F., 1982, Astron. Zh. 59, 92
Torbett, M.V., Campbell, B., 1986, Bul. Am. Astr. Soc. 18, 912
Viotti, R., Ricciordi, O., Ponz, D., Giangrande, A., Friedjung, M., Cassatella, A., Baratta, G.B., Altamore, A., 1983, Astron. Astrophys. 119, 285
Viotti, R., Altamore, A., Baratta, G.B., Cassatella, A., Friedjung, M., 1984, Astrophys. J. 283, 226
Viotti, R., 1987, private communication.

SPECTRAL VARIATIONS OF AG DRA BETWEEN 1981 AND 1985

C.C. Huang, Y.F. Chen
Purple Mountain Observatory, Academia Sinica

L. Chen
Nanjing Astronomical Instrument Factory, Academia Sinica

After a long quiet phase AG Dra underwent an outstandingly active phase with two outbursts in 1980 Nov. and 1981 Nov.(Viotti et al, 1984). Since then a new quiet phase has followed. In this work we analyse two spectra of AG Dra, of which, one was taken in 1981 by C.C. Huang at the Haute-Provence Observatory using the Marly spectrograph with a dispersion of 80 A/mm at the 1.2 m telescope, the other was obtained by Dr Y.Andrillat in 1985 with the same instrument.

Figure 1 shows the spectral variations of AG Dra between 1981 and 1985. The main features of the emission line spectrum are not much different between the two spectra, except that in 1985 there was a new wide weak emission line at 3488 A possibly due to FeII. There were a lot of strong emission lines due to H, HeI, HeII and OIII in both spectra. The spectrum of the late-type component was much more obvious in 1985 than in 1981. In 1985 the lines of CaII K, CaIλ4227 and the G-band of CH were quite strong. In addition we measured a lot of absorption lines due to FeI, SrII and TiII on the 1985 plate. The Balmer continuum emission and the blue contiuum were enhanced in 1981. On the 1981 plate the stellar spectrum in ultravio-let can be traced beyond 3200 A and the blue continuum heavily veiled the spectrum of the cool component. In 1985 they were much weaker.

Table 1. Relative intensities and FWHM of emission lines

Lines	Plate QA 1165 (1985 June 18)		Plate CT 87 (1981 July 1)	
	I(max)/I(cont)	FWHM	I(max)/I(cont)	FWHM
H	6.0	3.6	11.8	4.7
HeII λ4686	5.4	3.1	7.6	3.3
HeII λ4541	1.3		1.4	
HeI λ4471	1.3		1.8	3.3
H	5.2	2.4	6.7	3.6
H	4.7	3.2	6.2	3.9
H	11.1	2.4	7.3	3.2

J. Mikolajewska et al. (eds.), The Symbiotic Phenomenon, 205–207.

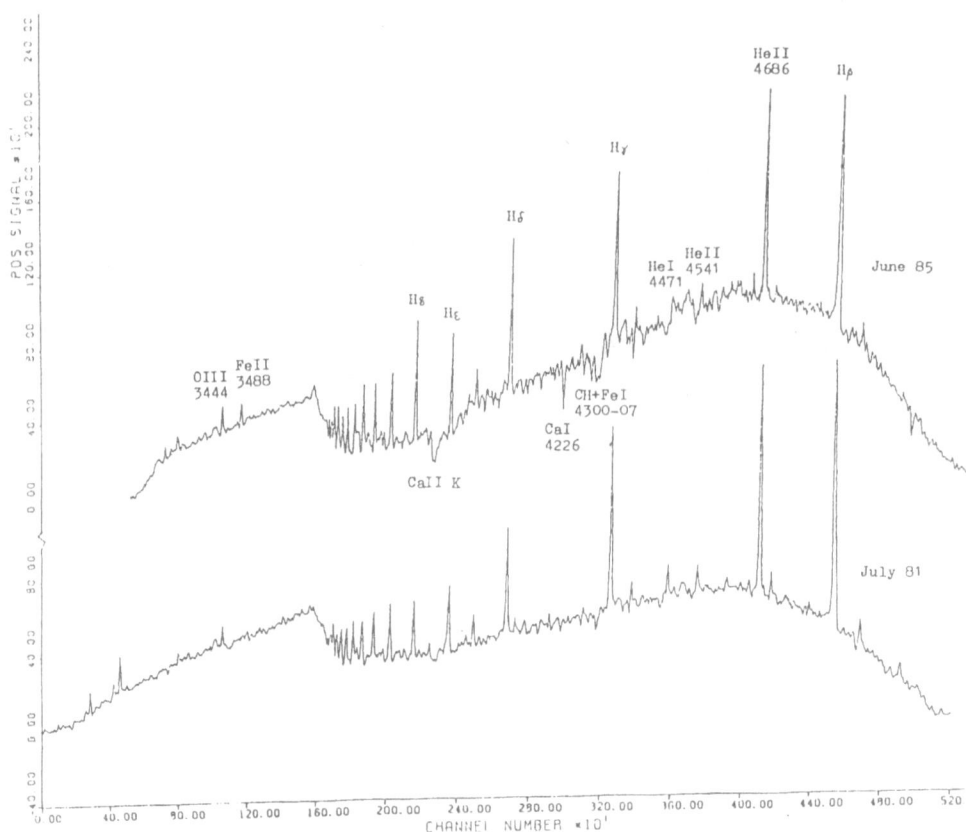

Figure 1. Density tracings of the spectrum of AG Dra.
Ordinates are Dx800 above fog.

Table 1 lists the relative intensities I(max)/I(cont) and FWHM in A of some emission lines. In 1981 the relative intensities of all the emission lines, besides Hε , were much greater than in 1985. The great relative intensity of Hε in 1985 is possibly caused by a greater decrease of the blue continuum in that spectral region. The line HeIλ4471 was a little stronger than the line HeIIλ4541 in 1981, while it was as strong as the latter in 1985. The FWHM of the emission lines were all larger in 1981 than in 1985.

Table 2. Radial velocities for AG Dra

| | 1985 | | 1981 | | 1985 | | 1981 | |
| | Emission Lines | | | | Absorption Lines | | | |
	V	n	V	n	V	n	V	n
H	-142.1±7.0	13	-146.0±2.8	20	-171.1±12.1	6	-157.9±7.2	5
HeI	-130.2±6.6	7	-154.4±3.0	15				
HeII	-140.1±0.8	2	-164.8±4.9	5				

Redial velocities were measured for emission and absorption lines with the 1985 plate and are summarized in Table 2. For comparison in this table we also list the radial velocities of 1981 taken from our previous measurements (Huang, 1982). The radial velocities of emission lines all decreased in 1985. This is in agreement with the possible periodic variations of radial velocity for emission lines of AG Dra (cf. Huang 1982, Fig. 2). Besides, the variations of radial velocity of HeI and HeII are much larger than that of H. The radial velocities of absorption lines probably continued to increase in 1985 (cf. Huang 1982, Table 1).

From the above results it can be seen that during the active phase in 1981 the emission lines and the Balmer continuum, as well as the blue continuum, were enhanced. However there was almost no diference in line excitation between the 1981 active phase and the quiet phase of 1985. This corresponds well with the results of the IUE observations of Viotti et al.(1984). As well, during the quiet phase in 1985, when the blue continuum faded, the spectrum of the cool star, which has a spectral type of K0Ib according to Huang (1982), appeared again. Finally, the difference in the radial velocities and their different variations between HeI, HeII and H indicate that they are from different regions.

Our optical data seem to support a binary model of AG Dra, but more radial velocity measurements with higher accuracy for both emission and absorption lines are required to clarify the nature of this interesting star.

We would like very much to express our thanks to Dr. Y.Andrillat for her valuable spectrum of AG Dra.

References

Huang, C.C.: 1982, IAU Colloquium No. 70, p.185.
Viotti, R., Altamore, A., Baratta, G.B., Cassatella, A., Friedjung, M. : 1984, Astrophys. J., <u>283</u> , 226.

CH CYG : TEN YEARS OF ACTIVITY

P.L. Selvelli
Astronomical Observatory
Via Tiepolo 11
Trieste - Italy.

ABSTRACT. The results of multi-frequency observations of CH CYG during the last outburst are reviewed, and the time-sequence of the variations related to the jets formation is discussed.

1-INTRODUCTION

CH Cyg was classified as a normal M6III semiregular variable until 1963, when spectroscopic observations made by Deutsch (1964) revealed the presence of an "extra" hot continuum and of emission lines of H, HeI, OI, [FeII], [OIII] etc., which were superimposed onto the M6III spectrum, thus producing a combination spectrum. These features constitute the distinguishing signature of a symbiotic star, and CH Cyg was included in this class of variables, although the absence of HeII emissions in its spectrum has cast some doubt on its classification as a "classical" symbiotic star.
The first activity episode lasted until Aug. 65 and was followed by another one with similar characteristics which started in June '67 and lasted until the end of 1970.
A description of the main characteristics of these first two outbursts can be found in Hack and Selvelli (1982) and in Kenyon (1987).
The third , last episode started around Aug. 1977 and is now probably close to its end. Unlike in the two previous outbursts, the behavior of CH Cyg during this last outburst has been monitored in a wide frequency range, from the radio to the x-ray. These observations, although generally not simultaneous or coordinated, have provided valuable data and set strict physical constraints on the nature of the system.
        Certainly, a very important step for understanding the physical model of CH Cyg was the discovery by Yamashita and Maehara (1979) of its duplicity with P $\sim$ 5700 days.
The M6III absorption lines vary in antiphase with respect to the absorption lines of once ionized metals, which are asso-

209

J. Mikolajewska et al. (eds.), The Symbiotic Phenomenon, 209–217.
© 1988 by Kluwer Academic Publishers.

ciated with the hot component. The Balmer lines absorption components vary also with the same trend (Mikolajewsky and Biernikovicz, 1985). The smallness of the amplitude in the r.v. curves and the presence of a large scatter around the mean points has led some researchers to suggest different mass ratios in the system. Thus, Tomov and Luud (1984) give M(hot)/M(giant)~0.3-0.5, while Hack et al. (1986) using the [Fe II] emissions give a ratio of about 1.
In any case, the length of the period implies that the giant does not fill its Roche lobe, and, therefore the transfer of material from the giant to the companion does not occur via Roche lobe overflow.

2-THE FIRST PHASES OF THE OUTBURST (from mid '77 to mid '81)

The first phases of the outburst were quite similar to the previous episodes, with the appearance of emission lines of H, HeI, FeII, OI, and a blue continuum veiling the giant's absorption lines. The once ionized metals show weak, broad emission bands in Sept. 1977, stronger emissions in 1978, and inverse P Cyg profiles since 1980. H_β and H_γ are double peaked and show generally V/R>1, with a true absorption core which goes below the continuum level. Very few absorption lines are observed in the visible by June 1980 (Wallerstein, 1981). H_α has a peak intensity of ~10 - 30 times the local continuum (Anderson et al., 1980) and wings which reach 400 km s^{-1} in 1980. Inverse P Cyg profiles in the Balmer lines started appearing already in 1977 (Faraggiana, 1980). It is noteworthy that in the 1967 outburst the Balmer lines showed only direct P Cyg profiles.

A flickering activity was detected in 1978 by Slovak and Africano (1978) with changes of ~0.1 mag. on a time-scale of ~20 min. , and of 0.02-0.04 mag. on a time scale of~ 5 min.. A similar behavior was observed also in 1967-1969 and in 1974 (Luud, 1980). The flickering increases generally toward the blue; 2% at λ7000, 30% at λ3200 .
Piirola (1982) also detected a variable linear polarization which increases toward the UV and reaches values of~1% .
Taranova and Yudin (1984), Ipatov et al. (1984) performed extensive IR observations from 1978 to 1982.
They found that the IR radiation had semiregular fluctua- tions coinciding in period and phase with the quiescent variability. Brightness variations occur with an amplitude of 0.25 mag. and a period of approximately 1000 days.
The near IR was interpreted as consistent with a 2600°K black body. An "excess" attributable to silicate dust is present longward of 3μ and is greater by about a factor of ten at λ > 20μ . The "excess"is correlated with the changes in the hot source , and is probably due to the heating of a silicate dust shell by the hot source. IRAS observations

at 10 μ have confirmed the excess.

The first UV observations were made soon after the l
launch of IUE by Hack (1979). Emission lines of low ionization character , mainly FeII and MgII , are present , mostly
in the λ 2000-3200 range , while at shorter λ absorptions
lines of once-ionized metals , mainly FeII and NiII,dominate
the spectrum (Hack and Selvelli, 1982a). On sept. 1980 the
continuum flux reached the value of $3 \cdot 10^{-12}$ erg cm^{-2}A^{-1}s at λ
~ 1400 Å . A comparison of the observed flux distribution
with the Kurucz's models indicate that no one of them agrees
with the observations, but the slope of the continuum toward short λ suggests a temperature less than 10^4 °K .

3-THE "SHELL" PHASE (from mid '81 to mid '84)

During this phase , the absorption spectrum of once-ionized
metals developed a more specific "shell" character.Between
June and Nov. 1981 sharp deep absorptions of FeII, TiII, and
CrII became clearly evident and these variations were accompanied by a weakening of the [FeII]lines and by a brightening
by 0.5 mag..The inverse P Cyg profiles were still clearly
evident in the hydrogen and FeII lines.
Wallerstein (1983) reported the presence of two absorption
shells in Oct. 1981.
The optical shell was studied by Luud and Tomov (1984) and
by Rodriguez (1984) who found T$_{exc}$ of the order of 9000-
10000°K, and log Ne $\sim 12.5 - 12.9$. These conditions are close
to those of the atmosphere of a AlIa star. The dilution factor W was estimated as $10^{-1}-10^{-3}$ by Luud and Tomov. The 1981-
1982 spectrum was studied also by Yoo (1984) who derived
from the FeII lines a dilution factor W $\sim 10^{-3}$.
IUE observations made in Dec.1981 (Hack et al., 1982b) showed a wealth of FeII, NiII, and CrII absorption lines. Persic
et al.(1982) estimated a colour temperature of 9000°K and
log g $\sim 1-2$, in agreement with the optical data.
The UV continuum reached its maximum intensity (of the
order of 10^{-11} erg cm^{-2}s^{-1}A^{-1} at 1400 Å)in Dec. 1981 ,
but the distribution was still similar to that of 1979 and
1980.
The optical spectrum between 1982 and 1984 has been studied
also by Mikolajewsky and BierniKowicz (1985), who have described the complex changes which occurred in the Hβ profile.
Since autumn 1983 the old shell of 1981 had become weaker
and sharp lines of neutral and ionized metals have appeared,
e.g. CaI 4227. This new shell is cooler than the previous
one and resembles that of an F-type supergiant.

Absolute spectrophotometry made in 1981 and 1982 by
Kaler et al. (1983) shows a clear correlation between the Hβ
flux and the ground U flux.

Ipatov et al.(1984) found a maximum in U during the spring of 1982, while a maximum in the optical occurred in 1983. It is noteworthy that the flickering became redder in 1982 (Spiesman et al.,1984). The amplitude was $\sim$20% of the total UBV in 1983, and showed a weaker dependance on λ (Reshetnikov and Khudyakova, 1984). No flickering was detected in the near IR, but again, flickering was detected in the H and K bands.

In Jan. 1984 (Selvelli and Hack,1985), high resolution IUE observations still showed the presence of a shell of once-ionized metals, together with the few emissions of low excitation (FeII, OI, MgII) which were present since the beginning of the outburst. The continuum was still almost as strong as in Nov. 1981 (with a flux at $\lambda\sim$1400 Å of $7\cdot10^{-12}$ erg cm^{-2} s^{-1} A^{-1}) and the distribution was quite similar.

4-THE JET FORMATION AND THE LAST OUTBURST PHASES

The last outburst phase started around July 1984, when dramatic changes were observes at all wavelengths. The veiling of the MIII absorption bands decreased substantially, at the same time as there was a drop by 1.2 mag. in the optical brightness. Emission lines of [OIII] and of [NeIII] were observed in August 1984, simultaneously with a strong increase in the [FeII] emissions (Tomov,1984).
VLA observations at λ=2 cm made after mid 1984 detected a strong radio outburst , and high resolution images showed a rapidly expanding jet . The map taken in Nov. 8, 1984 showed the presence of two jets expanding at 1".1 yr^{-1}, and originating in the star position. Other VLA observations in Jan. 22 1985, and May 3, 1985 confirmed this trend.
The total ejected mass was estimated by Taylor et al.(1986) as larger than $2\cdot10^{-6}$ M$_{\odot}$, and the expansion velocity was on the order of 1000/sin i km s^{-1} , if the system is at $\sim$ 400 pc.

The first UV observations after the formation of the jet were made in Jan.25,1985 and revealed dramatic changes in the UV spectrum (Selvelli and Hack,1985a). The UV continuum was much weaker than in Jan. 1984 (the flux at $\lambda\sim$1400 Å was less than 10^{-12} erg cm^{-2} s^{-1} A^{-1}) and very flat , and only emission lines were present, in a wide ionization range, from OI to HeII, CIV, and NV.
The most striking feature was the appearance of a wide Lyman alpha emission (Fig. 1),with an asymmetrical profile and an enhanced red wing , typical of formation in an accelerating outflow. It is remarkable that Johanssos and Jordan (1984) have reported the presence of a similar profile in high resolution spectra of cool giants and supergiants.
The peculiar shape of the CIV λ1550 emission, which shows a composite structure suggesting formation in an inhomogeneous expanding medium, is also remarkable (Fig. 2).

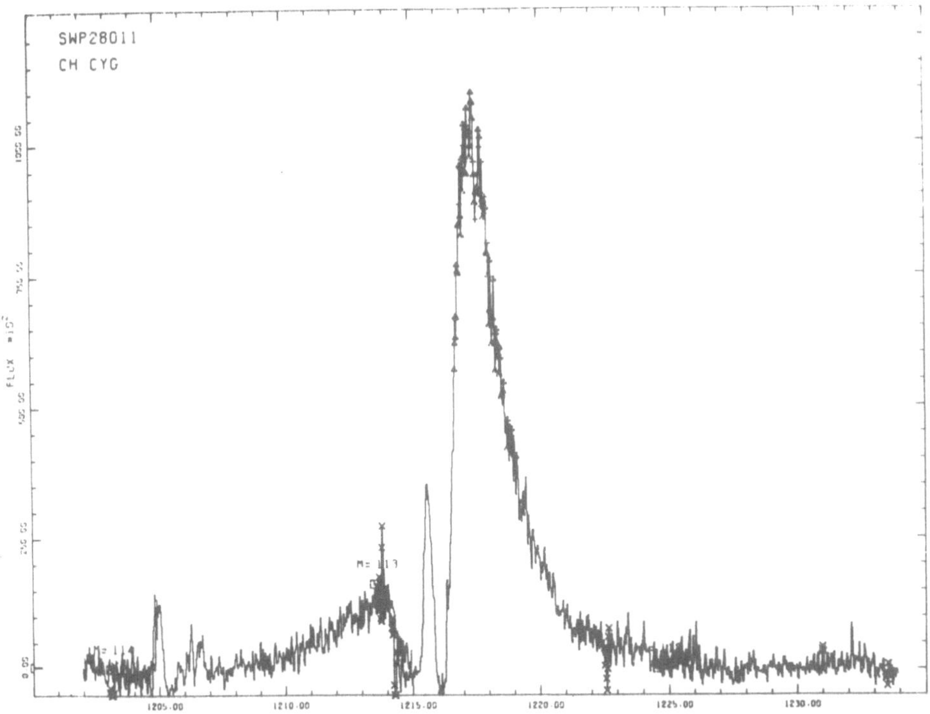

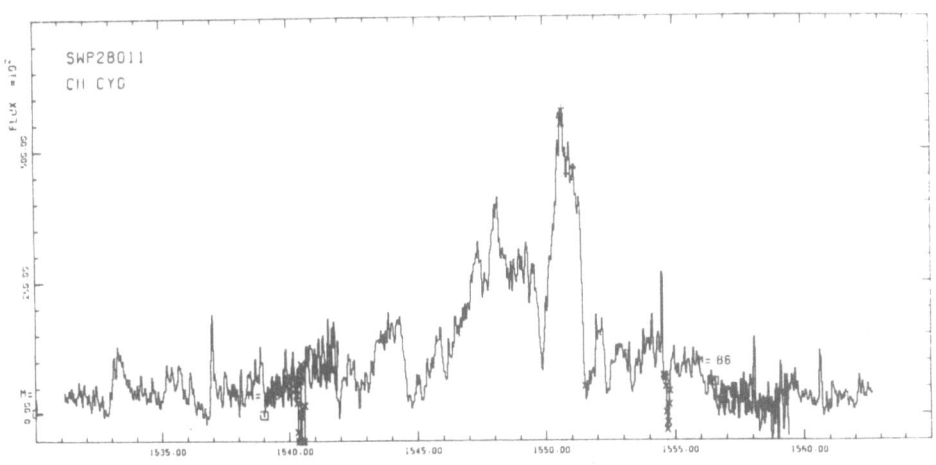

Figure 1. The Lyman-alpha emission (March 1986).

Figure 2. The CIV λ 1550 doublet (March 1986).

The FeII emission lines showed a great increase in number and intensity since 1985. The total FeII UV luminosity was of about 10^{34} erg s^{-1} in Jan. 1985. Marsi et al. (1987) applied the self absorption curve (SAC) method (Friedjung and Muratorio, 1987) to determine the physical parameters of the FeII emitting region. The slope of the SAC is in agreement with a model of formation of FeII in a wind region.

Mikolajewska et al.(1987) have recently made a systematic study of the variations in the UV continuum and in the UV emission lines using all the available low resolution IUE spectra, taken from 1978 to Dec. 1986. They found that, in general, the UV continuum can be matched with a combination of a Kurucz model atmosphere with $T \sim 9000\,°K$ and log g 1-2, and hydrogen b.-f. plus f.-f. emission with $Te \sim 10^{4} K$. The bulk of the high excitation UV emission lines (which appeared in 1985) are formed in a region with $10^{7} \leqq Ne \leqq 10^{9}$ cm^{-3} and electron temperature $10^{4} \leqq Te \leqq 2.10^{4}\,°K$. The typical emission measure is 10^{55} cm^{-3}. The emission measure of a weak hydrogen continuum observed during the 1985 minimum ($\sim 3 \cdot 10^{57}$ cm^{-3}) is the same as the radio emission measure and suggests that it is formed in the jets. It is remarkable that at all epochs the observed UV luminosity was far below the Eddington luminosity and ranged from $\sim 200\ L_\odot$ (Nov. 1981) to $\sim 2\ L_\odot$ (Oct. 1986).

A drop in the U light was observed at the beginning of May 1985 , and the plateau in the minimum was reached around mid May (JD 2446200) and lasted until September. The IUE observations confirmed this trend, with a flux at $\lambda \sim 1400$ Å of the order of $2.5 \cdot 10^{-13}$ erg cm^{-2} s^{-1} A^{-1}. Mikolajewsky et al. (1986) pointed out that a similar drop in the U light occurred about 5750 days before , and suggested that the drop was due to an eclipse. It is notable that the maximum in the radio flux (Taylor et al.,1986) occurred near May 1985 and that on May 24 1985 CH Cyg was detected by EXOSAT as an x-ray source with a 0.02-2.5 keV luminosity of $5 \cdot 10^{32}$ erg s^{-1}. The UV luminosity at nearly the same epoch was $4.6 \cdot 10^{33}$ erg s^{-1}. The x-ray data indicated a variability on a time scale of a few minutes , at a 94% confidence level. The emission measure was $2 \cdot 10^{55}$ cm^{-3} , smaller than the radio one by a factor of 10^{2}. It is also remarkable that previous x-ray observations with HEAO-2 gave only a marginal detection at 2-20 keV, while observations with EINSTEIN gave no detection after 8644 sec. of integration time (Wallerstein, 1981).

Several arguments in favour of the eclipse suggestion (e.g.: the single peak shape of the Balmer emission lines in July 1985, the minimum of the UV flux, the absence of optical flickering, etc.) have been listed by Szczerba et al. (1987). On the other hand , the x-ray flickering cannot be easily explained if the x-ray emission originates , as suggested in the boundary layer of the hot component. Moreover, in Sept. 1986 the UV flux fell to values lower than those observed during the "eclipse" ; this decrease , however, could be just the consequence of a general "decay" in the accretion disk (if any!) , in the last phases of the out-burst. If the eclipse was indeed real, a careful study of the changes in the continuum and emission lines at that epoch would be of paramount importance for the determination of the regions of formation of the continuum and of the emission lines, and of their sizes.

Recent radio observations (Taylor et al.,1987)show that the star is still active in the radio and suggest the presence of a secondary maximum at the end of June 1986.

Spectroscopic observations made by Solf (1987) at high spatial and spectral resolution in Sept. 1986, have re-vealed a very compact emission nebulosity in the [OIII] lines This feature is located at an angular distance of 1".1, north-west of the star , and has been identified with a component of the jets detected in the radio in Nov. 1984. The velocity of the flow has been estimated of the order of 800 km s^{-1} , and the direction of the jet axis is almost per-pendicular to the line of sight. It is noteworthy that the nebulosity has not been detected in H-alpha.

Very recent IUE observations made in June 1987 by Selvelli et al.(1987), indicate a decrease in the conti-nuum and emission lines intensity. Also ground-based obser-vations (Hack et al., 1987)indicate that the activity phase is almost over ; only a few, rather weak emissions are still present and the [OIII] and [NeIII] lines are below the detection level.

5.-COMMENTS AND SPECULATIONS

Observationally,the last outburst phases were characterized by a substantial drop in the optical and UV continua and by the formation of a jet, which, in turn, was mainly responsi-ble for the radio emission and the optical and UV high exci-tation emission lines, which started to appear in the second half of 1984. The x-ray detection of May 1985 is also cer-tainly related to this set of phenomena.
The time correlation among the various events which took place in the last outburst phases (from June 1984 to June 1987) suggests a causal, physical relation among them and also between them and the end of the activity
It is remarkable that "mutatis mutandis" the above reported variations (drop in luminosity, appearance of high excitation emission lines, radio emission) closely resemble the "obscu-ration events" (O.E.) described for R Aqr by Willson et al. (1981) , and more generally found in symbiotic Miras (White-lock,1987) , although no detection of jets has been reported

for this last class of objects. The obscuration events" have
been tentatively attributed to an "eclipse" of the mira's
IR emission by an extended accretion disk, or by clouds, or
alternatively, to spontaneous mass-loss from the mira.
The possible occurrence of an eclipse in CH Cyg around the
mid of 1985 has been claimed by Mikolajewsky et al.(1986).
It is possible that the eclipse was due to the orbital motion and occurred, by chance, close in time to the above -
mentioned sequence of severe multispectral variations. From
a more speculative viewpoint, a physical connection between
the eclipse occurrence and the other variations can be considered. Unless nature is indeed capricious, it seems unlikely that the whole set of events, eclipse included, had
occurred, just by chance, at nearly the same epoch, close
to the last phases of the outburst.
        A comprehensive model of CH Cyg has to take into
account all these observational constraints and determine
the physical processes which have led to the formation of the
jet contemporaneously with the decline in luminosity, the appearance of high excitation emission lines, radio emission,
x-ray emission, and the "eclipse",shortly before the end of
the outburst.
Systematic multi-frequency observations of other symbiotic
stars during their last outburst phases will cast light on
the extension of these phenomena to objects other than CH Cyg
and might also give significant constraints on the elusive
physical mechanismms which are responsible for the formation
of astrophysical jets .

REFERENCES

Anderson C.M.,Oliversen N.A.,Nordsieck K.H.,1980,Astrophys.
        J.242,188.
Deutsch A.J.,1964,Ann.Rep.Mt.Wilson and Palomar Obs. 11.
Faraggiana R. 1980,Astron.Astrophys.84,366.
Friedjung M.,Muratorio G.,1987,Astron.Astrophys.,in press.
Hack M.,1979,Nature 279,305.
Hack M.,Selvelli P.L.,1982a,"The Nature of the Symbiotic
        Stars" eds.M.Friedjung,R.Viotti,Reidel,p.131.
Hack M.,Selvelli P.L.,1982b,Astron.Astrophys.107,200.
Hack M.,Persic M.,Selvelli P.L.,1982 ESA SP 176,p.133.
Hack M.,Rusconi L.,Sedmak G.,Aydin C.,Engin S.,Yilmaz N.,
        1986,Astron.Astrophys.159,117.
Hack M.,Rusconi L.,Sedmak G.,Engin S.,Yilmaz N.,Boehm C.,
        1987,Astron.Astrophys.Suppl.Series, in press.
Ipatov A.P.,Taranova O.G.,Yudin B.F.,1984,Astron.Astrophys.
        135,325.
Johansson S.,Jordan C.,1984,M.N.R.A.S. 210,239.

Kaler J.B.,Kenyon S.J.,Hickey J.P.,1983,Publ.A.S.P. 95,1006.
Kenyon S.J.,1987,"The Symbiotic Stars",Cambridge Univ.Press.
Leahy D.A.,Taylor A.R.,1987 Astron.Astrophys.,in press.
Luud L.,1980,Astrofizika 16,443.
Luud L.,Tomov T.,1984a,Astrofizika 20,419.
Luud L.,Tomov T.,1984b,Sov.Astron.Lett.,10,360.
Marsi C.,Hack M.,Selvelli P.L.,1987, IAU Coll. 94,in press.
Mikolajevska J.,Selvelli P.L.,Hack M.,1987,Astron. Astrophys.
        in press.
Mikolajevska J.,Tomov T.,Mikolajevsky M.,1986,IAU Coll.93
        p.362.
Mikolajevski M.,Biernikowicz R.,1985,ESA SP 236,p.105.
Persic M.,Hack M.,Selvelli P.L.,1984,Astron.Astrophys.140,317
Piirola V.,1982,"The Nature of Symbiotic Stars",eds. M.
        Friedjung, R. Viotti, Reidel p.139.
Reshetnikov V.P.,Khudyakova T.N.,1984,Sov.Astron.Lett.10,281.
Rodriguez M.H.,1984,Astrophys.Space Science,102,103.
Selvelli P.L.,Hack M.,1985a,ESA S.P.236,p.263
Selvelli P.L.,Hack M.,1985b,Astron.Express 1,115.
Selvelli P.L.,Mikolajevska J.,Hack M.,1987,in prep.
Slovak M.H.,Africano J.,1978,M.N.R.A.S. 185,591.
Solf J.,1987,Astron.Astrophys.180,207.

Szczerba R.,Mikolajevski M.,Tomov T.,1987,these proceedings.
Taranova O.G.,Yudin B.F.,1984,Sov.Astron.28,299.
Taylor A.R.,Seaquist E.R.,Mattei J.A.,1986,Nature 319,38.
Taylor A.R.,1986, "RS Oph and the recurrent Nova Phenomenon
        ed. M.F. Bode, VNU Science Press, Utrecht,p.247.
Taylor A.R.,Seaquist E.R.,Kenyon S.J.,1987,these proceedings.
Tomov T.,1984,IBVS 2610.
Tomov T.,Luud L.,1984,Astrofizika 20,99.
Wallerstein G.,1981,Publ.A.S.P. 93,577.
Wallerstein G.,1983,Publ.A.S.P. 95,135.
Whitelock P.,1987,these proceedings.
Willson L.A.,Garnavich P.,Mattei J.A.,IBVS 1961,1981.
Yamashita Y.,Maehara H., Publ.Astron.Soc.Japan,31,307.
Yoo K.H.,Ann. Tokyo Astron. Obs.,20,75.

CH CYGNI HALF A CENTURY AGO – CHANGING ACTIVITY OF THE COOL COMPONENT

Władysław   Dziewulski[*],
Marek   Muciek, Maciej   Mikołajewski
Institute   of   Astronomy,
ul. Chopina   12/18 ,   PL-87100   TORUŃ,   Poland

CH   Cyg   was   observed   visually   by   Professor   Władysław Dziewulski   in   the   years   **1930-1940**   at   the   Vilno   Observatory, with   the   15   cm   refractor   using   Argelander's   method.   As   a skillful   and   experienced   observer   he   achieved   the   accuracy of   his   single   measurement   of   about   0.05   mag.   The   series   of his   652   estimates   of   CH   Cyg   has   been   recently   reduced   and is   discussed   in   this   paper.

Apparently   in   those   years   the   star   was   in   relatively quiet   state   (like   in   the   years   1970-1977),   when   in   the visual   region   the   light   of   *only   the   red   giant   component   of this   binary   is   visible.*   We   deduce   it   from:

a)   absence   of   any   emission   lines   in   spectra   taken   by
    A.   Joy   in   1930   (Joy   1942)
b)   visual   brightness   of   the   star   almost   never
    exceeding   7.0   mag.   what   is   typical   for   the   quiet
    state   of   pure   M6III   spectrum
c)   red   color,   resulting   from   comparison   of   our   visual
    brightnesses   and   the   photographic   ones   published
    by   S.   Gaposchkin   (1952)

Our   first   conclusion   from   the   analysis   of   our lightcurve   (which   cannot   be   shown   here   in   its   full   lenght but   will   be   published   elsewhere)   is   that   contrary   to   many previous   reports   *no   stable   periodicity*   can   be   seen. Sometimes   only   quasiperiods   of   100   day   or   400   day   seem   to excite   for   a   cycle   or   two   and   disappear.

And   yet   there   is   a   general   sense   in   this   chaotic   image. *The   amplitude* of   those   rather   erratic   light   variations   *seem to   change   with   the   orbital   phase* of   the   system.   As   can   be seen   in   Fig.1   the   red   giant   looks   the   most   active   near   one spectroscopic   conjunction   and   the   most   quiet   near   the   other.

The   solid   line   on   the   Fig.1   is   the   cosine   of   the   angle: hot   companion   –   red   giant   –   observer,   fitted   to   all   the points   except   of   the   last   six.   The   orbital   parametrs necessary   for   computing   this   angle   are   taken   from Mikołajewski,   Szczerba   &   Tomov   (this   volume)

219

J. Mikolajewska et al. (eds.), The Symbiotic Phenomenon, 219–220.

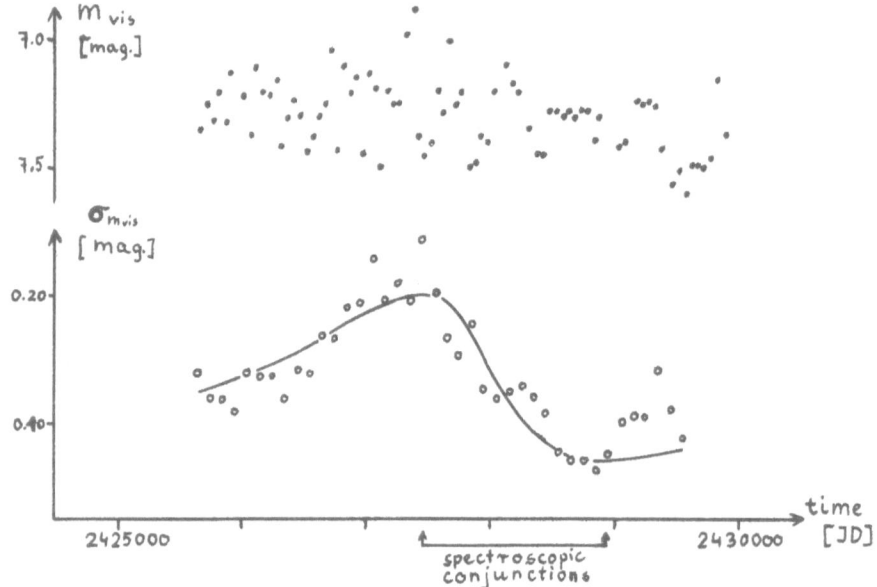

Fig.1 The time variation of the **visual brightness** averaged over 50-day intervals (top) and **dispersion** computed for the 300-day intervals, with step of 100 days (bottom).

It seems clear that the observed amplitude of light variations of the red giant follows a function of visibility of the hemisphere facing hot companion. It is so for most of the time covered by observations.

Finally, we suggest that in the quiet state of CH Cyg *the hemisphere of the red giant facing the hot companion is more active than the rest of the star.* The irradiation by hot star induces higher activity on the cool component. What kind of activity could it be? Spots? Flares? Nonradial pulsations?

References
1. Gaposchkin S. 1952 Ann. Harvard Obs. 118, 158
2. Joy A. 1942 Astroph. J. 96, 344

*Biographical note
Władysław Dziewulski - professor of astronomy at the Vilno University (1919-1939), later in Toruń, the Founding Father of the Nicolaus Copernicus University. He died on Feb.5,1962.

SPECTROSCOPIC ORBIT OF THE ECLIPSING SYMBIOTIC STAR CH Cyg

M.Mikołajewski[1], R.Szczerba[2], T.Tomov[3]
[1] Institute of Astronomy, UMK, Toruń, Poland
[2] N. Copernicus Astronomical Center, Toruń, Poland
[3] Rozhen Observatory, Bulgarian Academy of Sciences, Bulgaria

Mikołajewski et al.(1987) quoted convincing arguments that CH Cygni is eclipsing binary system:

1. Photometric minima in U light were separated by 5700÷5780 days which is very close to the spectroscopic period (P=5750±250d) found by Yamashita and Maehara (1979).
2. Variations of Hβ and Hγ profiles during both minima can be succesfully interpreted as an eclipse of a rotating disk or envelope from which these profiles originate (see also Fernie et al.,1986).
3. The flickering activity was absent during the 1985 minimum while observed before and after the minimum.
4. Only forbidden lines were unaffected by the minimum.

Recently, additional arguments for the eclipsing nature of CH Cygni have been found:

5. Deep minimum in the IUE integrated flux was coincided with the minimum detected in U light (Mikołajewska et al.,1987a).
6. Changes of Balmer decrement and Balmer continuum emission measure suggest a presence of relatively small dense region occulted during the minimum and extended low density region coinciding with the radio jets (Mikołajewska et al.,1987b).
7. Radial velocity of the jet spatially resolved by Solf (1987) in the [OIII]5007 line suggests the jet is practically perpendicular to the line of sight.

Yamashita and Maehara (1979) basing on 54 radial velocity measurements for the M giant found the first solution for spectroscopic orbit (line 1 in Table 1). Mikołajewski et al.(1987) estimated preliminary orbital elements taking into account eclipses. Simultaneously, they showed that radial velocities of the "shell" absorption lines of ionized metals roughly reflect the orbital motion of the hot component. However, strong variability of these lines as well as the fact that they could be observed only during the active phase led to a formal solution for the orbit inconsistent with the eclipses. Therefore, in present study, we have determined the accurate orbit only for the M giant basing on 170 available data points (Fig.1). The mean velocity curve was analyzed by the method of Lehmann-Filhes. The orbital elements thus obtained, adopting the period P=5700 days, give the time of eclipse

J. Mikolajewska et al. (eds.), The Symbiotic Phenomenon, 221–222.

only 60 days (!) later than observed (table line 2) and strongly argue for the eclipsing nature of CH Cygni. Taking into account observed time of eclipse (T_{ecl}=JD2446275) and adopting P=5700d we have calculated in each iteration ω from Kepler´s equation basing on the remaining elements which have been obtained in the standard way. Such procedure is strongly convergent and gives orbital elements with significantly improved accuracy consistent with the observed eclipses (table line 3, Fig.1).

Table 1. Orbital elements derived for the M giant in CH Cygni.

	N_0	P [d]	V_0 [km/s]	K [km/s]	e	ω [°]	T_0 [JD]	Δ
1	54	5750±250	-58.4±0.3	6.8±0.5	0.29±0.07	201±15	2445773±250	+146
2	170	5700	-57.7±0.3	4.9±0.5	0.48±0.08	136±12	2445032±129	- 60
3	170	5700	-57.7±0.3	4.9±0.5	0.47±0.07	142± 5	2445086± 55	0

here: $\Delta = T_{ecl}(obs) - T_{ecl}(calc)$; N_0- number of observations

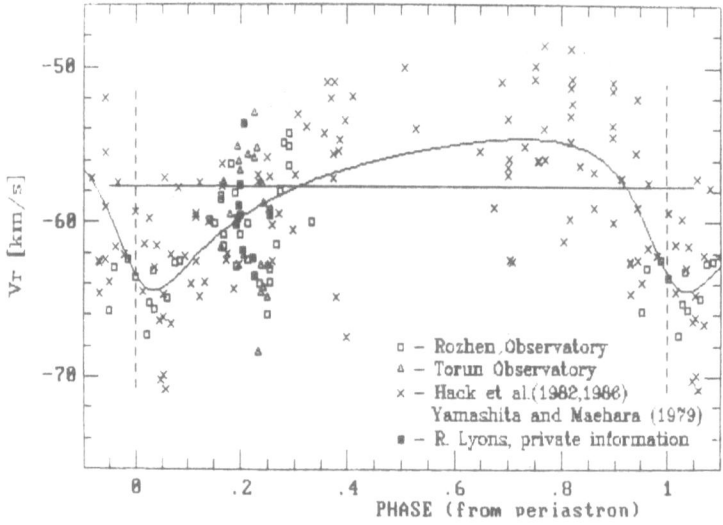

Fig.1 The radial velocity curve for the M giant of CH Cygni consistent with the observed eclipses (Table 1 line 3).

REFERENCES:
Fernie J.D., Lyons R., Beattie B., Garrison R.F.:1986, IBVS No.2935
Mikołajewska J., Selvelli P.L., Hack M.:1987a,
        submitted to Astron. Astrophys.
Mikołajewska J., Mikołajewski M., Biernikowicz R., Selvelli P.L.,
        Turło Z.:1987b, in proceedings of IAU Symp. No. 122,
        Reidel Co.,p.487
Mikołajewski M., Tomov T., Mikołajewska J.,:1987,
        Astrophys. Sp. Sci.131 ,733
Solf J.:1987, Astron. Astrophys.,180,207
Yamashita Y., Maehara H.:1979, Publ. Astron. Soc. Japan 31 ,307

SLOW AND RAPID CHANGES OF THE RADIAL VELOCITIES IN THE SYMBIOTIC BINARY CH CYGNI

A. Skopal

Astronomical Institute Slovak Academy of Sciences,
059 60 Tatranská Lomnica, Czechoslovakia

In the period 1982-1986, new 98 radial velocities (RVs) were determined (at the figures denoted by □). All spectrograms were obtained at the Toruń Observatory. RVs were measured at the Skalnaté Pleso Observatory. The main results are summarized in the figures 1-3. The phase is determined from the middle position of the hot component eclipse (JD 2 446 272) and from the minima difference observed in U light curve (5700d) (Mikolajewski et al., 1987).

Slow changes of the RVs of the absorption components of ionized metals reflect the orbital motion of the hot component only between the phases 0.7 and 0.0. The rapid changes are observed mainly between the phases 0.8 and 0.9. They are created by combination of the emission red wing with the absorption component (Skopal et al., 1987). RVs of the cool component are in antiphase (Tomov and Luud, 1984).

Slow changes of the RVs of emission Fe II and [Fe II] lines reveal similar behaviour (RVs Fe II lines are shifted about -5km/s). The RVs maximum of the [FeII] lines is shallow with scatter about 15km/s and shifted to the RVs maximum of the ionized metals approximatelly about 0.1P. The rapid changes of the RVs of the Fe II lines arise between the phases 0.8 and 0.9 too, where the 25km/s deep minimum is observed.

Figures 1-3 follow to conclusion that the RVs of emission forbidden and permitted lines of iron do not coincide with the orbital motion neither the hot nor the cool component. Slow changes of the RVs can be explaine by creating of these lines in stellar wind. The change of the radial component of the stellar wind velocity vector strongly depends on the spectroscopic elements. The rough estimate of the spectroscopic elements given by Mikolajewski et al. (1987) shows the possibility of the creation of the emission Fe II and [Fe II] lines in stellar wind around M component.

The deep minimum of the RVs of the emission Fe II lines coincide with the arising of the S-wave component (in detail: Skopal et al., in preparation). This fact indicates that the origin of these components of the Fe II lines is located either in small region near the hot spot in accretion disk or in shock front on the

J. Mikolajewska et al. (eds.), The Symbiotic Phenomenon, 223–224.
© *1988 by Kluwer Academic Publishers.*

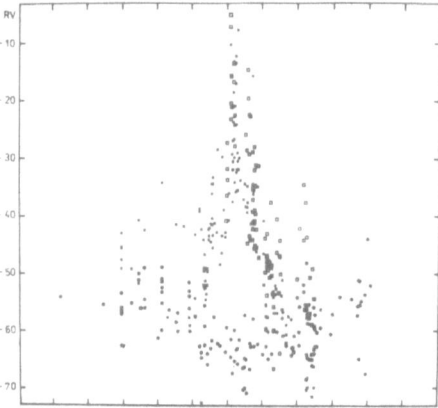

Figure 1. Radial velocities of the ionized metals (■,□) and of the M component (•)

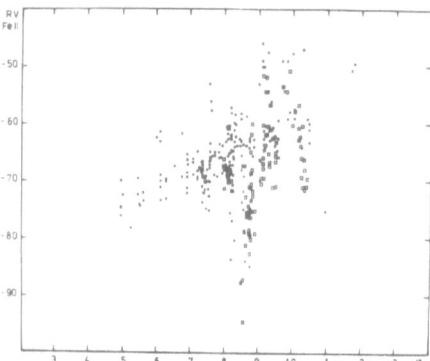

Figure 2. Radial velocities of the emission Fe II lines

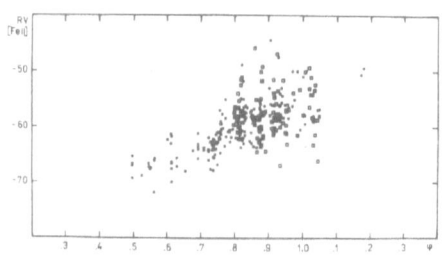

Figure 3. Radial velocities of the emission [Fe II] lines

cone surface, if an accretion from a stellar wind onto a compact component is assumed. Time of the passage through periastron (in 1982) and a position of both components ($-50° <$ omega$_{hot} < -30°$) (Mikolajewski et al., 1987) make possible both interpretations. These possibilities can be distinguished by determination of the Roche lobe size of cool component, which depend on the orbital eccentricity and on the ratio of rotational angular velocity to orbital frequency (Hadrava, in press).

References

Hadrava, P.: Bull. Astron. Inst. Czechosl., in press.
Mikolajewski, M., Tomov, T., Mikolajewska, J.: 1987, Astrophys. Space Sci. 131, 733.
Skopal, A., Mikolajewski, M., Tomov, T.: 1987, Astrophys. Space Sci. 131, 747.
Skopal, A., Mikolajewski, M., Biernikowicz, R.: Bull. Astron. Inst. Czechosl., in preparation
Tomov, T., Luud, L.: 1984, Astrofizika 20, 99.

RADIAL VELOCITIES OF CH CYGNI JUST AS THE JETS APPEARED

Ewa JANASZAK
Planetarium and Astronomical Observatory, Olsztyn, Poland
Maciej MIKOŁAJEWSKI
Institute of Astronomy,N.Copernicus University, Toruń, Poland

The period of maximum brightness of CH Cygni which started in mid
1981 finished at the end of July 1984, when sudden drop by about 1   mag
was observed (Fig.1). At the same time a radio outburst has occured and
the jet structure has developed (Taylor et al.1986). The Toruń Observatory has recorded the spectra of CH Cygni just at this moment.   Significant spectral variations, especially development of wide wings in Balmer
emission lines have been found (Mikołajewski and Tomov 1985).
     Mikołajewski et al. (1987) showed that the "shell" absorption lines
roughly reflect the orbital motion of a hot companion of the M giant.
Figure 1 presents radial velocities of these absorption lines of ionized
and neutral metals (mainly TiII, FeI) together with the v.rad. followed
from possible orbital solution suggested by Mikołajewski et al. About

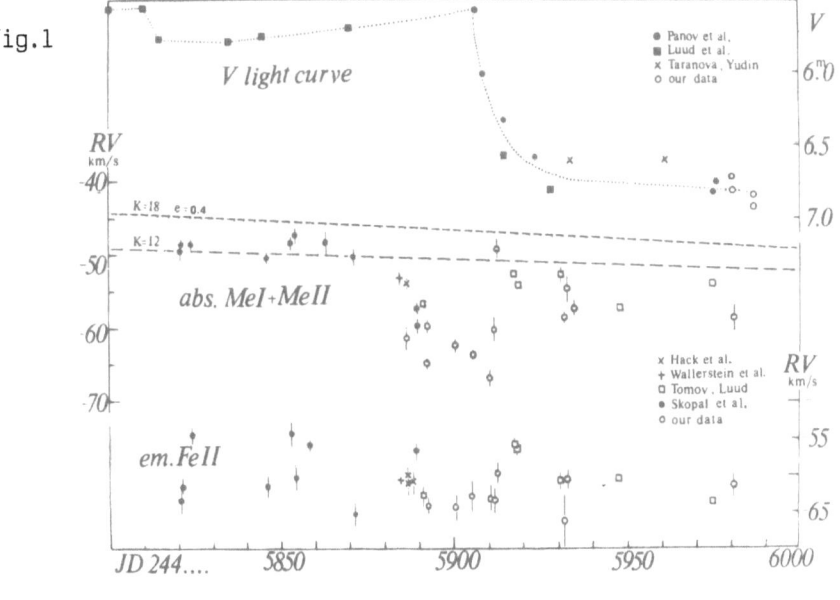

Fig.1

J. Mikolajewska et al. (eds.), The Symbiotic Phenomenon, 225–226.
© 1988 by Kluwer Academic Publishers.

Figure 2.

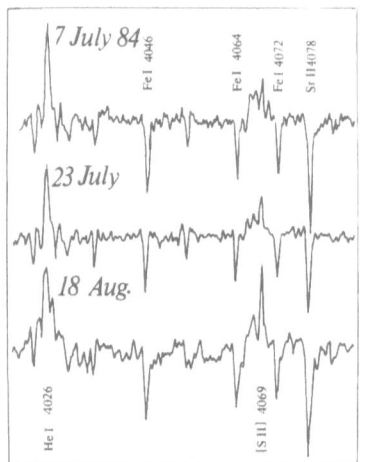

3-4 days after the drop started the RVs of the "shell" absorption lines very quickly (during 3 days) got out of a deep local minimum ($\Delta V_r \sim$ 15-20 km/s) preceding the drop by a mounth. The FeII emission lines, which are formed in the outer regions, do not show so large radial velocity variations (Fig.1). Simultaneously, we have not found any changes of the absorption line's profiles (Fig.2) which could account for the minimum of RVs, although they seem to be symmetrical only just before the drop (i.e. on July 23). There is also worth noticing the appearance of wide structure and red component of the HeI 4026 emission line.

The drop in brightness was not accompanied with any changes of B-V and U-B colours. Simultaneously, the "shell" absorption spectrum observed after the drop indicates similar excitation temperature. This suggest that the drop was due to a contraction of the emitting surface at nearly constant effective temperature. The increase of RVs observed at the time of drop of brightness seems qualitatively support this hypothesis. However, it is necessary to emphasize that the values of RVs during the minimum are always below the values followed from the orbital solution (Fig.1). This rather suggest a gradual expansion of the pseudophotosphere untill the time of drop, and then an ejection of the expanding matter perpendicularly to the line of sight.

References:

Hack,M.,Rusconi,L.,Sedmak,G.,Aydin,C.,Engin,S.,Yilmaz,N., 1986
      Astron.Astrophys.**159**,117
Luud,L.,Tomov,T.,Vennik,J.,Leediarv,L.,1986,Pisma v Astron.Zhu.**12**,870
Mikołajewski,M.,Tomov,T.,1986,MNRAS **219**,13
Mikołajewski,M.,Tomov,T.,Mikołajewska,J.,1987,Astrophys.Space Sci.**131**,733
Panov,K.P.,Kovachev,B.,Ivanova,M.,Geyer,E.H.,1985, Astrophys.
      Space Sci.**116**,355
Skopal,A.,Mikołajewski,M.,Biernikowicz,R.,1987 submitted to BAC
Taranova,O.G.,Yudin,B.F.,1985,Astr.Circ.**1370**,7
Taylor,A.R.,Seaquist,E.R.,Mattei,J.A.,1986,Nature **319**,38
Tomov,T.,Luud,L.,1987,preprint Tartu Observatory, in press
Wallerstein,G.,Bolte,M.,Whitehill-Batles,P.,Mateo,M.,1986,PASP **98**,330

THE SPECTRUM OF CH CYGNI - A SEARCH FOR RAPID LINE VARIATIONS

T. Tomov
National Astronomical Observatory Rozhen,
4700 Smoljan, PB 136, Bulgaria

D. Raikova
Department of Astronomy, Bulgarian Academy of Sciences,
1784 Sofia, blv. Lenin 72, Bulgaria

The spectacular changes of CH Cyg spectrum during the last period of activity and the flickering in short wavelength light suggest the possibility of rapid line variations (Walker et al.,1969). In November 1982 a series of spectral observations has been accomplished at Rozhen Observatory to investigate the subject.

In that period the spectrum of CH Cyg is dominated by type A shell lines. They are supposed to originate in an accretion disk surrounding the hot star of the binary system. The flickering in the optical wavelengths is characteristic for CH Cyg during periods of activity . Evidently it is related with the mass transfer. The investigation of its origin and relations with the other observational characteristics of the star could elucidate the accretion mechanism in this system.

On November 3, 1982 in 3 hours 11 spectra of CH Cyg have been taken at the coude spectrograph of the 2-m telescope in Rozhen Observatory. The original dispersion is 18 A/mm and the spectral range covered 3600-4900AA. The exposure time is 12-15 min.

The behaviour of some representatives of the different line systems has been studied. Some of them are shown in Fig.1. On the top the U - light curve of CH Cyg for the same time is presented. The photometric observations have been accomplished simultaneously with the spectral ones at Rozhen with 60-cm telescope (Chochol et al.,1984). There is well pronounced flickering of amplitude about 0.4^m and on a time scale of minutes or an hour.

The line profiles change in details. The violet emission peak of $H\beta$ exhibits noticeable intensity variations, the red one being practically constant. So does the red emission of $H\delta$. The absorption line ScII 3630.74A on some spectra is nearly equal to CrII 3631.72A and on others it is considerably deeper. CaI 4226.73A shell line is well outlined on some spectra and hardly discernable on others.

The equivalent widths of some representative lines have been measured: the absorptions FeI 4045.82A, FeI 4071.74A, CaI 4226.73A for neutral atoms, the resonance line SrII 4077.71A, two lines of ScII 3630.74A and 3651.80A exhibiting no emissions, the [FeII] emissions at 4359.34A, 4728.07A and 4814.55A. The absorption line variations are small and rather random with a slight tendency to increase after 18h 10m. The in-

J. Mikolajewska et al. (eds.), The Symbiotic Phenomenon, 227–228.

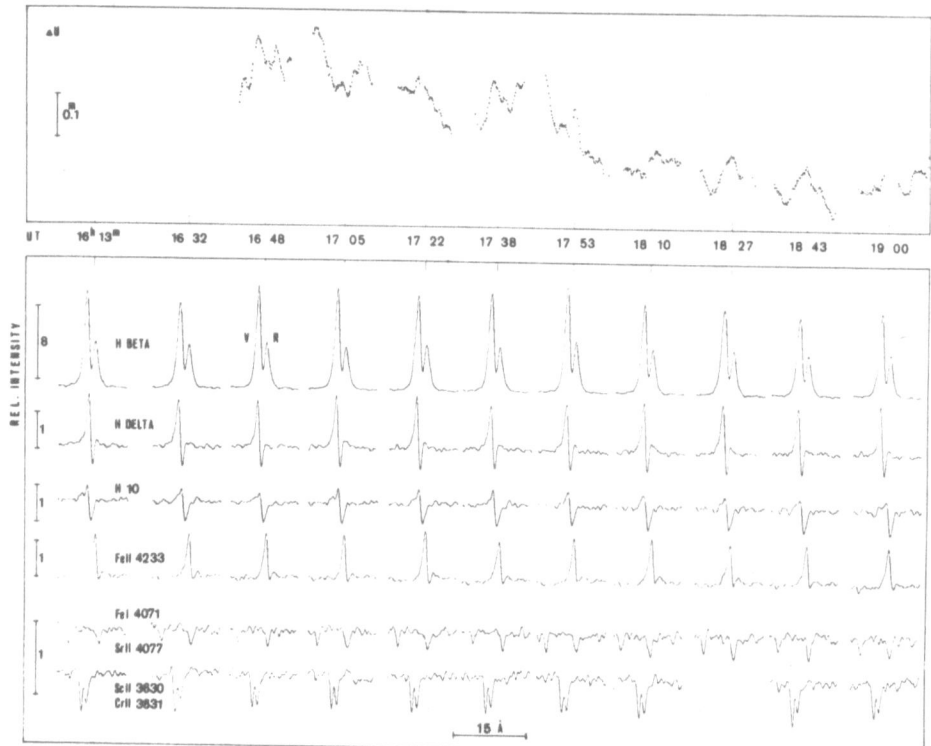

Figure 1. Relative intensity line profiles of CH Cyg spectrum on November 3, 1982. Top: U-light flickering at the same time. The UT is marked by the moments of mid-exposure.

tensity variations of the forbidden lines are evidently accidental.

The mean radial velocities from all the spectra are: -32.2 ± 0.6 km/s measured from 11 FeI absorption lines on each spectrum; -35.5 ± 0.3 km/s from 23 lines of ionized metals (ScII, TiII, CrII); -63.0 ± 0.4 km/s from 8 lines of [FeII]; -71.7 ± 0.3 km/s from 12 FeII emission lines. Their variations hardly exceed $2\sigma_i$ by amplitude (σ_i – the mean error of Vr from a given spectrum).

Our study showed no considerable rapid variations of the line spectrum of CH Cyg on the time scale of the flickering. The small profile changes do not correlate with the light variations. That suggests that the flickering does not originate in the regions where the studied line systems are formed.

REFERENCES

Chochol,D.,Hric,L.,Skopal,A. and Papousek,J.,1984,Contr. Astron.
        Obs. Scalnate Pleso,12,261
Walker,G.,A.,H.,Morris,S.,C. and Younger,P.,E.,1969,Ap. J.,156,
        No.1,Part1

POLARIZATION IN CH CYGNI DURING QUIET AND ACTIVE PHASES

M.H.Rodriguez

Main Astronomical Observatory, Ukrainian Academy
of Sciences, Kiev-127, U.S.S.R.

Our wide-band polarimetric observations of CH Cyg began in
1974 and continued in 1976 and 1984. During the first two
observational seasons the star was in quiet phase between
two outbursts, in 1984 the latest outburst that began in
1977 was declining. Figure shows mean values of the degree
of polarization P and the position angle θ for the different
observational periods. Standard errors calculated by photon
noise statistics are 0.02-0.04% for P and 1-5% for θ .
        In 1974 (October 12 - December 1) polarization
parameters did not change significantly. Sharp increase of
the degree of polarization in the shortwave region of the
spectrum and change with wavelength of the degree of polarization are typical for many cool luminous stars and
indicate existence of an asymmetric circumstellar envelope
around CH Cyg. Interstellar polarization in the direction
of CH Cyg should be quite small [1,2] ,as its distance is
330 pc and colour-excess E_{B-V} does not exceed 0.07^m.

        In 1976 (September 22 - October 10) wavelength
dependence of the degree of polarization remained the same
as it was in 1974, but the value of P in the shortwave region was larger.Orientation of the position angle differs
also from that in 1974, but there is no rotation of θ .
After October 1, 1976 the degree of polarization diminished
in all the filters, specially at shorter wavelengths, though
the brightness of the star in the same spectral interval
remained constant at 0.01^m level. Mean values of P and θ
for the both periods - before and after October 1 - are
shown in Fig. (curves 1976a and 1976b respectively).
Observations by Khudyakova [1] make it possible to prolong
the degree of polarization curve into near infrared.
It comes out that for the quiet state of the star P was at
the level of 0.5% in the interval from ~600 to 900 nm.
        In the course of the period of activity set in
after the 1977 outburst polarization in CH Cyg had quite

J. Mikolajewska et al. (eds.), The Symbiotic Phenomenon, 229–230.
© *1988 by Kluwer Academic Publishers.*

different nature.According to Piirola's data (2,3) that refer to the initial stage of star's activity, the degree of polarization varied within a large range - from 0.04 to 0.7% in red and near-IR, and from 0.3 to 1.6% in the UV. On an average it was lower th than in the quiet state, position angle changed al so. According to our data, in 1984 (Sept.22-Oct.8) the degree of polarization diminuished yet more, its wavelength distribution did not display the rise in the UV observed earlier, position angle was practi cally the same in all the wavelengths (see Fig.); statistically reliable va riations in the value of P from night to night were observed.

Drastic decrease of the degree of polarization in the UV in 1984 compar- ing with 1977-1979 should be related to weakening of blue continuum. This agrees with the observed correla- tion between UV flux and UV polarization (2) and indi- cates that a significant part of polarization at active stage is caused by additional hot radiation source. Nevertheless during the quiet stage, before the outburst, high degree of polarization in the short wave region obviously had different nature, and, as it is in the case of other cool giants, that polarization should be produced by light scattering on dust particles in the circumstellar envelope around CH Cyg.

REFERENCES

1. Khudyakova, T.N. 1987, Astrofizika 26, 153.

2. Piitola, V. 1981, Astrophys. Lab. Univ. Helsinki Report 2/1981, 127.

3. Piirola, V. 1983, in Cataclysmic Variables and Related Objects, IAU Colloquium 72, 211.

CONTINUED RADIO ACTIVITY FROM CH CYGNI

A.R. Taylor
Department of Physics, University of Calgary

E.R. Seaquist
Department of Astronomy, University of Toronto

S.J. Kenyon
Harvard–Smithsonian Center for Astrophysics

ABSTRACT. Following the discovery of the radio outburst and jet formation from CH Cygni in 1984, continued radio monitoring has been carried out with the Very Large Array. We present the preliminary results of these observations, including a high resolution image of the jet in 1986 – 1.5 years after the initial outburst. The star has remained active in the radio, undergoing at least one additional outburst. The radio jet has continued to expand and shows evidence of multiple episodes of ejection along the same axis.

1. INTRODUCTION

Since 1964 CH Cygni has been in an active state, exhibiting a high degree of optical variability and the combination spectrum of late-type continuum and emission lines characteristic of symbiotic stars. In July/August 1984 the optical continuum declined abruptly by more than a magnitude while higher excitation emission lines and very broad Balmer wings appeared (Tomov 1984, Selvelli and Hack 1985). At the same time radio observations revealed an outburst at centimeter wavelengths and the production of a jet of radio emitting gas expanding at a rate of 1.1 "/yr (Taylor et al. 1986). In this paper we report on continued radio observations of CH Cygni during 1985 and 1986.

2. OBSERVATIONS AND RESULTS

Observations at frequencies of 1.5, 5 and 15 GHz have been obtained with the National Radio Astronomy Observatory Very Large Array at intervals of about three months. Figure 1 shows the flux densities at these frequencies for the three year period April 1984 to April 1986. The curves delineate at least two distinct outbursts that evolve on a time scale of about 400 days. The time scale and spec-

J. Mikolajewska et al. (eds.), The Symbiotic Phenomenon, 231–232.

tral evolution of the outbursts are similar to the radio outbursts of novae. The beginning of a third outburst is suggested by our last observation.

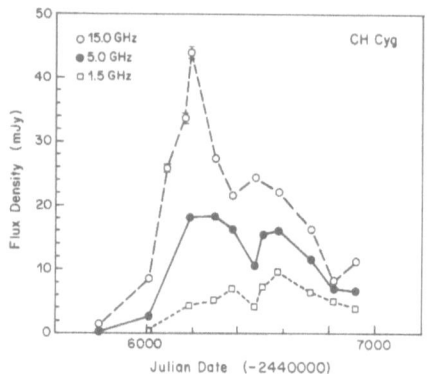

Figure 1. Radio Flux Densities of CH Cygni at three frequencies

In February 1986 the VLA returned to the highest resolution A configuration. The radio image at $\lambda = 2$ cm is shown in figure 2. The total extent of the jet is 1.5", roughly consistent with the expansion rate measured in 1984. The jet also shows additional radio components, indicating that a further episode of material ejection has occurred. The spacing of the components suggests an interval of about 200 days between ejection events.

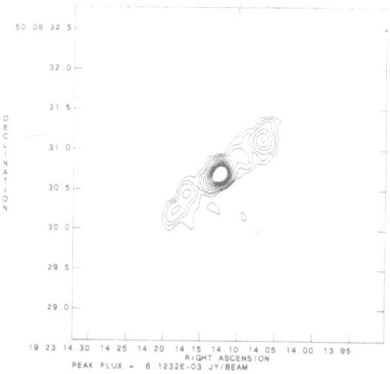

Figure 2. Radio image of CH Cygni from February 1986 at $\lambda = 2$ cm

References

Selvelli, P.L. and Hack, M. 1985 *Astronomy Express*, **1**.
Taylor, A.R., Seaquist, E.R. and Mattei, J.A. 1986, *Nature*, **319**, 38.
Tomov, T. 1984 *Inf. Bull. of Var. Stars*, No. 2610.

AN "ACCRETOR-PROPELLER" MODEL OF CH CYGNI

M. Mikolajewski & J. Mikolajewska
Institute of Astronomy,
Nicolaus Copernicus University,
Chopina 12/18 , PL-87100 Torun, Poland

The most spectacular episode in the history of CH Cyg was undoubtedly a sudden drop of brightness (by about 1 mag) during the last weak of July 1984, coinciding with the ejection of two collimated jets (Taylor *et al.* 1985) perpendicular to the line of sight (Solf 1987) and the orbital plane (Luud *et al.* 1986). Supercritical accretion cannot be responsible for the jet ejection mechanism, because the luminosity of the accreting component was always about 2 orders of magnitude lower than the Eddington limit (Mikolajewska *et al.* 1987). This mechanism is also inconsistent with the rapid drop of brightness coinciding with the ejection of material. Simultaneously, it is conspicuous that the increase of brightness to optical maximum was also rapid ($\sim$1 mag from July to September 1981; e.g. Kaler *et al.* 1983).

We propose a preliminary model of the activity of CH Cyg based on the theory of an oblique rotator (Lipunov 1987) and assuming an accreting white dwarf with magnetic axis nearly parallel to the rotational axis as the energy source of the hot component. The luminosity of a rotator due to dissipation of rotational energy and release accretion energy is redistributed by the outer layers of material accreted from the M6 giant wind and observed as the 9000K pseudophotosphere ($R\overset{\sim}{=}10\text{-}2R_\odot$). We also assume that both the rapid increase of brightness in 1981, as well as its sudden drop in 1984, was due to a transition through *"the catastrophic equilibrium"*, in which a small variation in the accretion rate made the rotator pass from the *"propeller state"* into the *"accretor state"* (1981) and vice versa (1984). The luminosity of the rotator abruptly changes as a result of these transitions.

The luminosity of the *accretor* just before the drop in brightness and the *propeller* just after the drop in luminosity (Lipunov 1987) is:

$$L_{before}=\dot{M}_c GM/R_0+\mu^2\omega/3R_c^3 \; ; \quad L_{after}=\dot{M}_c GM/R_A+\mu^2\omega/3R_A^3$$

where $\dot{M}_c$ is the "catastrophic" accretion rate and M, R_0, $\mu,\omega=2\pi/P$ are the mass, radius, magnetic dipole moment, and angular rotational velocity of the white dwarf, respectively. The Alfven radius (R_A) is equal to the corotation radius (R_C) at the time of the transition ($R_A(\dot{M}_c)=R_C$), so:

$$R_A=(\mu^2/\dot{M}_c(2GM)^{1/2})^{2/7} \; ; \qquad R_C=(GM/\omega^2)^{1/3}$$

J. Mikolajewska et al. (eds.), The Symbiotic Phenomenon, 233–234.
© *1988 by Kluwer Academic Publishers.*

The solution of this system of 4 equations yields:

$$R_c/R_0 = (2/9)^{1/2}[(L_{before}/L_{after}) - 1] ; \qquad \omega = (GM)^{1/2}R_c^{-3/2}$$

$$\dot{M}_c = (L_{before} - L_{after})(1/R_0 - 1/R_A)/GM ; \qquad \mu^2 = \dot{M}_c(2GM)^{1/2}R_A^{7/2}$$

The spectroscopic orbit (Mikolajewski *et al.*, *this Proceedings*) and the mass ratio $q \simeq 3.5$ (Mikolajewski *et al.* 1987) implies $M = 1 M_\odot$. The photometric and IUE data suggest $L_{before} = 240 L_\odot$ and $L_{after} = 70 L_\odot$ (see also Mikolajewska *et al.* 1987). Taking $R_0 = 10^9$ cm we have: $R_c \simeq 5 R_0$, $\mu \simeq 10^{33}$ Gcm^3, $\dot{M}_c \simeq 10^{-7} M_\odot$/yr and $P \simeq 200$s. The derived value of the magnetic field is close to that expected for magnetic white dwarfs (King 1985). The rotational period of the white dwarf would be difficult to detect in the case of CH Cyg because of the optically thick envelope.

In the proposed model, the jets represent material from accretion columns blown off along the rotation axis at the time of the transition from *accretor* to *propeller* ($A \rightarrow P$). It is very striking that the theoretical collimation angle for a jet, $\theta = (R_0/R_A)^{1/2} = 0.45$ (Lipunov 1987), is comparable to that observed in CH Cyg (20^0–30^0; Taylor *et al.* 1985). Adopting the mass in the jets, $2 \times 10^{-6} M_\odot$ (Taylor *et al.* 1985), the work performed by the magnetic field during the transition $A \rightarrow P$, $W = GM_{jets}M(1/R_0 - 1/R_A) \simeq 10^{44}$ ergs, is comparable to the kinetic energy of the jet motion. In CH Cyg, about 1% of the cool giant wind can be captured by the white dwarf (see Mikolajewski *et al.* 1987). To mantain the "catastrophic" accretion, $\dot{M}_{wind} \simeq 10^{-5} M_\odot$/yr is required, which is unacceptable for the M6 giant. This suggests that material must be accumulated around the white dwarf for a long time before accretion is possible. Using formulae given by Livio & Warner (1984) and the known orbital parameters of CH Cyg an accretion disk with a radius 0.5–$2.5 \eta^2 R_\odot > R_c \simeq 5 R_0$ can be formed, (η is dimensionless parameter of order 1). We estimate that about $10^{-6} M_\odot$ was accreted during the active period 1963–1987 and that a comparable amount of material was ejected during the 1984 event. Therefore, at least 100–1000 years are necessary to reconstruct the disk-envelope around the white dwarf for an assumed, reasonable mass-loss rate from the giant, $\dot{M}_{wind} \simeq 5 \times 10^{-7}$–$5 \times 10^{-8} M_\odot$/yr. In fact, no activity of the hot component was detected before 1963 (*e.g.* Dziewulski *et al.*, *this Proceedings*.).

Finally, the accreted mass required to initiate a hydrogen shell flash on a white dwarf is about $5 \times 10^{-6} M_\odot$ (*e.g.* Prialnik 1987); thus we may expect a nova like eruption in each new activity cycle. In fact, R Aqr, the symbiotic system most closely related to CH Cyg, also seems to be an ancient nova (Kafatos & Michalitsianos 1982)

REFERENCES:
Kafatos, M. & Michalitsianos, A.G., 1982, *Nature* 298, 540
Kaler, J.B., Kenyon, S.J. & Hickey, J.P., 1983, *P.A.S.P.* 95, 1006
King, A.R., 1985, in *Recent Results on Cataclysmic Variables*, ESA SP-236, 133
Lipunov, V.M., 1987, *Astrophys. Space Sci.* 132, 1
Livio, M. & Warner, B., 1984, *Observatory* 104, 152
Luud, L., Tomov, T., Vennik, J., Leedjarv, L., 1986, *Pisma v Astr. Zhu.* 12, 870
Mikolajewska, J., Selvelli, P.L, Hack, M., 1987, submitted. to *Astron. Astrophys.*
Mikolajewski, M., Tomov, T., Mikolajewska, J., 1987, *Astrophys. Space Sci.* 131, 733
Prialnik, D., 1987, *Astrophys. Space Sci.* 131, 431
Solf, J., 1987, *Astron. Astrophys.* 180, 207
Taylor, A.R., Seaquist, E.R. & Mattei, J., 1985, *Nature* 359, 38

A REVIEW OF THE R AQUARII SYSTEM

A.G. Michalitsianos and M. Kafatos
NASA Goddard Space Flight Center
Laboratory for Astronomy and Solar Physics
Greenbelt, Maryland 20771 USA

ABSTRACT. The spatially resolved nebula that characterizes the D-type symbiotic R Aquarii has afforded investigators a unique opportunity to probe the extended emission line regions. Its extensive and complex radio morphology, that includes SiO emission from the only symbiotic associated with maser emission, has provided important clues concerning the mass expulsion process in interacting binary radio stars. Infrared, radio, optical, UV and X-ray observations of the system are discussed in context with models which have been proposed to explain the appearance of the brilliant jet.

1. INTRODUCTION

1.1 Early Spectroscopic Observations

Of the modest number of known D-type symbiotic systems, R Aquarii (M7e) is unique among the group, having been studied extensively over a wide range of wavelengths. Its relatively close distance of d = 180 to 300 pc has afforded investigators an opportunity to probe the emission properties of the circumstellar-meniscus nebula which characterizes the system at optical, and more recently, at radio continuum wavelengths (Hollis et al. 1987). Since the early observations of Lampland of the 1920's, R Aquarii is known to be associated with an extended filamentary nebula which surrounds a 387^d period Mira. The presence of nebular forbidden lines in optical spectra was identified by Merrill (1934,1940 and 1950), who measured their radial velocities and observed the rise of the continuum from the hot source of the system. Between 1922 to 1933, the continuum of the hot subdwarf actually dominated the strong TiO absorption features associated with the Mira; the hot companion achieved a visual magnitude m_v ~8, which rivaled the Mira even at maximum light. Merill also recorded the presence of P-Cygni structure in the Balmer lines during the spectroscopic outburst of the 1920. Since that time R Aqr has seemingly returned to a quiescent state, in which the

J. Mikolajewska et al. (eds.), The Symbiotic Phenomenon, 235–243.

nebular emission lines of [N I], [N II], [S II], [O II], [O III], [Fe II], [Fe III], He I, He II and the Balmer lines are superimposed on the strong TiO features of the cool M star. The possible cyclic nature of the blue continuum source in R Aqr proposed by Merrill (1950), in which the hot source would again brighten in the mid-1960's, was not observed (Jacobsen and Wallerstein 1975).

The extended nebula consists of a filamentary structure that is comprised of two broad, and almost symmetric arcs, whose intersecting points form a double meniscus-shaped nebula, that is ~90-arcsec in EW extent. Nearly perpendicular to the EW nebula, a NS nebula of approximately 1-arcmin extent is also present. Both NS and EW nebulae are centered on a spatially unresolved HII region which surrounds the Mira and hot companion. Based upon the expansion velocity of the EW filaments, Merrill (1950) deduced that the outer nebula was created during a nova outburst which he estimated took place about 600 years ago. However, from ancient chronicles it has been suggested that the nova may have actually taken place about 1000 years ago (Kafatos and Michalitsianos 1982), or more precisely as a "guest star" of 1073 AD, as noted by Li (1985). Alternatively, Solf and Ulrich (1985) propose the EW and NS nebulae were formed in two distinct ejection events, in which the EW "ring" was formed 640 years ago, while the NS nebula was created in a more recent outburst, about 185 years ago.

1.2 Radio Continuum and SiO Maser Emission

Gregory and Seaquist (1974) detected the first radio emission from R Aqr at 10.5 GHz. Their observations indicated the source was variable on a timescale of ~1 month, over which time the flux density varied by a factor ~5. In addition, they deduced an upper limit to the size of the radio emitting region of ~1 arcsec, while the radio flux densities obtained at 10.5, 8.085 and 2.695 GHz suggested a spectral index of α ~-0.1, ($S \propto \nu^{\alpha}$), consistent with optically-thin bremsstrahlung emission. Lepine, LeSqueren and Scalise (1978), Zuckerman (1979) found SiO maser emission in R Aqr from the collisionally pumped vibrationally excited J = 2 - 1, v = 1 transition. Zuckerman (1979) finds that the SiO lines at 43.12203 and 86.24327 GHz indicate radial velocities -28 km s^{-1} (LSR), which is close to the radial velocities of the K I and Ca I lines obtained by Wallerstein and Greenstein (1980), but is not consistent with most of the absorption features associated with the Mira, which indicate -20 km s^{-1} as the most commonly measured systematic velocity for the star. Emission lines formed at lower densities, such as [O III] and [N III], indicate larger radial velocities, that range up to -60 km s^{-1}, and suggest the velocities in the nebula increase outwardly (Wallerstein and Greenstein 1980). The discrepancy between the SiO maser velocity and optical lines from the Mira may be quite important, because SiO maser emission associated with M giants and M supergiants is believed formed in the extended atmospheres of these stars, where column densities are sufficiently high that SiO can be

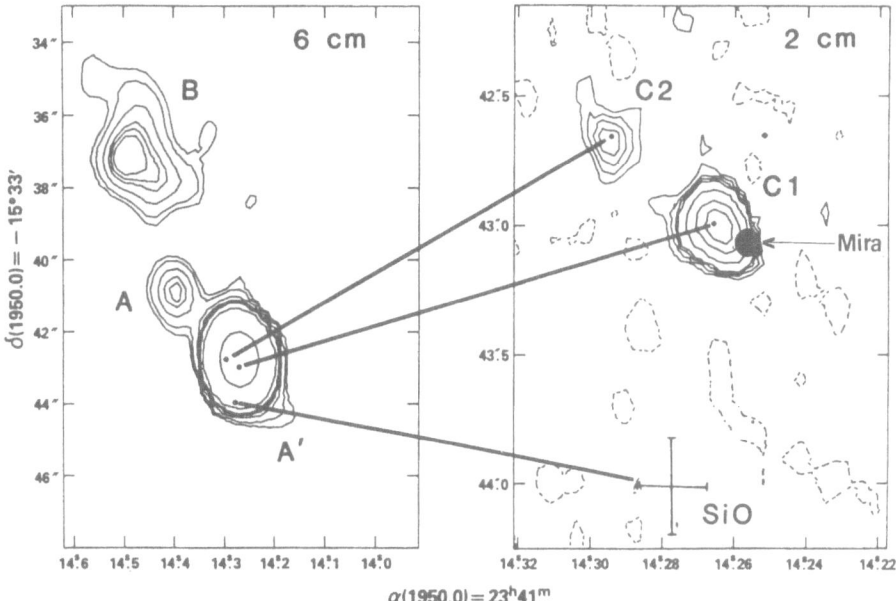

Figure 1: VLA 6 and 2-cm radio continuum maps showing position of the SiO Maser in R Aqr, and the astrometric optical Mira Position

efficiently pumped by collisions (Elitzur 1980). Accordingly, the SiO maser should be found in close proximity to the HII region, in which the Mira and hot subdwarf are embedded.

However, sub-arcseond radio continuum observations obtained with the Very Large Array (VLA) and SiO observations obtained with the Hat Creek Radio Interferometer demonstrate that the SiO maser is far removed from the HII region (Hollis et al. 1985). The centroid of SiO maser emission is ~ 1-arcsec south ($4.5x10^{15}$cm at d = 300 pc) of the HII region, which corresponds to the spatially resolved feature C1, one of the two radio features that comprise the central HII region. (Fig. 1). Moreover, astrometric observations indicate the Mira is located within feature C1 (Michalitsianos et al. 1987), consistent with the Hat Creek observations which indicate the maser is not in the near vicinity of the long period variable. This result will very likely impact models concerned with SiO maser emission in single late-type M giants as well. It may be that shocks formed several hundred Mira radii from a high velocity wind could achieve sufficient column densities that SiO maser emission can be excited through collisions (Elitzur, private communication). This question remains completely open for further investigation.

2. THE NATURE OF THE RADIO/OPTICAL/UV JET

2.1 Radio Continuum Observations

Wallerstein and Greenstein (1980) first noticed the appearance of a brilliant optical "spike" or "jet" in 1977 protruding about 6".5 - NE from the central star. From their spectra they concluded the "spike", as well as other points in the nebula are generally lower in density ($n_e \sim 10^2 - 10^3 cm^{-3}$) compared with the HII region, based upon the relative intensities of the [O II] and [S II] lines. Subsequently, RG1 and UG1 images of the jet from Lick 3-m direct plates obtained by Herbig (1980) indicated that the jet had probably brightened, and rivaled the Mira at minimum light ($m_V \sim 11$ mags.). A comparison of the 1980 plates with those obtained by Herbig in 1970 indicated that the jet had probably appeared within the decade. Radio maps obtained with the VLA at 6-cm (with ~4-5-arcsec spatial resolution; Sopka et al. 1982) closely matched the morphology of the Lick-UG1 images of the jet.

Higher spatial resolution (~ 1-arcsec) observations at 6-cm with the VLA (Kafatos et al. 1983) further resolved the jet and HII region in three discrete regions of emission, consisting of a spatially resolved feature B located about 6".5 from R Aqr at a position angle (p.a.) 29°, feature A, at 2".5 and p.a. = 45°, and emission from a compact HII region that corresponds to the position of the binary (Figure 2, VLA map). The extended contours SE of the R Aqr (labeled A') provide bipolar symmetry with feature A. Mauron et al. (1985) obtained high resolution optical images in near UV light confirming the radio structure, from which they concluded that features A and A' comprise the jet, while feature B is not dirtectly associated. However, given that features A and B have nearly the same p.a., and are relatively strong radio sources, it is unlikely that feature B not associated with physical processes that created A and A'. The spatial extent of feature B enabled Hollis et al. (1985) to obtain its spectral index $\alpha \sim -0.1$ from 20, 6 and 2-cm VLA maps. This index is consistent with optically-thin bremsstrahlung emission also found by Gregory and Seaquist (1974), but from the integrated radio emission of R Aqr at 10.5, 8.085 and 2.695 GHz. However, Hollis et al. (1986) also found that $\alpha \sim +0.6$ in the HII region, consistent with a spherically symmetric, optically-thick wind (Wright and Barlow 1975). The combined index of the jet and HII region is $\alpha \sim +0.3$, which cautions against surveys of unresolved stellar radio systems, in which the combined spectral indices of optically-thick and -thin emitting regions can introduce systematic errors (cf. Seaquist et al. 1984), if the jet nebulosity of R Aqr is typical of the contributions which extended radio structure can make in symbiotic stars. Additionally, a comparison of the 6-cm maps obtained between 20 Sept 1982 and 3 Feb 1984 indicate the integrated flux density of the map ~11.793 mJy had not changed over this period. If the radio flux from the system is highly variable (Gregory and Seaquist 1974), the system can have periods of prolonged radio quiescence as well.

Sub-arcsec 2-cm VLA maps of by Hollis et al. (1986) revealed the presence of another radio feature in the system C2 (Figure 1, right panel) at 0".5 and p.a. = 55°, whose flux density at 2-cm of 1.37 mJy is weaker compared with the HII region (feature C1), which is 8.52 mJy. In all, the radio components we associate with the jet, therefore, consists of features C2, A and B, with A' possibly indicating counterjet activity. The oblate contours that define feature C1 suggests a prolonagtion in the direction of C2, and indicates the HII region is spatially resolved. Astrometric observations of the luminous Mira indicate it is located within the contours of C1 (Michalitsianos et al. 1987), consistent with the model that the suspected ~44 year binary orbit (Willson et al. 1981) is surrounded by a compact HII region that is photoexcited by the intense radiation field of the hot subdwarf and/or accretion disk.

Finally, VLA observations obtained by Hollis et al. (1987) at 6-cm revealed radio emission from the outer ~2' - EW nebula. Typical electron densities derived from the radio emission of the extended EW nebulosity are $n_e \sim 10^2$ to $10^3 cm^{-3}$, consistent with estimates from optical nebular lines (cf. Wallerstein and Greenstein 1980; Solf and Ulrich 1985). Curiously, wide-field 6-cm maps also revealed a cluster of a dozen radio sources, most of which are ~3' south and/or west of R Aqr. Together with 20-cm data, these observations suggest limits of the index α, which indicate they are thermal, while other sources in the field with $\alpha < 0$ may be extragalactic background sources. If the cluster of sources is associated with R Aqr, they may provide the first direct evidence of a prehistoric outburst of the system (Hollis et al. 1987).

2.2 Far-UV, Optical Emission and Extinction in the System

The complex nature of the circumstellar extinction associated with R Aqr makes it difficult to obtain n_e and T_e from the emission measure. From optical spectra acquired in 1977, Wallerstein and Greenstein (1980) obtained an E(B-V) = 0.67 mags. for the inner nebula, while extinction in the vicinity of the jet appeared to be almost negligible. Similarly, Kaler (1981) obtained an E(B-V)=0.65 from the normalized fluxes of lines calculated from the recombination theory of Osterbrock (1974), approprite to $n_e \sim 10^6 cm^{-3}$, and $T_e \sim 10^4 K$. Whitelock et al. (1983) found that the peculiar variability of Hα indicates that Wallerstein and Greenstein's value of 0.67 could be questionable. In order to explain these discrepancies, Whitelock et al. (1983) propose that an opaque dust cloud partially obscured the Mira between 1975 to 1978. From the Balmer lines, Hβ, Hγ, Hδ, and He I lines, (Brugel et al. 1984) have concluded that the extinction varied from E(B-V)=0.4 in 1979 to E(B-V)=0.05 in 1982, which together with Wallerstein and Greenstein's value of 0.67 in 1977, indicates that central star was partially obscured by a dust-cloud in 1977. Probably a combination of variable extinction and geometrical changes strongly affect the nature of the line and continuum emission from the source. The partial obscuration of R Aqr by an opaque dust cloud

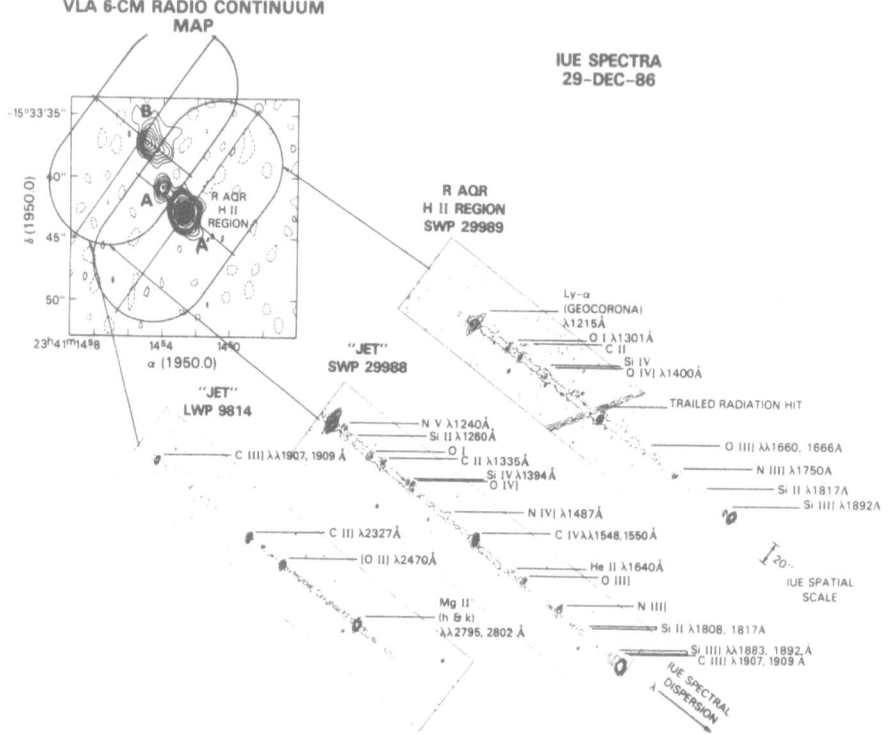

Figure 2: IUE 10x20" entrance slit shown on VLA 6-cm radio continuum map. Emission features are indicated for two slit positions; SWP 29988 & LWP 9814 centered on jet feature B, SWP 29989 on HII region.

is also suspected for another similar system, RX Puppis, that is based upon variations in J and V magnitudes over a timescale of ~2 years (Whitelock et al. 1984).

The spatial extent of the jet has afforded investigators an opportunity to examine the far–UV emission of the jet separately from the HII region. Low resolution IUE spectra ($\Delta\lambda \sim 6$Å resolution) indicate considerable differences exist in excitation between these regions (Kafatos et al. 1986). In addition to most of the resonance and intercombination lines of the jet exhibiting quasi-periodic variations in line strengths on timescales of $\gtrsim 1.5$ years, a systematic increase in line intensity is observed over 4 to 5 years; while the line intensities appear essentially constant in the HII region. The appearance of N V $\lambda\lambda 1238,1242$, and strengthening of He II $\lambda 1640$ emission in the jet, indicates that higher excitation conditions prevail at the jet offset position, compared with the HII region, where N V is absent, and He II emission is weak (Michalitsianos et al. 1986; Kafatos et al. 1986) (Fig. 2). The appearance of N V in the jet in 1985 is consistent with the first unambiguous detection of soft X-ray emission in the 0.25 to 1 KeV energy range with EXOSAT by Viotti et al. (1987). Because higher excitation conditions prevail in the jet, soft X-rays are likely to be more intense in the vicinity of features A and B, where N V and He II are prominent. EXOSAT does not have sufficient spatial resolution to determine where X-rays could be concentrated in the system.

Additionally, Paresce et al. (1987) obtained narrow band filter images with the ESO 2.2m coronograph telescope in Hα, and [N II], where they detected the appearance of a new feature, located at 8".4 and p.a. = 20°. They suggest this new feature may be an extension of feature B, and provides evidence for shock excitation. The appearance of this new feature sometime between 1984 and 1986 is contemporaneous with the detection of N V at the jet offset position, and with the detection of soft X-rays from the system.

3.0 MODELING R AQUARII AND ITS JET

A number of models have been proposed to explain the nature of the complex nebulosity and temporal emission from this object. It seems reasonable that the emission line spectrum of the star can be explained by a compact HII region with characteristic densities and temperatures of $\gtrsim 10^6 cm^{-3}$ and T_e ~15,000K (Michalitsianos 1980). The intense radiation field emanating from the hot subdwarf and/or accretion disk, photoionizes the HII region (Kafatos et al. 1986), but also powers a stellar wind. Evidence for an optically-thick stellar wind in the system is indicated by the spectral index α ~ +0.6 obtained by Hollis et al. (1985) in the radio, and from the first successful IUE SWP-HIRES spectrum of the central HII region, in which the anomalous C IV doublet intensity ratio of $I(\lambda 1548)/I(\lambda 1550)$ ~ 0.6. This ratio being less than the optically-thick limit of unity can explained if a hot optically-thick wind, with an expanion velocity at least equal to the doublet separation of $\Delta\lambda = 2.6Å$, or $\gtrsim$ 500 km s^{-1}, results in P-Cygni profile absorption of blue component photons by the red doublet component (Michalitsianos et al. 1987). This will enhance emission of the $\lambda 1550$ line relative to $\lambda 1548$ line, because the source function $S_{\nu,R}$ of the red doublet now depends on nonlocal values of $S_{\nu,B}$ of the blue doublet (cf. Castor and Lamers 1979; Olson 1982).

The appearance of the jet and the mechanism of its continued and sustained excitation is under dispute. Solf and Ulrich (1985) have shown that the velocities associated with features A and B (relative to the star) are characteristically $\lesssim$100 km s^{-1}. Solf and Ulrich (1985) suggest that feature B was part of the major outburst which formed the NS inner nebula about 185 years. The appearance of feature B in 1977 is the result of collisions with differentially moving material in the expanding nebula, which decelerates the parcel, resulting in shock heating and a sudden brightening of the parcel which took place in late 1970's.

The morphology of features A, B and C2 from high resolution VLA maps, however, suggests that the parcels which comprise the jet indicate systematic, organized flow from the system. As such, the radio features define an arc on the small-scale (up to ~7 arcsec from the central star) which is similar to the broad filamentary arcs that form the EW and NS nebula (on the 1-arcmin scale), as evident in narrow-band [O III] images (Michalitsianos et al. 1987). This morphology could suggest that each parcel was ejected sequentially in repeated outbursts, while the central "cannon" precesses or rotates (Kafatos et al. 1986). Thus, there is evidence for an extended S-pattern in the NS bipolar nebula, which could reflect the historical sequence of outbursts from the system.

Also evident from IUE spectra, the continued increase in excitation at the jet offset position is not consistent with the cooling timescale of ~2-years for nebular material, if shock excitation were the only means of heating ejected parcels of gas in a low density gas, where $n_e \sim 10^2$ to $10^3 cm^{-3}$. Kafatos et al. (1986) suggest that the illumination of perviously ejected parcels by an intense cone of ionizing radiation created by a thick-accretion could explain morphology of the radio jet features, and the modulations of UV emission lines found in IUE spectra of feature B, if the broad radiation cone of the disk slowly precesses while also varying in intensity. This modulation could be the result of a variable mass accretion rate. A sufficient body of evidence is not presently available which favors either of these models. Quite probably, a combination of shock and photoionization models might eventually prove to be correct.

R Aqr could constitute a prototype of directed mass loss or jet activity in symbiotic systems. RX Pup and HM Sge also show similar structure in the C IV line profiles, indicating that these systems may also posses hot star winds. Moreover, the close proximity of R Aqr to earth affords us a unique opportunity to study the physics of jet formation. The suspected accretion disk in R Aqr could be visible with HST. The low velocities prevalent in the R Aqr jet have been explained as a result of ejection from the outer, cool regions of the giant accretion disk that surrounds the hot star. Grain opacity would be high in this region, resulting in efficient acceleration of gas parcels, even at luminosities much less than the Eddington limit (Kafatos et al. 1986). The existence of a hot stellar wind is only evident in the C IV profiles of the HII region, while the bulk of the ejected gas moves at much lower velocities ($\leqslant 100$ km s^{-1}). The jet of CH Cyg may be related to jet activity in R Aqr, in the sense that high ejection velocities observed of $\gtrsim 1000$ km s^{-1} are achieved because ejection takes place closer to the hot subdwarf (Taylor et al. 1986).

References

Allen, D.A. and Wright, A.E. 1987, in Proc. IAU Colloq. 103, The Symbiotic Phenomena (D. Reidel Pub. Co.-Holland), in press.
Brugel, E.W., Cardelli, J.A., Szkody, P. and Wallerstein, G. 1984, Ap.J., **98**, 78.

Castor, G.L. and Lamers, H.J.G.L.M. 1979, Ap.J., **39**, 481.

Elitzur, M. 1980, Ap.J., **240**, 553.

Gregory, P.C. and Seaquist, E.R. 1974, Nature, **247**, 532.

Herbig, G. 1980, IAU Cir. No. 3535.

Hollis, J.M., Kafatos, M., Michalitsianos, A.G. and McAlister, H.A. 1985, Ap.J., **289**, 765.

Hollis, J.M., Michalitsianos, A.G., Kafatos, M., Wright, M. and Welch, W.J. 1986, Ap.J. (Letters), **309**, L53.

Hollis, J.M., Kafatos, M., Michalitsianos, A.G., Oliversen, R.J. and Yusef-Zadeh, F. 1987, Ap.J. (Letters), in press.

Kafatos, M. and Michalitsianos, A.G. 1982, Nature, **298**, 540.

Kafatos, M., Michalitsianos, A.G. and Hollis, J.M. 1986, Ap.J. Supp., **62**, 853.

Kaler, J.B. 1981, Ap.J., **245**, 568.

Lepine, J.R.D., LeSqueren, A.M. and Scalise, E. Jr. 1978, Ap.J., **225**, 869.

Li, Jing 1985, Chin. Astron. and Astrophys., **9**, 322.

Mauron, N., Nieto, J.L., Picat, J.P., Lelievre, G. and Sol, H. 1985, Astron. and Astrophys., **142**, 413.

Merrill, P.W. 1934, Ap.J., **81**, 312.

_____ .1940, Spectra of Long Period Variable Stars (Chicago Univ. Press), pp. 82-89.

_____ . 1950, Ap.J., **112**, 514.

Michalitsianos, A.G., Hollis, J.M. and Kafatos, M. 1986, Canadian J. of Phys., **64**, 523.

Michalitsianos, A.G., Oliversen, R.J., Hollis, J.M. and Kafatos, M. 1987, A.J., in press.

Michalitsianos, A.G., Kafatos, M., Fahey, R.J., Viotti, R., Friedjung, M., Cassatella, A., Piro, L. 1987, Ap.J., submitted.

Olson. G.L. 1982, Ap.J., **255**, 267.

Osterbrock, D.W. 1974, Astrophysics of Gaseous Nebulae (San Francisco: Freeman).

Paresce, F., Burrows, C. and Horne, K. 1987, Ap.J., submitted.

Seaquist, E.R., Taylor, A.R. and Button, S. 1984, Ap.J., **284**, 202.

Solf, J. and Ulrich, H. 1985, Astron. and Astrophys., **148**, 274.

Sopka, R.J., Herbig, G., Kafatos, M. and Michalitsianos, A.G. 1982, Ap.J. (Letters), **258**, L32.

Taylor, A.R., Seaquist, E.R. and Mattei, J.A. 1986, Nature, **319**, 38.

Viotti, R., Piro, L., Friedjung, M. and Cassatella, A. 1987, Ap.J. (Letters), in press.

Wallerstein, G. and Greenstein, J.L. 1980, Pub. Ast. Soc. Pacific, **92**, 275.

Willson, L.A., Garnavich, P. and Mattei, J. 1980, Inf. Bull. Var. Stars, nor. 1961-1963.

Whitelock, P.A. Feast, M.W., Catchpole, R.M., Carter, B.S. and Roberts, G. 1983, MNRAS, **203**, 351.

Whitelock, P.A., Menzies, J.W., Evans, T. Lloyd and Kilkenny, D. 1984, MNRAS, **208**, 161.

Wright, A.E. and Barlow, M.J. 1978, MNRAS, **170**, 41.

Zuckerman, B. 1979, Ap.J., **230**, 442.

THE ULTRAVIOLET SPECTRUM OF RX PUPPIS

M. Kafatos and A.G. Michalitsianos
NASA Goddard Space Flight Center
Laboratory for Astronomy and Solar Physics
Greenbelt, MD 20771 USA

ABSTRACT. The UV spectrum of the peculiar star RX Puppis has afforded
symbiotic star investigators a wealth of information for unraveling its
mysteries. RX Pup and R Aqr, both being of the D-type variety, are now
better understood as result of an extended coverage of observations at
different wavelengths including radio observations using the VLA. These
stars present challenges to our understanding of the symbiotic phenomenon
and clues to other astrophysical phenomena like jets. Resolution of the
question whether RX Pup has a jet system and an associated system of
rings/extended disk or, alternatively, a colliding winds region will be
resolved by high resolution radio observations or future observations
using the Hubble Space Telescope.

1. INTRODUCTION

RX Puppis is a peculiar, emission line object related to slow novae
(Swings and Allen 1972). It contains a 580^d period Mira (Whitelock et
al. 1984) with an effective temperature of ~2400 K (Barton, Phillips and
Allen 1979). It, therefore, is a D-type symbiotic. It is not just interesting in itself but also since it may be a link between D-type symbiotics and slow novae.
    In the 1940's (Swings and Struve 1941) it displayed a rich high-
excitation emission spectrum (cf. Kafatos, Michalitsianos and Feibelman
1982). Night to night variations in the Balmer line intensities were
found in 1975 (Swings and Klutz 1976). Multicomponent P-Cygni profiles
of the Balmer lines and the stronger Fe II lines were detected in 1976
(Klutz, Simonetto and Swings 1978) with sharp blueshifted absorptions in
the Balmer lines ranging up to -1100 km/s. There is an indication of
broad P-Cygni structure in the C IV profiles seen by the "International
Ultraviolet Explorer" (IUE) although no direct detection of it (Kafatos
et al. 1982; Kafatos, Michalitsianos and Fahey 1985). The far UV spectrum
of RX Pup obtained with the IUE shows a number of prominent emission lines and a weak continuum. High resolution (HIRES) structure of the strongest UV lines is complex and variable (Kafatos et al. 1985). High excitation lines become present when the star becomes fainter (Allen and

245

J. Mikolajewska et al. (eds.), The Symbiotic Phenomenon, 245–247.

Wright 1987). In 1982 the UV emission lines became stronger (Kafatos et al. 1985). The multicomponent structure of the C IV profiles seen with IUE is not unique: Recently, observations of the central H II region around R Aqr (Michalitsianos et al. 1987) reveal similar structure as well as HM Sge (Mueller and Nussbaumer 1985). Since R Aqr contains a prominent jet system, there remains the possibility that RX Pup and other D-type symbiotics contain similar structures. RX Pup is a prominent radio source (Seaquist 1977; Hollis et al. 1986; Seaquist and Taylor 1987) and reveals extended radio structure (Hollis et al. 1986) but due to its greater distance compared with R Aqr it does not afford us as much detail as that revealed in high resolution VLA maps of the latter object. The IR properties of RX Pup are complex. The presence of the Mira is only revealed from J and L photometry (Whitelock et al. 1983). The decline in the visual light cannot easily be attributed to variations in either the hot or the cool component (Whitelock et al. 1984).

2. MODELS OF RX PUPPIS

The 1548/1550 C IV intensity ratio has been observed to be always less than unity (Kafatos et al. 1985). Typical velocity differences between individual components are 40 - 50 km/s. The anomalous doublet ratio of the C IV profiles has been attributed to the existence of a hot wind with $700 \lesssim v_w \lesssim 1000$ km/s (Kafatos et al. 1985; Michalitsianos et al. 1987). Michalitsianos et al. (1987) plot the 1548/1550 ratio versus the IUE FES magnitude or versus the C IV line intensity and find an inverse relationship. This they dub the "CIV doublet ratio-intensity effect". It is also found in HIRES profiles of the central H II region surrounding R Aqr. The mass loss in the hot wind in both stars is very small, $M \lesssim 10^{-11}$ $M_\odot$/yr, much less than the cool wind emanating from either the Mira or the outer regions of a cool disk. The colliding wind model for RX Pup (Allen and Wright 1988) would meet then with difficulties in the sense that the hot star wind is negligible in terms of dynamics and is only discernible in the C IV profiles. To date two models have been proposed for RX Pup:

Kafatos et al.(1985) proposed a model involving an extended disk which has broken up into rings. Their conclusion is based on the fact that the kinematic properties of the C IV profiles defy a simple colliding wind model and that the components are remarkably constant in velocity space. The rings shroud the hot star but allow photoionizing radiation to escape perpendicular to the equatorial plane. Kafatos et al. (1985) offer the ring model as the most viable since very long orbital periods ($\gtrsim 50$ yr) are not known for any D-type symbiotic including RX Pup (Whitelock 1987). The system of rings is embedded in the hot wind. A cool wind probably engulfs the Mira and the hot star.

Allen and Wright (1988) have offered a colliding wind model for RX Pup. They argue that the viscosity parameter $\alpha \sim 0.01$ required by the ring model is too low. While the viscosity parameter is low compared to what is suspected for dwarf novae (Pringle 1981) it is still not unreasonably low. As in the Kafatos et al. (1985) work, Allen and Wright (1988) assume the presence of the hot star wind. Their mass loss rates are in the range 10^{-5} - 10^{-4} $M_\odot$/yr, in disagreement with the value that one obtains from the presence of the absorption trough in the $\lambda 1548$ profile (Michalitsianos et al. 1987). Moreover, the orbital period of the two stars in the Allen and Wright (1988) model is thousands of year, unreasonably high for the symbiotic phenomenon. Finally, Allen and Wright conclude that the hot star would be radiating $L \sim 5 \times 10^4 \; L_\odot$, more than the $10^4 \; L_\odot$ luminosity of the Mira (Kafatos et al. 1982). Without the disk/ring system, the hot component in RX Pup should be fully exposed and produce an intense continuum both at optical and UV wavelengths. Such a hot star continuum has never been observed.

Both models have some common properties (the hot star wind) and it is obvious that RX Pup continues to provide many challenges as to its nature. One should perhaps accept the possibility that no simple model can explain all the observed features of RX Pup. We have to emphasize that the models of Seaquist and Taylor (1987) and Allen and Wright (1988) depend on the assumption of the presence of spherically symmetric optically thick winds. If radio observations of RX Pup reveal instead extended structure which implies jet activity, those models will become untenable. Due to the large distance of RX Pup any jet structure in this star similar to R Aqr will have sub-arcsec features. VLA observations and future Hubble Space Telescope observations provide the only available means to discern the structure of this most interesting star. It is obvious that long-term observations will help establish the binary properties and outburst mechanism of RX Puppis.

3. REFERENCES

Allen, D.A. and Wright, A.E. 1988, Proc. IAU Coll. 103, "The Symbiotic Phenomenon", ed. M. Friedjung, R. Viotti (D.Reidel Publ. Co.).

Barton, J.R., Phillips, B.A. and Allen, D.A. 1979, M.N.R.A.S.,187, 813.

Hollis, J.M., Oliversen, R.J., Kafatos, M. and Michalitsianos, A.G. 1986, Ap.J., 301, 877.

Kafatos, M., Michalitsianos, A.G. and Fahey, R.P.1985, Ap.J.Sup.59, 785.

Kafatos, M., Michalitsianos, A.G. and Feibelman, W.A. 1982, Ap.J., 257, 204.

Kafatos, M., Michalitsianos, A.G. and Hollis, J.M. 1986, Ap.J.Sup., 62, 853.

Klutz, M., Simonetto, O. and Swings, J.P. 1978, A&A, 66, 283.

Michalitsianos, A.G. et al. 1987, Ap.J., submitted.

Mueller, B.E.A. and Nussbaumer, H. 1985, A&A, 145, 144.

Pringle, J.E. 1981, Ann.Rev.Ast.Ap., 19, 137.

Seaquist, E.R. 1977, Ap.J., 211, 547.

Seaquist, E.R. and Taylor, A.R. 1987, Ap.J. (in press).

Swings, J.P. and Allen, D.A. 1972, Pub.A.S.P., 84, 523.

Swings, J.P. and Klutz, M. 1976, A&A, 46, 303.

Swings, J.P. and Struve, O. 1941, Ap.J., 94, 291.

Whitelock, P.A. 1987, Pub.A.S.P., 99, 573.

Whitelock, P.A., Catchpole, R.M., Feast, M.W., Roberts, G. and Carter, B.S. 1983, M.N.R.A.S., 203, 363.

Whitelock, P.A., Menzies, J.W., Lloyd Evans, T. and Kilkenny, D. 1984, M.N.R.A.S., 208, 161.

A MODEL FOR RX PUPPIS

David Allen *Anglo-Australian Observatory*
PO Box 296, Epping, NSW 2121, Australia
and
Alan Wright *C.S.I.R.O. Division of Radiophysics*
PO Box 76, Epping, NSW 2121, Australia

The dust-rich (D-type) symbiotic stars appear always to comprise a mira variable and a star of temperature $\sim 10^5$ K which ionizes the mira's circumstellar envelope. On timescales of decades they change little, save for the slow pulsation of the mira at infrared wavelengths. RX Puppis is a striking exception to this generalization, however, since it undergoes extraordinary changes.

RX Pup drifts between two extremes, exhibiting intermediate states. Table I summarises the known characteristics in the extreme states. Because of its variability, models of RX Pup are more highly constrained, giving us a chance to understand one of the D-type symbiotics.

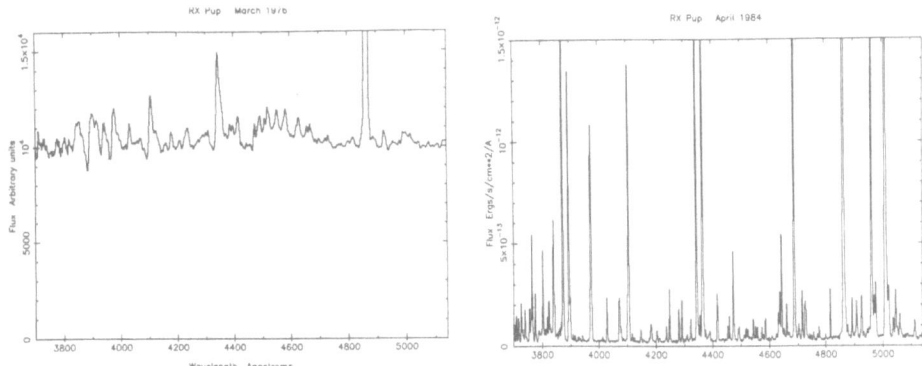

Optical spectra of RX Pup from the Anglo-Australian Telescope. The breadth of the lines in 1976 may be due to lower spectral resolution.

We have developed a model that accounts for *all* the major observed properties of RX Pup in both of its states and during transitions. The key to our model is the wind from the hot star, which is strong enough to prevent it accreting from the mira's outflow.

In the low-excitation state the mass-loss rate from the hot star is so large that its wind becomes optically thick, generating a continuum at about 20,000 K. We estimate that $\dot{M} \sim 10^{-4}$ $M_\odot$ yr^{-1}. In the high-excitation state $\dot{M}$ falls by about one order of magnitude, the photosphere recedes, and the temperature exceeds 100,000 K. The

249

J. Mikolajewska et al. (eds.), The Symbiotic Phenomenon, 249–250.

high mass-loss rate strongly suggests that the hot star is close to its Eddington luminosity, and is thus most credibly a white dwarf undergoing a shell flash.

TABLE I   The extreme states of RX Puppis

	Low excitation	High excitation
Dates	1960-75	1905?, 1940, 1981-present
V	8.5	13
Optical emission	H, Fe II, [Fe II]	Range of emission to [Ne V], [Fe VII]
Radio spectrum	Flat, ~ 20 mJy	$S_\nu \propto \nu^{0.8}$, rising to > 500 mJy
Mira reddening	Very low	$A_V > 5$ mag
Dust emission	Weaker	Stronger
Evidence for wind	P Cyg, 1100 km s^{-1}	> 800 km s^{-1} from C IV doublet
Ultraviolet	?	Complex emission; strong C IV

Different radio spectra are seen in the two states. In its high-excitation phase RX Pup has a spectrum which is optically thick ($S_\nu \propto \nu^{0.8}$) to a frequency between 100 and 400 GHz. The separation of the stars can be inferred to be $\sim 100 A.U.$, so the orbital period is many centuries. The dwarf must have accreted gas very slowly from the mira's outflow. Slow accretion is, in fact, necessary to produce a shell flash of duration approaching one century. The radio spectrum in the low-excitation state is flat and weak, and attributed to remnant free-free radiation from low-density regions of the mira's wind where recombinaton has yet to take place.

A standing shock forms where the wind pressures from the two stars balance. At times of high mass-loss from the white dwarf this is so close to the mira's surface that over much of the hemisphere facing the dwarf no dust can form in the mira's wind. When Ṁ falls, the shock recedes and the dust shell grows to surround the mira. We currently view the system from the hemisphere that contains the white dwarf, which explains the higher reddening observed to the Mira in its high-excitation state.

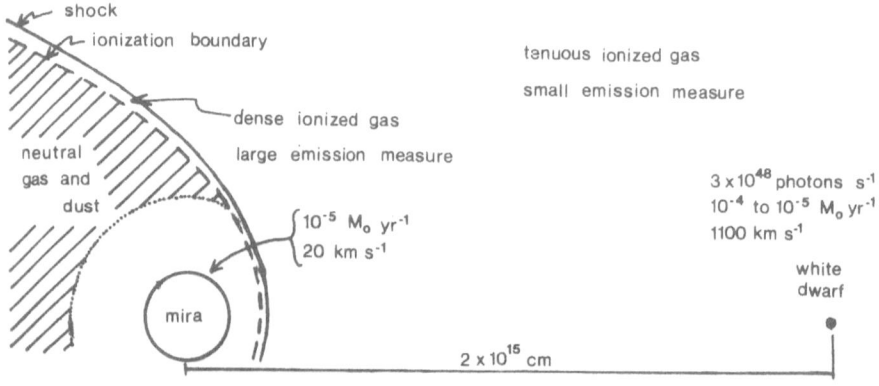

Cross section of RX Pup.

SYMBIOTIC STAR AG PEGASI - RETROSPECT AND PROSPECTS

D. Chochol
Astronomical Institute, Slovak Academy of Sciences
059 60 Tatranská Lomnica, Czechoslovakia

Symbiotic star AG Peg (HD 207757), which is still declining from the outburst in the last century, appears to
be related to the very slow RT Serpentis type novae: binary stars consisting of a cool M giant and a hot component,
which underwent outburst.

1. PHOTOMETRY

AG Peg was ninth magnitude star before the year 1850. Then
it slowly rised in brightness and reached fifth magnitude
in 1885. After the sudden drop of brightness in 1892 due
to the dust formation in expanding envelope, the star returned to the sixth magnitude and then it was continuously
decreasing in brightness (Lundmark, 1921; Rigollet, 1947;
Belyakina, 1968). Nowadays the decrease of brightness is
0.021 mag/yr (Fernie, 1985).
Belyakina (1968) found 0.3 mag variations in brightness. The variations follow the 800 days spectroscopic
(orbital) period discovered by Merrill (1929). Belyakina
(1970) showed that the variations of brightness could be
caused by the reflection effect: ultraviolet radiation of
the hot component heats up the facing hemisphere of the
cool component. Hutchings et al. (1975) suggested that the
hot component rotates rapidly and ejects material which
streams toward the cool component. A bright region on the
M star responsible for 0.3 mag light variation is heated
by collisions from the impinging stream. Many authors tried
to improve the period: P = 827 days (Meinunger, 1981),
P = 820 days (Belyakina, 1985), P = 816.5 days (Fernie,
1985), P = 813 days (Paul and Luthardt, 1987).
Infrared photometry shows that AG Peg contains a normal M giant without any evidences of variability. AG Peg
lies near the locus of non-variable M giants in (J-K),
(K-L) diagram (Feast et al. 1983).
Ultraviolet broad band photometry on the Orbiting As-

J. Mikolajewska et al. (eds.), The Symbiotic Phenomenon, 251–255.

tronomical Observatory 2 showed that the hot component is luminous UV source (Gallagher et al. 1979).

2. SPECTROSCOPY

Since Mrs. Fleming´s discovery of emission hydrogen lines in the spectrum of AG Peg taken at Harvard College Observatory in 1893 the spectroscopic behaviour of AG Peg has been progressively more complex. On the first spectra AG Peg resembled a peculiar Be star with P Cygni profiles of hydrogen and faint emission lines of singly ionized metals. In 1915 He I appeared in absorption (Merrill, 1916), in 1920 He I went from absorption to emission. The TiO bands belonging to the cool component appeared in 1920 and became more conspicuous as a blue continuum faded. From the begining of 1922 the spectrum has been gradually developing features characteristic of symbiotic stars evolving to Wolf-Rayet type. In 1931 He II emission appeared and the level of ionization represented by emission lines continuously increased (Merril 1929, 1932, 1942, 1951a, 1959). The narrow absorption lines of the M giant have been gradually emerging and from 1943 many of them could be accurately measured to obtain radial velocity curve of the M giant. By about 1950 P Cygni profiles vanished from hydrogen lines. In 1960s N V lines were detected.

Merrill in many papers pointed out that emission lines exhibited variations both in radial velocities and intensities with 800 days period, but relative phasing of different lines did not coincide. In 1920s He I lines were ≈ 160 days out of phase with H I lines, in 1950s the lines nearly coincided. The mean velocity derived from hydrogen lines decreased from + 16 km/s in 1915 to - 27 km/s in 1952. Merrill (1951 b) found that He I 3888 A exhibits multiple absorption components with velocities from - 56 km/s to - 428 km/s. Kolotilov (1975) found absorption component in the line He I 10830 A at velocity - 460 km/s.

Periodic radial velocity variations of narrow absorption lines of M giant were used for determination of spectroscopic orbit. The spectroscopic elements determined by different authors are in Table I. Assuming reasonable estimate of orbital inclination ($40° < i < 60°$) and cool star mass 3-4 $M_\odot$ the f(m) derived by Hutchings et al. (1975) yields the mass ≈ 1 $M_\odot$ for the hot component.

AG Peg is the brightest symbiotic star in UV region with well developed UV continuum and bright emission lines. Broad lines with P Cygni profile resemble a Wolf-Rayet wind. They arise close to the hot component. Narrow semi-forbidden emission lines are formed near the cool M giant. UV spectra are discussed in the papers: Keyes and Plavec (1980 a,b), Slovak (1982), Penston and Allen (1985) and in the work Chochol et al. (1987a) in this book.

TABLE I

	Cowley and Stencel (1973)	Hutchings et al.(1975)	Slovak (1987)	Garcia and Kenyon (1987)
P (days)	830	820	818.7	796
K_1(km/s)	5.6	5.5	4.85	6.4
γ(km/s)	-16.8	-16.3	-16.23	-14
e	0.25	0.23	0.235	-
ω	251°	222°	154°	-
T_o (JD)	2439045	2440928	2444514.6	-
f (m)	0.014 $M_\odot$	0.013 $M_\odot$	0.009 $M_\odot$	0.022 $M_\odot$

Spectrophotometric investigation of AG Peg made by Boyarchuk (1966) showed that the hot component is a WN6 type star, the cool one M3 III star. Gaseous clouds, which surrounds the system have T_e = 17 000 K, n_e = 7×10^6 cm^{-3}, R $\approx$ 10^5 $R_\odot$ and mass M $\approx$ 0.001 $M_\odot$. Keyes and Plavec (1980 a, b) classified the cool component as M1.7 III star with an effective temperature 3570 K. A distance of 500 pc and a radius of cool component 56 $R_\odot$ were derived supposing the luminosity class III. UV continuum could be fitted by a model of stellar atmosphere with T_{eff} = 30 000 K and log g = 4.5 (reddening E_{B-V} = 0.12) or 100 000 K black body with superimposed Balmer continuum. Kenyon and Webbink (1984) showed that UV colour-colour diagnostic indicated a hot stellar source with T_{eff} = 57 500 K (reddening E_{B-V} = 0.15). Kenyon and Fernandez-Castro (1987) classified the cool component from TiO and VO indices as M3.0 $\pm$ 0.4 III star.

3. RADIO OBSERVATIONS

AG Peg was one of the first symbiotic stars detected in the radio. VLA observations show three sources of radio emission: a nebular sphere with a diameter of 1".5, that has had an average expansion velocity of 15 km/s since about 1850 when ejection began, a point source, and a condensation of material inside the large nebula (Hjellming, 1985).

4. BINARY MODELS

There is no doubt that AG Peg is interacting binary. According to generally accepted models the cool star loses the matter by Roche lobe overflow or by stellar wind and this material is accreted onto main sequence star or white dwarf. The cool star in AG Peg is substantially smaller than corresponding Roche lobe radius (285 $R_\odot$ according to Hutchings et al. 1975), so that cool M giant can lose the matter only by stellar wind. The wind could be accreted on a cool white dwarf. Outburst observed in the last century could be ex-

plained by a degenerate flash, which led to an expansion
to A-F supergiant (Kenyon and Truran, 1983). Described
spectroscopic and photometric development after the outburst is in agreement with the model.
     Hutchings et al. (1975) surprisingly found that the
matter in AG Peg is transferred from the hot rapidly rotating component towards the cool one. They left the puzzle
of the origin of the outburst a century ago. Penston and
Allen (1985) tried to explain this puzzle in the frame of
thermonuclear event model. The hot component loses the mass
by the high velocity wind. The ionization of the hot component extends close to the surface of M giant and there
gives rise to the lower-velocity emission. Ablation tail
of this emission gives the illusion of a stream of gas flo-
wing from the hot component to the cool one.
     Chochol et al. (1987ab) showed that UV observations
could be explained in agreement with Hutchings et al.(1975)
model. The model seems to be supported also by the compu-
tation of spectroscopic orbit for the cool component for
periods 813 - 827 days using the RV data published by Cowley and Stencel (1973) and Hutchings et al. (1975). The
photometric minimum does not occur in spectroscopic conjuction but precedes the conjuction from 0.03 (period 813
days) to 0.09 (period 827 days) of the period (from 25 to
75 days) indicating the presence of a bright spot on M
giant due to the collision of the stream from the hot component with M giant.
     AG Peg is still declining from the outburst in last
century. The conservation of angular momentum when the star
shrinks could lead to the rotational shedding of mass. An
envelope of hot star could be transferred to the cool one.
As it was shown by Kříž (1982) the star rotating at the
critical velocity forms the inner excretion disk which en-
circling the star. Excretion disk takes away the mass, the
angular momentum and the rotational energy from the central
star. The disk could extend to the corresponding Roche-lobe,
where the transported matter overflows to the companion.
In AG Peg the inner excretion disk is supported not only
centrifugally but also by the wind from the cool component.
Some kinds of disk instabilities could cause the outburst
in last century. From evolutionary point of view the hot
component of AG Peg could be a helium subdwarf in "post
case B" of mass transfer.

REFERENCES-

Belyakina, T.S.: 1968, Astron. Zh. 45, 139.
Belyakina, T.S.: 1970, Astrofizika 6, 49.
Belyakina, T.S.: 1985, Inform. Bull. Var. Stars No. 2697.
Boyarchuk, A.A.: 1966, Astron. Zh. 43, 976.
Chochol, D., Komárek, Z., Vittone, A.: 1987a, in The Sym-

biotic Phenomenon, eds. J. Mikolajewska, M. Friedjung, S.J. Kenyon and R. Viotti, D. Reidel, Dordrecht, in press.

Chochol, D., Komárek, Z., Vittone, A.: 1987b, in Proc. of the 10. European Regional Astronomy Meeting, Prague, August 1987, in press.

Cowley, A., Stencel, R.: 1973, Astrophys. J. 184, 687.

Feast, M.W., Catchpole, R.M., Whitelock, P.A., Carter, B.S., Roberts, G.: 1983, Mon. Not. Roy. Astron. Soc. 203, 373.

Fernie, J.D.: 1985, Publ. Astron. Soc. Pacific 97, 653.

Gallagher, J.S., Holm, A.V., Anderson, C.M., Webbink, R.F.: 1979, Astrophys. J. 229, 994.

Garcia, M.R., Kenyon, S.J.: 1987, in The Symbiotic Phenomenon, eds. J. Mikolajewska, M. Friedjung, S.J. Kenyon and R. Viotti, D. Reidel, Dordrecht, in press.

Hjellming, R.M.: 1985, in Radio Stars, eds. R.M. Hjellming and D.M. Gibson, D. Reidel, Dordrecht, p. 97.

Hutchings, J.B., Cowley, A.P., Redman, R.O.: 1975, Astrophys. J. 201, 404.

Kenyon, S.J., Fernandez-Castro, T.: 1987, Astron. J. 39, 938.

Kenyon, S.J., Truran, J.W.: 1983, Astrophys. J. 273, 280.

Kenyon, S.J., Webbink, R.F.: 1984, Astrophys. J. 279, 252.

Keyes, C.D., Plavec, M.J.: 1980a, in The Universe at Ultraviolet Wavelengths, ed. R.D. Chapman, NASA CP-2171, 443.

Keyes, C.D., Plavec, M.J.: 1980b, in IAU Symposium No. 88: Close Binary Stars: Observations and Interpretations, eds. M.J. Plavec, D.M. Popper and R.K. Ulrich, D. Reidel, Dordrecht, p. 365.

Kolotilov, E.A.: 1975, Astr. Tsirk. No. 865.

Kříž, S.: 1982, Bull. Astron. Inst. Czechosl. 33, 302.

Lundmark, K.: 1921, Astron. Nachr. 213, 93.

Meinunger, L.: 1981, Inform. Bull. Var. Stars, No. 2016.

Merrill, P.W.: 1916, Publ. Michigan Obs. 2, 71.

Merrill, P.W.: 1929, Astrophys. J. 69, 330.

Merrill, P.W.: 1932, Astrophys. J. 75, 413.

Merrill, P.W.: 1942, Astrophys. J. 95, 386.

Merrill, P.W.: 1951a, Astrophys. J. 113, 605.

Merrill, P.W.: 1951b, Astrophys. J. 114, 338.

Merrill, P.W.: 1959, Astrophys. J. 129, 44.

Penston, M.V., Allen, D.A.: 1985, Mon. Not. Roy. Astron. Soc. 212, 939.

Paul, H., Luthardt, R.: 1987, in The Symbiotic Phenomenon, eds. J. Mikolajewska, M. Friedjung, S.J. Kenyon and R. Viotti, D. Reidel, Dordrecht, in press.

Rigollet, R.: 1947, l'Astronomie, 61, 247.

Slovak, M.: 1987, in The Symbiotic Phenomenon, eds. J. Mikolajewska, M. Friedjung, S.J. Kenyon and R. Viotti, D. Reidel, Dordrecht, in press.

THE OBSERVATIONS OF AG PEG DURING 1985-87

A.A.Boyarchuk, T.S.Belyakina, A.E.Tarasov
Crimean Astrophysical Observatory
Nauchny, Crimea, 334413 U.S.S.R.
N.Tomov
National Astronomical Observatory
Rozhen, Bulgaria.

ABSTRACT. The light curves in UBVRI and profiles of the emission lines of H_α, HeII 5411 and HeI 5876, 6678 obtained during 1985-87. The radial velocities of the emission lines HeI 6678 and HeII 5411 demonstrated clear periodic variations in antiphase with the radial velocity defined for the absorption line.

J. Mikolajewska et al. (eds.), The Symbiotic Phenomenon, 257.
© *1988 by Kluwer Academic Publishers.*

THE PHOTOMETRIC PERIOD OF AG PEGASI

R. Luthardt
Central Institute for Astrophysics
of Academy of Sciences of GDR
Sonneberg Observatory
DDR-6400 Sonneberg

ABSTRACT. The photometric period of AG Pegasi has been determined to be 813 days on the basis of 165 photoelectrical observations from 1961 to 1986. When splitting the interval different periods were found.

For the symbiotic star AG Pegasi different periods ranging from 733 to 830 days have been published. Most of them are near 820 days, but two periods deviate very strongly from this value: Slovak found 733+30 days using visual estimations of the AAVSO between 1974 and 1982. A period near this value was also determined by the present author with 760 days using visual estimations of the AFOEV, the AAVSO and photoelectrical observations. A period change was suggested but there wasn't an explanation for this. In 1985 Fernie published a period of 816.5 days based on 89 photoelectrical data.

Because of these strong differences a new determination has been done on the basis of 165 photoelectrical observations in V from 1961 to 1986 using the method of Lafler and Kinman.

The lightcurve shows a decline both in the mean magnitude and in amplitude. This slope has to be corrected before period determination. Fernie corrected for the slope using the maxima of brightnes only. The following correction formula considers both the maxima and minima of light:

$$V_{cor} = -0.424 \cdot 10^{-4} \, (JD-2440000) + V$$

The V magnitudes corrected by this equation are the basis for period determination. For the whole interval we got a period of 813 days. But splitting the lightcurve into two parts and doing the same procedure separatly for each one different periods have been calculated:

259

```
              time interval            period

JD = 243 7000 - 244 7000       813 days
     243 7000 - 244 1000.       853
     244 1000 - 244 7000        803
```

    The period of the first subinterval is nearly 50 days longer than that of the second. If this effect is real, it might be explained by shifting the hot spot on the surface of the M giant. This causes a variable rotational period of this spot. But also the scattering of the measurements and the large gap in the course of the lightcurve may cause this effect.

    As a conclusion the short periods of 733 and 760 days have not been confirmed. The mean period of AG Pegasi has been determined to 813 days. This is in agreement with the determination of Fernie, but it seems difficult to get an exact period on the basis of photometric data because of the decline in mean brightnes and amplitude.

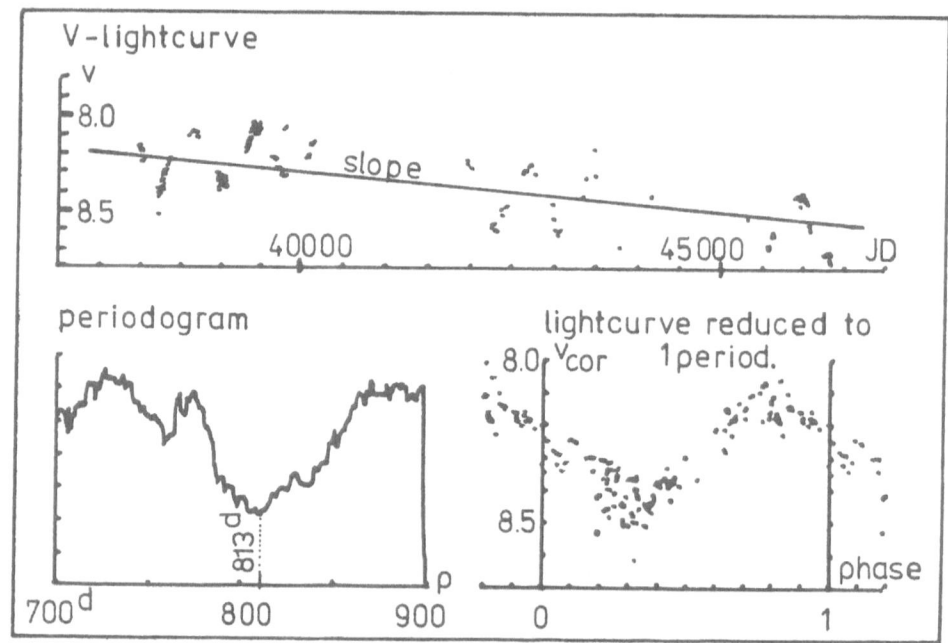

REFERENCES:
Chochol,D.C. 1987; Private communication
Fernie, J.D. 1984, Pub.A.S.P. 97,653
Lafler,J. and Kinman,T.D. 1965;Ap.J.Suppl.11,216
Luthardt,R. 1984; Inf.Bull.Var.Stars No.2495
Slovak,M.H. 1982, J.A.A.V.S.0.11,67

THE CAUSES OF THE LIGHT VARIATIONS IN AG Peg

B. F. Yudin

Sternberg State Astronomical Institute, Moscow,
U.S.S.R.

It can be shown that the long-term periodic variations in the U band, distinctly observed in AG peg by
UBV photometry (Belyakina 1970) from the early 1960's
with amplitude increasing, over since, have in fact two
principal reasons for their existence. One of these is
transparency of the envelope to the Lc-emission of the
hot component in AG Peg arisen in the 1950's and increasing since then. The second one is connected with the changes of the transparency of the compact part of the gas envelope located close to the hot component when in moves
along its orbit (Yudin 1987). It is interesting to note
that for the orbital solution with non-zero eccentricity
(Hunchings et al. 1975) the transparency minimum and consequently the flux emission maximum for the gas envelope
in AG Peg are observed in the moment when the hot component
is close to its apoastron. The amplitude of the periodic
variations in V increases due to the existence in AG Peg
of the effect of noticeable heating of the cool component's
hemisphere facing its hot companion. Using the TiO bands in
the red the cool component in AG Peg is classified as
slightly earlier than M2 from illuminated side and slightly
later than M3 from the opposite one (Ipatov and Yudin
1986).
Nowadays the radiation flux from the gas envelope,
that is its optical depth, changes from maximum to minimum
by about a factor of 2.5. In 1984, at the maximum of its
brightness, only about 37% of the total amount of the Lc-
quanta were absorbed in the gas envelope. At the same time,
the hot star's temperature and bolometric luminosity were
equal to about $7.9 \cdot 10^4$ K and $5.7 \cdot 10^3$ Lo respectively
(L(h,bol) 4 L(c,bol); Ipatov and Yudin 1986). A comparison of the hot component's bolometric luminosity at present with that derived for the moment of its visual brightness maximum show that the decline from the maximum have
not been accompanied by the noticeable luminosity changes

261

J. Mikolajewska et al. (eds.), The Symbiotic Phenomenon, 261–262.
© *1988 by Kluwer Academic Publishers.*

yet (Boyarchuk 1967).

The unique light curve of the AG Peg during its outburst, which has been continuing since 1850's, and its very long-lived W-R phase accompanied by the very high rate of mass loss by the hot component ($M>10^{-6}$ Mo yr^{-1}), having shed already more than 10^{-4} Mo from its hydrogen envelope, could be understood if we suppose that the present outburst of the hot component in AG Peg was connected with the helium (but not hydrogen) shell flash.

The differences in the light curves of symbiotic novae (AG Peg, V1016 Cyg, HM Sge) and classical symbiotic stars (Z And, CI Cyg, AG Dra) are due to a different role of the accretion process in these systems. In symbiotic novae, accretion is not a noticeable energy sources of the hot subdwarf. It can play only role of a supplier of hydrogen fuel to its surface when the hot subdwarf already is cooling down. In classical symbiotic systems, at least during the outbursts of their hot components, accretion becomes a predominant source of energy of the hot subdwarf.

The transition from a symbiotic nova to a classical symbiotic star takes place at the moment when the cool component's size is approaching the size of its Roche lobe, resulting in a sharp increase of the accretion rate of its matter onto the hot component (Yudin 1987). The nonstationarity of this process leads to the emergence of the nova-like outbursts on the light curves of symbiotic stars which are now called classical.

References

Belyakina, T. S.: 1970, Astrofizika 6, 49.
Boyarchuk, A. A.: 1967, Astron. Zh. 44, 1222.
Hutchings, J. B., Cowley, A. P. and Redman, R. O.:
        1975, Astrophys. J. 201, 404.
Ipatov, A. H. and Yudin, B. F.: 1986, Pis'ma Astron. Zh.
        12, 936.
Yudin, B. F.: 1987, Astrophys. and Space Sci., in press.
Yudin, B. F.: 1987, Astron. Tsirk., in press.

ULTRAVIOLET VARIABILITY OF THE SYMBIOTIC STAR AG PEG

D. Chochol[1], Z. Komárek[2], A. Vittone[3]
1 Astronomical Institute, Slovak Academy of Sciences,
  059 60 Tatranská Lomnica, Czechoslovakia
2 Public Observatory, 071 01 Michalovce, Czechoslovakia
3 Osservatorio Astronomico Capodimonte, Via Moiariello
  16, 80131 Napoli, Italy

Symbiotic star AG Peg consists of a hot subdwarf with a WN6
spectrum and a cool M3 giant, which is not filling its Roche lobe
(Boyarchuk 1967, 1985). A detailed study of profiles, equivalent
widths and radial velocities of emission lines in optical spectra
allowed Hutchings et al. (1975) to conclude that a hot subluminous
star approximately 1 $M_\odot$ rotates rapidly and ejects material which
streams towards the cool M giant with the mass 3-4 $M_\odot$. UV observations seems to support this model.

UV observations provided from the databank of the IUE satellite
were obtained in 1978-81 by different observers. The observational
material consists of 12 high dispersion SWP spectra and covers the
region 1200 - 2100 A. The spectra were reduced at Trieste observatory using standard IUESIPS package. The radial velocities of emission
lines were measured on tracings and corrected for the motion of Earth
and satellite. The orbital phases in our paper were computed according to Meinunger's (1981) ephemeris:

$$\text{Min.} = JD\ 2\ 428\ 250 + 827^d \times E\ ,\qquad\qquad (1)$$

derived from photometric minima of brightness.

UV spectrum of AG Peg shows a number of strong emission lines.
From the shapes of the line profiles it is possible to distinguish
three types of emission lines:
1. Very broad emission lines with width (FWHM) 600 - 1000 km/s:
   He II 1640 Å, N IV 1718 Å, N V 1238,1242 Å.
2. Narrow emission lines with width (FWHM) 40 - 70 km/s:
   O III] 1661,1666 Å, Si III] 1892 Å, C III] 1909 Å.
3. Combined broad emission lines with narrow component:
   N IV] 1486 Å, C IV 1548,1550 Å.

The broad emission lines are formed in higly ionized region near
the hot rapidly rotating object. The broad emission lines N V 1242 A,
N IV 1718 Å and C IV 1550 Å show broad absorption in the blue wings
of these lines, which indicate mass-loss wind (Penston and Allen,
1985). The proof that the wind is not isotropic comes from the observation of C IV 1548 Å line, which is in some phases depressed due
to blending with broad absorption component of broad emission line
C IV 1550 Å (Fig. 1). Fig. 2 shows that the wind is variable and

J. Mikolajewska et al. (eds.), The Symbiotic Phenomenon, 263-264.

seems to be more enhanced in the direction of the cool component. The wind can easily drive material ejected by the hot component towards the cool one.

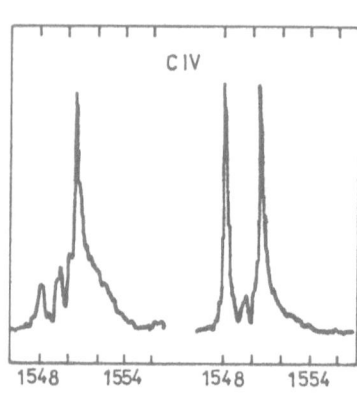

Figure 1. C IV lines in orbital phases 0.729 and 0.507.

Figure 2. Phase dependence of line flux of C IV 1548 Å (dots) and 1550 Å (crosses) lines.

Radial velocities of nebular emission lines O III], Si III], C III], N IV] fit radial velocity curve of the cool component indicating that these lines are formed near this component (Fig. 3).

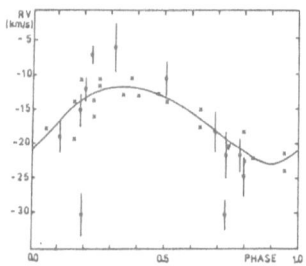

Figure 3. Radial velocities of nebular emission lines with r.m.s. errors. The data for RV curve of the cool component (crosses) are from the works of Cowley and Stencel (1973) and Hutchings et al. (1975).

REFERENCES

Boyarchuk, A.A.: 1967, Soviet Astr. 10, 783.
Boyarchuk, A.A.: 1985, In: ESA Workshop: Recent Results on Cataclysmic Variables, ESA SP-236, p. 97.
Cowley, A., Stencel, R.: 1973, Astrophys. J. 184, 687.
Hutchings, J.B., Cowley, A.P., Redman, R.O.: 1975, Astrophys. J. 201, 404.
Meinunger, L.: 1981, Inform. Bull. Var. Stars, No. 2016.
Penston, M.V., Allen, D.A.: 1985, Mon.Not.R.astr.Soc. 212, 939.

A New Absorption-Line Orbit for the Symbiotic Nova AG Pegasi

M. H. Slovak
University of Wisconsin, Madison, Wisconsin, USA

D. L. Lambert
University of Texas, Austin, Texas, USA

ABSTRACT.   Precise cross-correlation absorption-line velocities have been derived using high resolution near infrared coude' spectra for the symbiotic nova AG Pegasi. A revised weighted orbital solution is presented based on both extant photographic and the new absorption-line velocities.

1. Orbital Studies of AG Pegasi

The pioneering work of Merrill (1951) clearly illustrated the complex velocity variations of the rich emission-line spectrum of AG Pegasi = HD 207757 which followed an approximate period of 800 days. Cowley and Stencel (1973) presented the first orbital solution for AG Peg based on twelve velocities measured from coude' spectra near Hα and nine determinations from Merrill (1951, 1959). Adopting a mass ratio $q = 3$ and assuming an inclination i = 36 degrees, they concluded that the M giant secondary did not fill its Roche lobe. An additional analysis was performed by Hutchings, Cowley and Redman (1975) with eleven new absorption-line velocities. They adopted an orbital period P = 820 days and argued that the hot primary was ejecting a mass stream towards the cool secondary. A large phase shift of $\Delta\phi$ = 0.20P was found for the Balmer lines between their data and the earlier Merrill velocities, and similar phase shifts appeared for other emission-line velocities as well. In order to clearly address these interesting features, we have obtained a series of high resolution digital coude' spectra over a complete orbital cycle and have derived precise ($\delta v = \pm$ 1.2 km/s) absorption-line velocities from infrared spectra near 8400 A.

2. New High Resolution Data

2.1. Digital Reticon Spectroscopy

Over 60 high resolution (0.2 A) coude' spectra of AG Pegasi were obtained using the 2.7m McDonald Observatory Reticon spectrograph (Vogt, Tull and Kelton 1978) from 1977 to late 1980, covering one complete orbital cycle (820 days). Four spectral regions were chosen, centered on 4684 A, 5007 A, 6563 A, and 8410 A, which include both the prominent emission features of He II, [O III], Hα as well as strong absorption lines from Ti I and Fe I (Fig. 1). Simultaneous observations of selected late-type giants with a-quality velocities were also obtained and used as templates to derive cross-correlation velocities.

2.2. Cross-correlation Radial Velocities

An automated cross-correlation method was developed to derive high precision absorption-line velocities (Slovak 1982).   Discrete cross-correlation functions (DCCF) are calculated for each program-standard star pair and fit with numerical cubic splines rather than an imposed functional form. The first derivative of the spline yields the

J. Mikolajewska et al. (eds.), The Symbiotic Phenomenon, 265–267.

266

zero crossing, providing the maximum of the DCCF to a fraction of a diode. Cross-correlation velocities determined in this way were independently verified against conventionally measured velocities using Fe-Ne comparison spectra. The mean of the residual velocities between the two methods is -0.98 ± 0.71 km/s and is dominated by the uncertainties in the adopted velocities for the a-quality standards (±0.5 km/s).

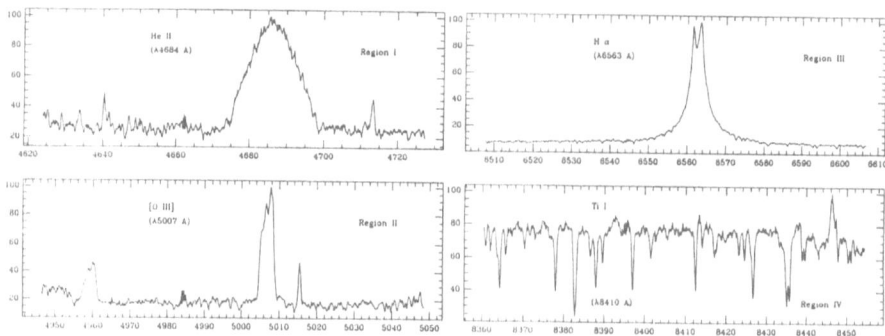

Figure 1. High resolution Reticon spectra of AG Peg. Strong emission lines dominate Regions I - III and the absorption lines are veiled. The majority of the absorption-line velocities were derived from Region IV data, clearly showing strong, uncontaminated absorption features of Fe I and Ti I.

3. Revised Orbital Solution

Using absorption-line velocities from Merrill (1959), Cowley and Stencel (1973), and Hutchings, Cowley and Redman (1975), in addition to the new data, orbital elements were calculated using a modified Wilsing-Russell code, which has been enhanced to accomodate large eccentricities and which performs differential correction of the preliminary elements. The best determined photometric period P = 816.5±0.9, based on an extensive analysis of photoelectric data (Fernie 1985), was adopted as the initial orbital period.

The observed velocities and the orbital solution are seen in Figure 2. The mean residual velocity of the data about the calculated curve is -0.21 km/s with a standard deviation of 2.24 km/s. The relatively large scatter reflects either intrinsic velocity variations or is due to the larger errors associated with the photographically determined velocities. Final orbital elements are presented in Table 1 and are compared to the previous solution from Cowley and Stencel (1973).

4. References

Cowley, A. , and Stencel, R. 1973, *Ap. J.*, **184**, 687.
Fernie, J. D. 1987, *Pub. A. S. P.*, **97**, 653.
Hutchings, J. B., Cowley, A. P., and Redman, R. O. 1975, *Ap. J.*, **201**, 404.
Merrill, P. W. 1951, *Ap. J.*, **113**, 605.
------------. 1959, *ibid,* **129**, 44.
Slovak, M. H. 1982, PhD Thesis (University of Texas: Austin).
Vogt, S. S., Tull, R. G., and Kelton, P. W. 1987, *Applied Optics*, **17**, 574.
Wilson, R. E. 1953, *General Catalogue of Stellar Radial Velocities* (Washington: Carnegie), Publication **601**.

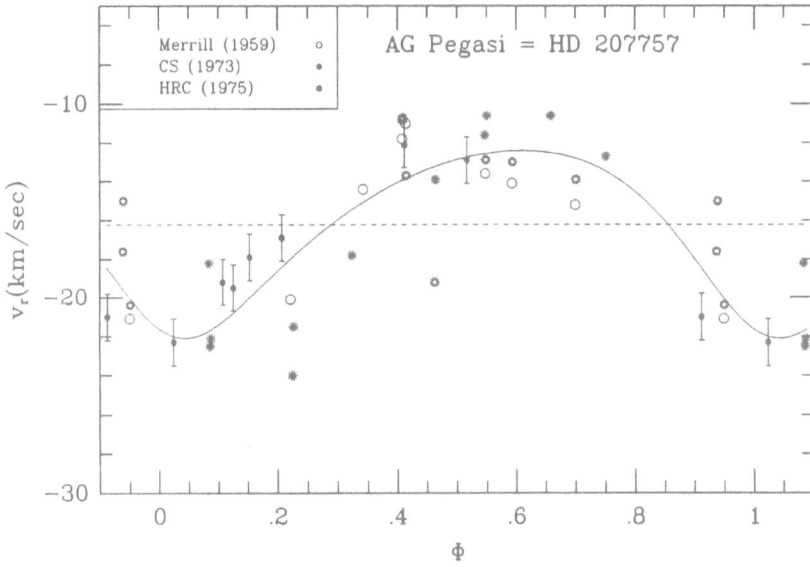

Figure 2.   Absorption-line radial velocities and calculated orbit (solid line) for AG Peg. The systemic velocity is also indicated (dashed line). Photographic velocities are shown as various symbols, identified in the legend. New cross-correlation velocities are seen as filled symbols with 1σ error bars.

Table I

Orbital Elements for the Symbiotic Nova AG Pegasi		
P (days)	818.72 ± 2.5[1]	830.14 ± 1.68[2]
K (km/s)	4.98 ± 0.37	5.6 ± 0.2
γ (km/s)	-16.15 ± 0.28	-16.8 ± 0.2
e	0.28 ± 0.08	0.25 ± 0.05
Ω (°)	149.9 ± 20.2	251.1 ± 3.2
T_0 (days)	2444507.8 ± 33.9	2439045.99 ± 24.98
a sin(i) (10^7 km)	5.42 ± 0.13	6.19
f (M) ($M_\odot$)	0.009	0.014

[1] Present solution.
[2] Cowley and Stencel (1973).

THE SYMBIOTIC NOVAE

Roberto Viotti
Istituto Astrofisica Spaziale, CNR, Frascati

ABSTRACT. A small number of symbiotic stars are characterized by a single nova-like outburst with a very slow time evolution and several years of permanence at maximum. We describe in detail RR Tel and discuss the main properties of these objects.

1. INTRODUCTION

Two kinds of nova-like outburst can be identified in the symbiotic phenomenon, the Z And-type with non-periodic recurrent maxima, and the RR Tel-type with only one single event recorded in their historical light curve. Allen (1980) listed seven stars having the character of very slow novae or symbiotic novae: AG Peg, RT Ser, RR Tel, V1016 Cyg, HBV 475 (V1329 Cyg), HM Sge and AS 239 (V2110 Oph). More recently a similar behaviour was observed in the symbiotic, nova-like variable PU Vul. Table 1 summarizes the main data on these eight symbiotic novae. With respect to classical novae, their light curve is characterized by a slow rise to maximum and an even slower decline (or no decline at all), and in some cases a small amplitude of the visual outburst. The light curves of symbiotic novae are shown in Fig.1. The oldest symbiotic nova is AG Peg which in the middle of the past century underwent a major nova-like outburst with a very slow increase of the visual luminosity, and a still longer decline to the present magnitude which is close to the pre-outburst luminosity (Fig.1). Presently AG Peg displays a typical symbiotic spectrum with a cool continuum showing strong TiO bands, characteristic of an M3 giant star, prominent emission lines of low and high ionization, and a hot continuum extending to the far-UV (e.g. Hutchings et al. 1975; Penston and Allen 1985). It should be noted that without the knowledge of the AG Peg light curve so many years back, its nova-like nature would have not been recognized on the basis of its present behaviour, and especially the many differences with other 'classical' symbiotic novae, such as RR Tel, V1016 Cyg and HM Sge. It is thus conceivable that several other symbiotic objects actually belong to the category of symbiotic novae, but, because of the long time scale involved, their main nova-like outburst has not been recorded.

J. Mikolajewska et al. (eds.), The Symbiotic Phenomenon, 269–277.

Table 1. The Symbiotic Novae (1)

star	To (a)	1"	b"	magnitude (b)	(c)	(d)	spectrum (e)	(f)	IR (g)	notes (h)
AG Peg	1855	69	-30	9	6	8.3	M3		S	2
RT Ser	1909	14	+10	>16	9.5	13	M	A8	S	2
V2110 Oph	1940:	05	+04	-	11:	22	>M3		D	2,4
RR Tel	1944	342	-32	14v	6	11	M	F5	D	h,2
V1016 Cyg	1964	75	+06	14	11	11	M	neb	D	h,5
V1329 Cyg	1966	78	-05	14v	11.5	13-14	M5	neb	S	6
HM Sge	1975	54	-03	>17	11	11	M	neb	D	h,5
PU Vul	1978	63	-09	15v	8.8	8.8	M4	A7	S	2

Notes to the table:
a. Beginning of the outburst. b. Pre-outburst magnitude. c. Maximum luminosity. d. Post-maximum (present) magnitude. e. Cool spectral component. f. Equivalent spectral type at maximum. g. Infrared type (S=stellar, D=dust). h. Periodic variations in the IR.
References: 1. Allen (1980). 2. Kenyon (1986). 3. Adams and Joy (1928). 4. Allen (1978). 5. Ciatti et al. (1978). 6. Grygar et al. (1979).

2. RR TELESCOPII

In this section we discuss in detail RR Tel which could be considered as the best representative of the category of symbiotic novae. Much of the following discussion also applies to other symbiotic novae. The star was discovered as variable by Mrs. Fleming (1908) many decades before its main outburst. Later Mrs. Mayall (1949) from the analysis of 600 Harvard observations from 1889 to 1947 found that during 1889 to 1930 RR Tel showed little evidence of periodic variability, the observed range being about 1.5 mag with maxima ranging from 12.5 to 14 mag. After 1930 the variations appeared clearly periodic with a mean period of about 387 days and an amplitude of about 3 mag. As discussed below, this periodicity with the same period is now observable in the IR (Feast et al. 1983). In late 1944 the periodicity stopped and the star rapidly brightened from mpg=14 to 10 in a few days, then rose to the seventh magnitude by mid 1945. The following years RR Tel remained at maximum luminosity until 1949, reaching the sixth magnitude during 1948, then gradually faded to the present V=10 in about 14 years (see e.g. Kenyon 1986). The outburst of RR Tel was discovered in 1949, three years after the outburst. It was a lucky chance that the star was still at maximum and that major spectral changes occured after the discovery. The early spectral evolution of RR Tel was followed at the Bosque Alegre and Radcliffe Observatories. We shall discuss the Bosque Alegre spectra on the basis on a reproduction of unpublished data kindly provided me by Jorge Sahade. The first spectrum was taken in April 1949 and showed a continuum with many absorption lines of singly ionized metals. Some absorptions are flanked at longer wavelengths by a weak emission compo-

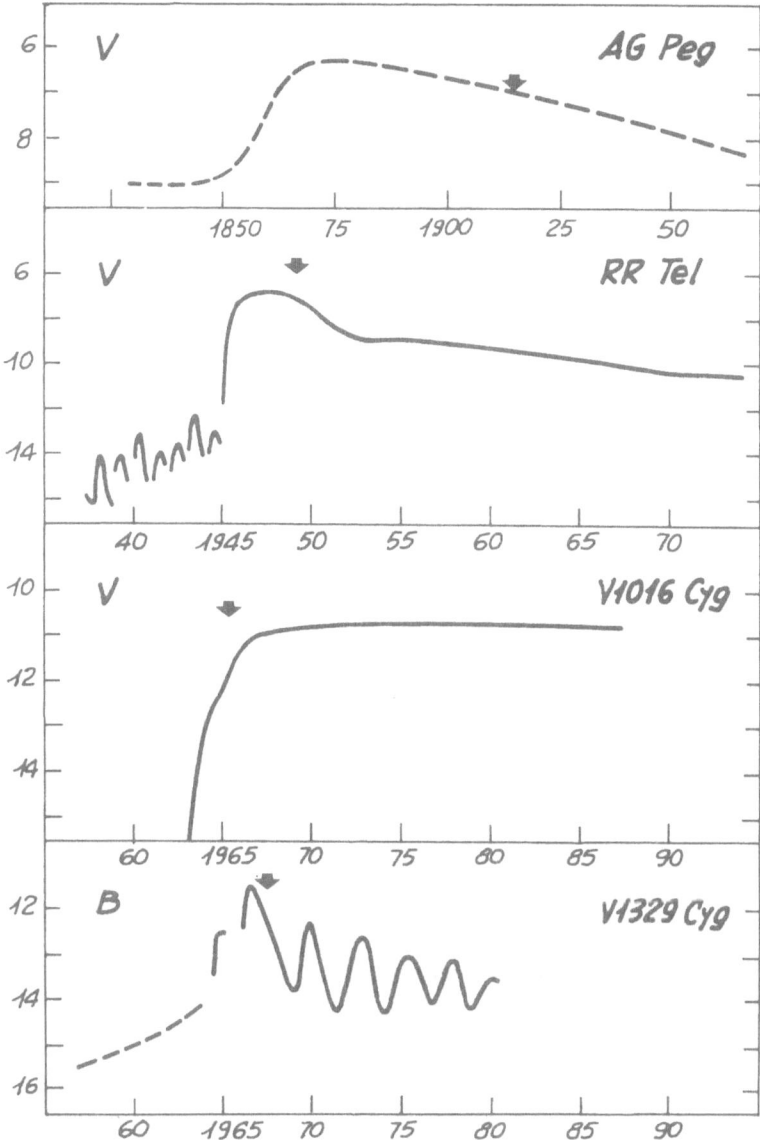

Fig.1 The light curves of four symbiotic novae. Note the different decline rates (in AG Peg the abscissa scale is contracted). The arrows indicate the time of the first spectroscopic observations.

nent indicating a P Cygni profile. This emission became more prominent in July 1949. By September 11 the continuum is weaker and the emission lines dominate the spectrum of RR Tel, although the mean line excitation is still low. The first spectra obtained at Radcliffe were discussed by Thackeray (1950) who found in June-August 1949 strong absorptions of CaII and hydrogen. Also TiII was present in absorption, while Hβ was absent, probably filled-in by emission. Mayall (1949) reports on low quality, low dispersion spectra taken when RR Tel was at maximum brightness. All these earlier observations agree in giving an F-super-giant spectral type (cF5, according to Thackeray 1950). In a later paper, Thackeray (1977) reports that the relative shift of the absorption lines was of -100 km/s. The remarkable spectral change occurred between August and September 1949 was first noted by Thackeray (1950) who found that in his spectra all the absorption lines disappeared and a rich emission line spectrum appeared with prominent hydrogen, CaII and especially FeII emission.

The spectral evolution of RR Tel in the following years is best described by Thackeray (1977). Since 1949 the star showed a gradual increase of the mean ionization of the emission line spectrum. HeII and NIII appeared around August 1950 (see also Pottasch and Varsawsky 1960). These authors also identified HeI absorption with a velocity of -685 km/s (in 1951) and -865 km/s (in 1952), indicating the presence of a high expansion velocity in RR Tel. Then between 1951 and 1952 [OIII] and [NeIII] flared while the permitted FeII lines faded (Thackeray 1977). The increase of the level of ionization continued through 1953 and 1954 and is best illustrated by the sequential appearence of higher and higher ionization stages of iron, from FeIII to FeVII (Fig.2). Presently RR Tel displays a very rich emission line spectrum both at optical and UV wave-lengths. A comprehensive discussion of its UV spectrum was given by Penston et al. (1983). RR Tel also has a strong IR excess which could be due to thermal dust emission. However, the near-IR appears variable with a period of 387 days - the same as that found during the pre-outburst phase - and with a large amplitude typical of that of Miras (Feast et al. 1983). The presence of a late type star is also supported by the marginal identification of TiO bands in the visible (Thackeray 1977). It should be noted that long period fluctuations of the visual luminosity of RR Tel was found by Heck and Manfroid (1985) and Kenyon and Bateson (1984). The period of 374 days is close to the above reported values and should be associated with the Mira pulsation, although the optical spectrum is dominated by the emission lines.

3. DISCUSSION

Symbiotic novae have the common caracteristic of having undergone one single major outburst, and to be symbiotic. Since among the few so far identified symbiotic novae there exists a variety of behaviour and of time scales, it would be important to understand whether they represent an extreme case of symbiotic stars or they are more related to the category of novae. Let us discuss the different aspects of the phenome-non.

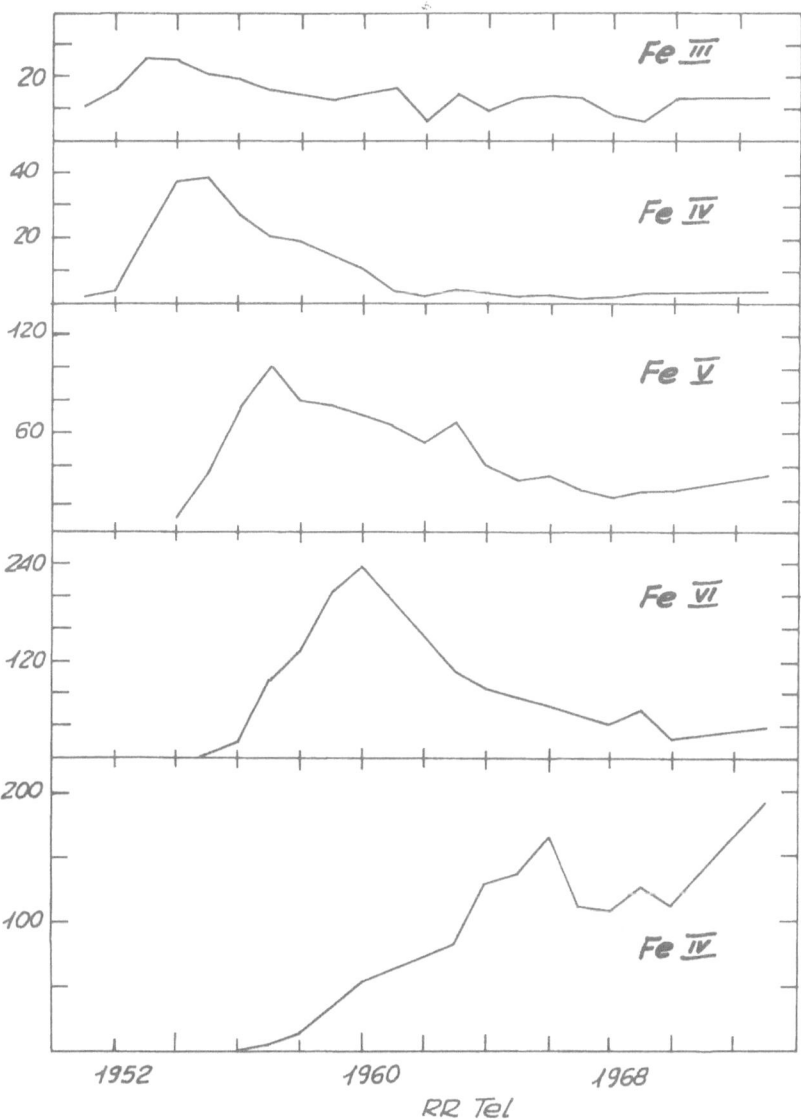

Figure 2. The sequential evolution of the iron emission
lines in RR Tel from 1951 to 1970 (from Thackeray 1977).
The gradual increase of the ionization of the emitting
envelope is indicated by the appearence of successive ionization stages.

The pre-outburst phase is rather well known only for RR Tel and V1329 Cyg. Both showed large amplitude, periodic variations which are attributed for RR Tel to Mira pulsation and for V1329 Cyg to eclipses of a binary system. For these stars and V1016 Cyg the spectrum at minimum was M. The M spectrum was also observed in PU Vul during its deep fading of 1980, and in AG Peg and RT Ser after decline from maximum. In V1016 Cyg and HM Sge which have not (yet) declined after the maximum, the M spectrum in the optical is hidden by the strong continuum and line emission from the circumstellar region. In these stars the presence of a late type spectral component is shown by the large amplitude, Mira type variations of the near-IR (Table 2). These variations are also observed in RR Tel and imply that the cool spectral component is the main contributor to the near-IR spectrum. The spectrum however does not peak in the near-IR as expected. Kenyon et al. (1986) have studied the IR energy distribution of three D-type symbiotics, including the symbiotic novae RR Tel and HM Sge, and concluded that their IR spectrum is consistent with those of highly reddened Mira variables. Thus they suggest that the cool, Mira-type component is embedded in a dense dust shell while the hot source lies outside the shell. In these systems, the evidence of a late type component is also indicated by the marginal presence of TiO bands in the optical, and by the steam absorption in the IR. The conclusion is that in these symbiotic novae (and in the other symbiotic stars as well) there is clear evidence for the presence of an M 'star'.

The observations and energy balance considerations also imply that there is another component which is responsible for the outburst. The presence of a binary system is without doubt in AG Peg and V1329 Cyg. In the other objects the radial velocity appears constant, but it is difficult to decide whether this is due to a very long orbital period, as suggested by Kenyon et al. (1986), or to the inclination of the orbit with respect to the line of sight. The strong UV radiation of the hot component should destroy the circumstellar dust, as it could have happened in AG Peg and V1329 Cyg which S-type symbiotics. The persistence of dust in the D-type symbiotics should be explained by a larger separation of the system. Note that Friedjung et al. (1984) have suggested that the deep minimum of PU Vul in 1980 could be explained by dust condensation and hidding of the warm, A-type component.

The rise to maximum was rather fast in RR Tel, V1329 Cyg and HM Sge (Table 2), but very slow in V1016 Cyg and especially in AG Peg and RT Ser. Thus this different behaviour is apparently not related to the other features of the symbiotic novae. The amplitude of the 'outburst' ranges from 3 to >6 magnitudes. In most cases this should be attributed to the presence of the M companion, which is bright (with respect to the peak luminosity) in AG Peg, and very faint in RT Ser, V2110 Oph and HM Sge. Yet this cannot explain the slowness of AG Peg and RT Ser.

The spectrum at maximum in another intriguing problem. RT Ser, RR Tel and PU Vul displayed an intermediate (A-F) spectral type, similar to that of classical novae (without however the large violet displacements of the absorption lines). No such spectrum was observed in the other symbiotic novae. Nevertheless this should be possibly caused by the lack of early spectral observations. For instance V1016 Cyg was first observed spectroscopically near the end of its long lasting rise phase. The first spectra of V1329 Cyg were taken one year after its light maximum (Fig.1). It should also be recalled that in RR Tel the A-type spectrum disappeared probably in a few days at the beginning of September 1949. Should this star have been spectroscopically observed since mid September instead of April-June 1949, we should have lost the A-type phase. Thus the absence of this phase in some symbiotic novae could be an observational effect. What is also important to remark is the presence of rapid spectral change also in these very slow novae.

In general in symbiotic novae emission lines are rather narrow in comparison with novae. But there are a number of interesting exceptions. Crampton et al. (1970) found in V1329 Cyg the [NeIII] and [OIII] lines with a multiple structure with peaks extending from -240 to +250 km/s. Crampton et al. also discovered on the continuum several broad and shallow emission features which can be identified with WN5-type spectral features with an expansion velocity of about 2300 km/s. In the UV spectrum of AG Peg the high ionization lines (NV, CIV, NIV, HeII) present a broad and complex structure resembling that of WR stars (Penston and Allen 1985) and implying radial velocities of several 100 km/s. Broad WR features were also observed in the otpical and UV spectrum of RR Tel (e.g. Thackeray 1977; Penston et al. 1983; Ponz et al. 1982). In this star, as discussed above, high velocity P Cygni profiles were observed during its decline phase. Thus in symbiotic novae there is evidence for the presence of hot high velocity regions, although most of the emitting region is generally characterized by a low velocity field. A 'stratification' is also suggested by the correlation between emission line width and ionization energy (e.g. Penston et al. 1983; Muratorio and Friedjung 1982).

Another puzzle of the symbiotic phenomenon is the light curve after maximum. Four objects displayed a gradual fading which took several years or decades. The e-folding decline time varied from 7-9 years for RT Ser and RR Tel to 40 years for AG Peg. In the case of V1329 Cyg the decline time was derived from a fit of the UV emission line flux variation, taking into account the 950 days periodicity (Nussabaumer et al. 1986). It should be mentioned that Allen (1981) and Willson et al. (1984) from the analysis of the X-ray flux in the three symbiotic novae, RR Tel, V1016 Cyg and HM Sge, suggested a very slow decline of the flux after the outburst. Willson et al. give an e-folding decay time of 5 to 50 years, which is in agreement with the visual decay time of RR Tel. Three recent symbiotic novae, V1016 Cyg, HM Sge and PU Vul, are still at maximum, which means that their decay time should be larger than 38 to 125 years (see Table 2).

Like in RR Tel discussed above the spectral evolution of symbiotic novae after outburst is in general similar to that of classical novae, except for the much longer time scale, with a gradual increase of the mean ionization of the emission line spectrum. This is best observed in RR Tel, V1016 Cyg, HM Sge and V1329 Cyg. In the first three stars the nebular spectrum appeared at maximum luminosity, and in V1016 Cyg and HM Sge the nova-like spectral evolution occurred at constant visual magnitude. (It should be noted that in these stars, and in most symbiotic objects as well, emission lines largely contribute to the broad band photometry, so that care should be taken in the interpretation of their

Table 2. Characteristic times scales

star	rise (a)	decay (b)	Mira (c)	orbit (d)	K (e)	ref.
AG Peg	16	40	-	816.5	5.1	1,2
RT Ser	14	7:	-	-	-	1
RR Tel	<0.3	9	374.2	-	-	1,3
V1016 Cyg	2-3	>125	472	-	n	4
V1329 Cyg	0.3	12-20	?	950	62	5
HM Sge	<0.4	>65	500-600	-	n	4
PU Vul	1	>38	-	-	n	1

Notes to the table: (a) Rise time (total) in years. (b) 1/e decay time (years). (c) Period (in days) of the Mira pulsation. (d) Orbital period (days). (e) Semi-amplitude (in km/s) of the radial velocity curve (n= no variation observed). References: (1) Kenyon 1986. (2) Hutchings et al. 1975. (3) Feast et al. 1983. (4) Ciatti et al. 1978; Taranova and Yudin 1983; Willson et al. 1984. (5) Nussbaumer et al. 1986; Grygar et al. 1979.

light curves). This is one main difference with classical novae. In the case of PU Vul the A-type maximum spectrum is still present eight years after the outburst, with the star still at maximum. Perhaps this could be related to the much lower expansion velocities observed in symbiotic novae.

We should finally recall the important results recently obtained in the radio which are described in the volume Radio Stars (Hjellming and Gibson 1985) and discussed by Viotti (1987). Of particular interest is the discovery of radio nebulae around some symbiotic novae with complex structures, such as jets, halos and bipolar nebulae which might be associated with the outburst and their binary nature. Also important is the time evolution of the radio emission of HM Sge (e.g. Kwok 1982) and of the 'radio nova' CH Cyg (Taylor et al. 1986). These phenomena are described elsewhere is this volume.

We may conclude that, in spite of the many important results so far obtained especially outside the optical region, the phenomenon of the symbiotic novae is still not well understood, and their relationship with novae open to discussion. Also the present list of symbiotic novae is far from being statistically significant. Fields which should require future work are the optical and UV emission lines observed at high resolution for the study of the circumstellar gas dynamics, high spatial resolution imagery in the UV, visual and IR, high resolution near-IR spectroscopy of the cool components, and, finally, a systematic search of nova-like events associated with symbiotic systems.

REFERENCES

Adams, W.S., Joy, A.H.: 1928, Pub.astr.Soc.Pacific 40, 252.
Allen, D.A.: 1978, Inf. Bull. Var. Stars No.1399.
Allen, D.A.: 1980, Mon. Not. R. astr. Soc. 190, 75.
Allen, D.A.: 1981, Mon. Not. R. astr. Soc. 197, 739.
Ciatti,F., Mammano,A., Vittone,A.: 1978, Astr. Astrophys. 68, 251.
Crampton, D., Grygar, J., Kohoutek, L., Viotti, R.: 1970, Ap. Lett. 6, 5.
Feast, M.V. et al.: 1983, Mon. Not. R. astr. Soc. 202, 951.
Fleming, W.P.: 1908, Circ. Harvard Coll. Obs. No.143.
Friedjung, M. et al. (1984), in The Future of Ultraviolet Astronomy
    Based on Six Years of IUE Research, NASA CP-2349, p.305.
Grygar, J., Hric, L., Chochol, D., Mammano, A.: 1979, Bull. astr. Inst.
    Czech. 30, 308.
Heck, A., Manfroid, J.: 1985, Astron. Astrophys. 142, 341.
Hjellming,R.M., Gibson,O.M.(eds.): 1985, Radio Stars, Reidel, Dordrecht.
Hutchings, J.B., Cowley, A.P., Redman, R.O.: 1975, Astrophys. J. 201, 404.
Kenyon, S.J.: 1986, The Symbiotic Stars, Cambridge University Press,
    Cambridge, Great Britain.
Kenyon, S.J., Bateson, F.M.: 1984, Publ. astr. Soc. Pacific 96, 321.
Kenyon, S.J., Fernandez-Castro T., Stencel, R.E.: 1986, Astron. J. 92,
    1118.
Kwok, S.: 1982, The Nature of Symbiotic Stars, M. Friedjung and R.
    Viotti eds., Reidel, Dordrecht, p.17.
Mayall, M.W.: 1949, Harvard Bull. No.919, 15.
Muratorio, G., Friedjung, M.: 1982, The Nature of Symbiotic Stars,
    Reidel, Dordrecht, p.161.
Nussbaumer, H., Schmutz, W., Vogel, M.: 1986, Astr. Astrophys. 69 154.
Penston, M.V., Allen, A.D.: 1985, Mon. Not. R. astr. Soc. 212, 939.
Penston, M.V. et al. 1983, Mon. Not. R. astr. Soc. 202, 833.
Ponz, D., Cassatella, A., Viotti, R.: 1982, The Nature of Symbiotic
    Stars, Reidel, Dordrecht, p.217.
Pottasch, S.R., Varsavsky, C.M.: 1960, Ann. d'Astrophys. 23, 516.
Taylor, A.R., Seaquist, E.R., Mattei, J.A.: 1986, Nature 319, 38.
Taranova,O.C., Yudin,B.F.: 1983, Astron. Astrophys. 117, 209.
Thackeray A.D.: 1950, Mon. Not. R. astr. Soc. 110, 45.
Thackeray, A.D.: 1977, Mem. R. astr. Soc. 83, 1.
Viotti, R.: 1987, Planetary and Protoplanetary Nebulae from IRAS to ISO,
    Frascati Workshop, A. Preite Martinez ed., Reidel, Dordrecht, p. 163
Willson, L.A., Wallerstein, G., Brugel, E.W., Stencel, R.E.: 1984,
    Astron. Astrophys. 133, 154.

SYMBIOTIC NOVA PU VUL (KUWANO-HONDA OBJECT): SOME RESULTS
OF COORDINATED INVESTIGATIONS

R.E.Gershberg and N.M.Shakhovskoj
Crimean Astrophysical Observatory
Crimea Nauchny 334413 USSR

Since 1979 a team consisting of T.S.Belyakina, N.I.Bondar',
K.K.Chuvaev, Yu.S.Efimov, R.E.Gershberg, V.I.Krasnobabtsev,
E.P.Pavlenko, P.P.Petrov, I.S.Savanov, N.I.Shakhovskaya,
N.M.Shakhovskoj, A.G.Shcherbakov and V.I.Shenavrin from
the Crimea, Dr. V.Piirola from Finland and Drs D.Chochol,
L.Hric and J.Grygar from Czechoslovakia are carrying out
a coordinated study of pecular Kuwano-Honda (PU Vul)
object, and this short report is made on behalf of the
team.
    PU Vul flared up from 14^m to 9^m in 1978 and up to
date maintains the high brightness. Our studies include
the optical and IR photometry, spectroscopy, spectrophotometry, polarimetry and UV observations with the space station Astron. The general character of brightness variations and time distribution of our observations are given in
Fig.1.
    The data analysis lead us to following conclusion:
PU Vul is a binary consisting of a normal M giant and
exploded component; in maximum (1983) the latter was not
distinguishable from a supergiant F5 in respect of absolute luminosity ($-6\overset{m}{.}3$), physical conditions and chemical
abundance of atmosphere. PU Vul is of about 5.3 kpc from
the Sun and about 800 pc above the galactic plane. The
deep minimum in 1980-81 was found to be due to an episode
of a dust envelope formation and subsequent dissipation
of the envelope around the exploded component similar to
typical slow novae. Such a model explains many observed
features of PU Vul in minimum: appearance of the TiO bands
but very small U-B and B-V indices, IR fluxes, wavelength
dependence of polarization.
    In 1983 the photospheric temperature of exploded
component reached a minimum, then the temperature began
to increase and the size of the component began to decrease. As a results, during 1979-86 PU Vul traced a closed
curve in the V, B-V plane - see Fig 2, plotted by Yu.S.

J. Mikolajewska et al. (eds.), The Symbiotic Phenomenon, 279–281.

Efimov. Brightness variations occurred not monotonously but as a set of discrete "bluerings" with a duration of about a month. In 1982-86 we did not find quasi-periodical brightness variations with a characteristic time of about 80^d that existed before the dust formation episode but we found irregular variations with amplitudes up to $\Delta U \approx 0^m\!.2-0^m\!.3$. In the same period we had significant variations of parameters of intrinsic polarization of PU Vul radiation with an amplitude up to 1% and with a characteristic time of several days. Variations of polarization parameters are accompanied with strong changes in a shape of $p(\lambda)$ and are weakly correlated with brightness variations, therefore it is a hard task to offer a simple model for polarization feature of PU Vul.

From 1983 to 1986 the energy distribution in the PU Vul spectrum changed from F5 to A2. Simultaneously an excess in the Balmer jump region appeared; hydrogen, Ca II IR triplet and Fe II lines showed emission of a complicate structure involving the P Cyg profiles. The curve-of-growth analysis showed that excitation temperature of the hot component of PU Vul growed up from 6300 K in 1979-82 to 8000 K in 1984. At the same time the electron density increased by ten times and became similar to that in the atmospheres of giants. We found some indications on appearance of chemical anomalies similar to that of Am stars: Ca and Sc deficit and excess of the Fe group elements (excluding Fe itself) and heavier increasing on the average with the atomic weight -see Fig 3, prepared by I.S.Savanov.

The M giant brightness maintains at a constant level within 10-15% during this history of PU Vul.

Our studies permit to attribute the Kuwano-Honda object to extreme slow novae of the RT Ser type. We found an excelent agreement between the observed features of PU Vul and characteristics of an accreting hot white dwarf in the stage of quasi-stationary surface thermonuclear burning that have been calculated by Dr I.Iben (Ap J 259, 244, 1982) -see Fig 4. We have practically coincidence in M_V and T_{eff} for the time of temperature minimum and in flaring up time to the last 5^m. The loop in Fig 2 is in agreement with a prediction in Fig 4.

Thus, observations of the Kuwano-Honda object give us unique possibility to look at very rare process of a slow evolution of an exploded white dwarf. Many aspects of such a process is still hardly obscured and require new observational data and theoretical efforts.

In more complete form our results have been published in Astron.Astrophys. 132, L 12-14, 1984; Bull. Crimean Ap Obs 72, 3-72, 1985; Commun Konkoly Obs N 86,351-354, 1986.

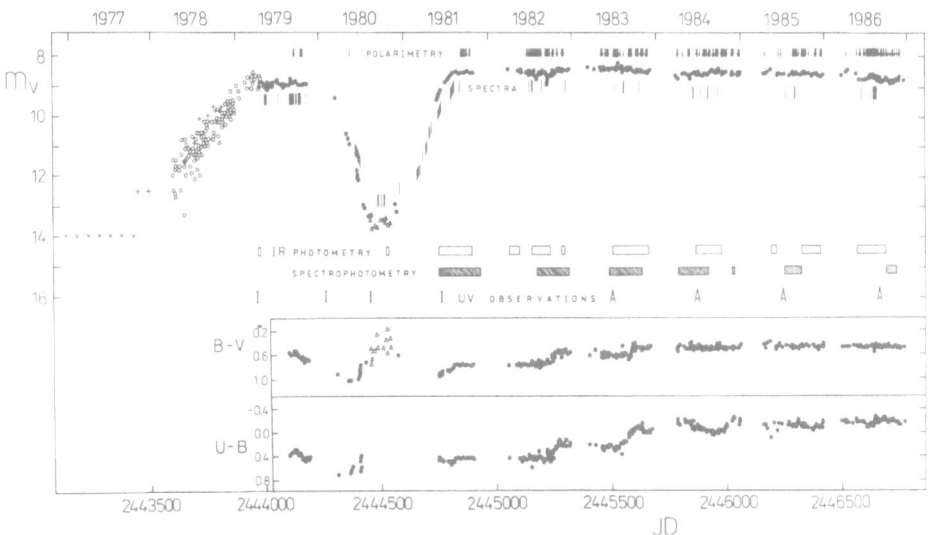

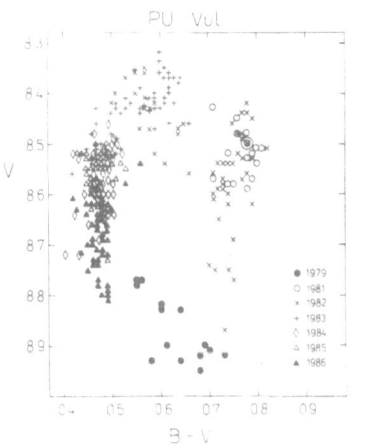

Fig.1. Light and colour curves and distribution of other observations of PU Vul.

Fig 2. The colour-magnitude diagram for 1979-1986 years.

Fig 3. Abundances of elements in the atmosphere of the exploded component of PU Vul.

Fig 4. Theoretical track of accreting white dwarf calculated by I.Iben and PU Vul's position at maximum (1983) in L,T plane.

Fig 2

Fig 3

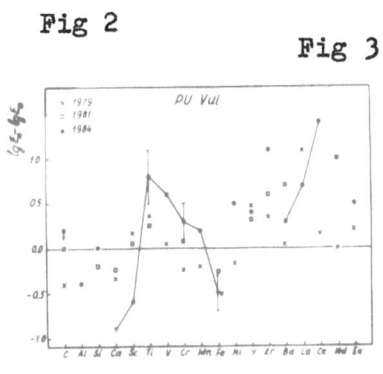

Fig 4

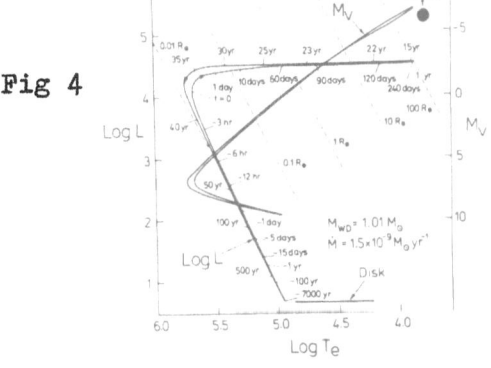

SPECTROSCOPIC VARIATIONS OF THE SYMBIOTIC NOVA PU VUL

L. Hric, D. Chochol
Astronomical Institute, Slovak Academy of Sciences
059 60 Tatranská Lomnica, Czechoslovakia
B. Kovachev
Bulgarian Academy of Sciences, National Astronomi-
cal Observatory Rozhen, 4700 Smoljan, Bulgaria

ABSTRACT. Evidence for formation and subsequent expansion
of an A-F supergiant envelope in the very slow symbiotic
nova PU Vul caused by quasiperiodic injections of matter
from M giant to the A-F supergiant is given.

INTRODUCTION

The very slow symbiotic nova PU Vul has been studied
intensively since the outburst which started in 1979 and is
still in progress. UBV photometry and spectroscopy of the
nova taken by several groups of observers around the world
support a suggestion that the binary system PU Vul consists
of a massive mass-losing M6 giant ($M \sim 7.5$ $M_{\odot}$) and a CO
dwarf with mass 1.01 $M_{\odot}$ accreting hydrogen-rich matter at a
rate 10^{-9} $M_{\odot}/yr$. This dwarf mimics in outer appearance an F
supergiant (cf. Kenyon, 1986; Chochol and Grygar, 1987 and
references therein). Chochol and Grygar (1987) computed the
elements of the spectroscopic orbit of the F supergiant.
An orbital period of 3200 d and an eccentricity of 0.64
were found. Radial velocities, especially after periastron
passage, display considerable oscillations. The aim of our
paper is to show that the oscillations connect with formation and subsequent expansion of the A-F supergiant envelope due to the quasiperiodic injections of matter from the
M giant to the A-F supergiant after periastron passage.

SPECTROSCOPIC DATA AND RESULTS

Spectra of PU Vul with a dispersion of 1.7 nm/mm were
obtained at the Coudé focus of the 2-m telescope at the
Rozhen Observatory on July 8/9, 1985, October 27/28, 1985
and September 20/21, 1986 exactly at the time when the
O - C deviations on radial velocities curve published by
Chochol and Grygar (1987) reached extremal values. The main

283

J. Mikolajewska et al. (eds.), The Symbiotic Phenomenon, 283–284.
© *1988 by Kluwer Academic Publishers.*

differences between the spectra are as follows: in the
spectrum from July 85 absorption shell lines of ionized metals (Mg, Si, Ca, Ti, V, Cr, Fe, Sr, Zr) are present. The
mean radial velocity is (22.0 ± 1.5) km/s. The O - C deviation is 20.7 km/s. The second spectrum, from Oct. 85, exhibits mainly neutral metals (Mg, Al, Si, Ca, Ti, Cr, Mn,
Fe). The mean radial velocity is $(- 18.1 \pm 1.5)$ km/s. The
O - C deviation is - 14.5 km/s. The large differences between the radial velocities clearly connect with the expansion of an A-F supergiant envelope. The expansion is best
visible in the profile of the Ca II - K line in Fig. 1. In

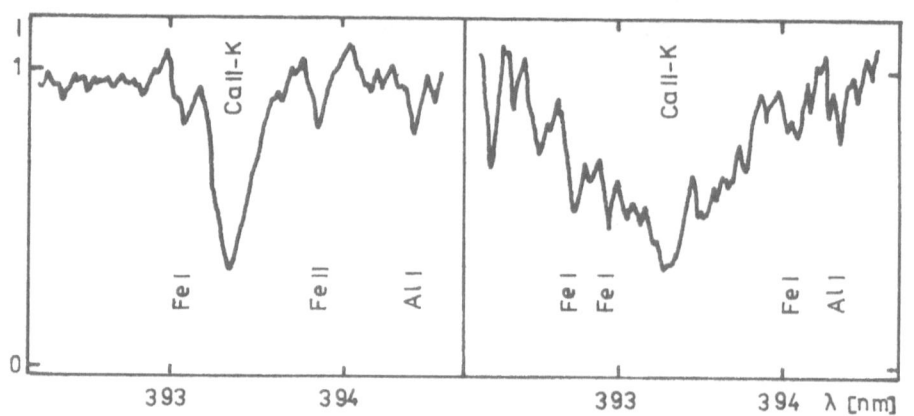

Figure 1. Ca II line on the spectra from July and Oct. 85.

the sp. from Sept. 86 both neutral and ionized metals are
present. The mean radial velocity from all lines is (- 44.3
± 2.9) km/s. The O - C deviation is - 53 km/s. The Ca II - K
line decrease in intensity but is even broader than in the
sp. from Oct. 85. Due to the fact that the radial velocities
of neutral and ionized metals differ only in the range of
errors we suggest that the expanding envelope is gradually
ionized. We tried to find a periodicity in the O - C deviations of the RV curve in above mentioned paper by method of
minimization of phase dispersion (Stellingwerf, 1978). A period of 263.5 d with statistical significance 0.692 was fo-
und. Observed RV changes clearly connect with physical processes in the interacting binary PU Vul and can be caused by
a more or less periodic increase of mass transfer from the
M6 giant to the A-F supergiant followed by an expansion of
the envelope of A-F supergiant, after periastron passage.

REFERENCES

Chochol, D., Grygar, J.: 1987, Astrophys. Space Sci. 131,487
Kenyon, S.J.: 1986, Astron. J. 91, 563.
Stellingwerf, R.F.: 1978, Astrophys. J. 224, 953.

Emission LINE ANALYSES OF HBV 475, V1016 CYG, AND HM SGE

S. Tamura
Astronomical Institute, Tohoku University
Aobayama, Sendai   980
Japan

ABSTRACT. Results of the high dispersion spectroscopic observations on
HBV 475, V1016 Cyg, and HM Sge are presented. Due to yearly observations
of HBV 475 since 1981, radial velocities which have been measured from
Hα and Hγ can be explained by a bipolar like non-spherical flow combined
with the rotation. Highly resolved profiles of Hα, [FeVII]λ6087, [OIII]
λλ4959,5007, and HeIIλ4686 are obtained from V1016 Cyg and HM Sge as
well as HBV 475. Characteristic differences of them can be seen among
these symbiotic stars.

1. Time Variation of Emission Line Profiles in HBV 475

Line profiles of Hα and Hγ are displayed in Fig. 1a,b and radial velocities of them are plotted in Fig. 2 superimposed on the previous result
(solid circles; Iijima et al 1981). To explain our results(open circles
and squares, crosses, asterisks which correspond to Hγ, Hα at FWHM, components of Hα, Hα at broad wings respectively), we introduce a bipolar
like non-spherical flow combined with the rotation (indicated by real
and broken lines). During these observations we noticed other characteristics on the strength and profile of HeIλ5016 line. It is available to
explain this strength based upon the optical depth effect in absorption
of λ537. The title of preprint is "Spectroscopic investigation of HBV
475 in optical regions". This work will be published elsewhere.

2. Gaussian fitting analyses on the dynamical structure of ionized
   regions of HBV 475, V1016 Cyg, and HM Sge

Calibrated profiles of Hα, [FeVII]λ6087, [OIII]λ5007, and HeIIλ4686 of
HBV 475, V1016 Cyg, and HM Sge are presented in Fig. 3a,b,c. It seems
that almost all of these lines are consisted of several components which
show different velocities among observed ions. This fact suggests various ionization stratification. A full paper is in preparation.

Reference

Iijima,T., Mammano,A., and Margoni,R. 1981, *Ap. and Space Sci.*, **75**, 237

J. Mikolajewska et al. (eds.), The Symbiotic Phenomenon, 285–286.

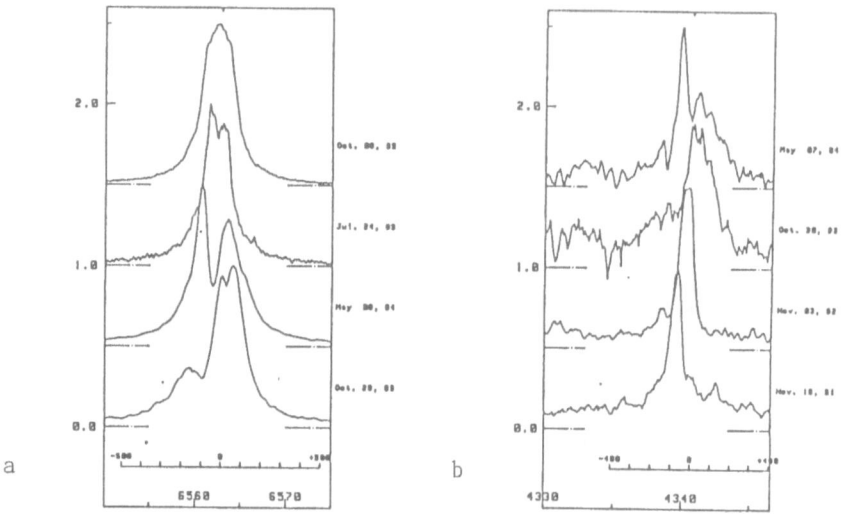

Figure 1. Time variations of line profiles in Hα(a) and Hγ(b). Abscissae are wavelengths and velocities in the heliocentric system.

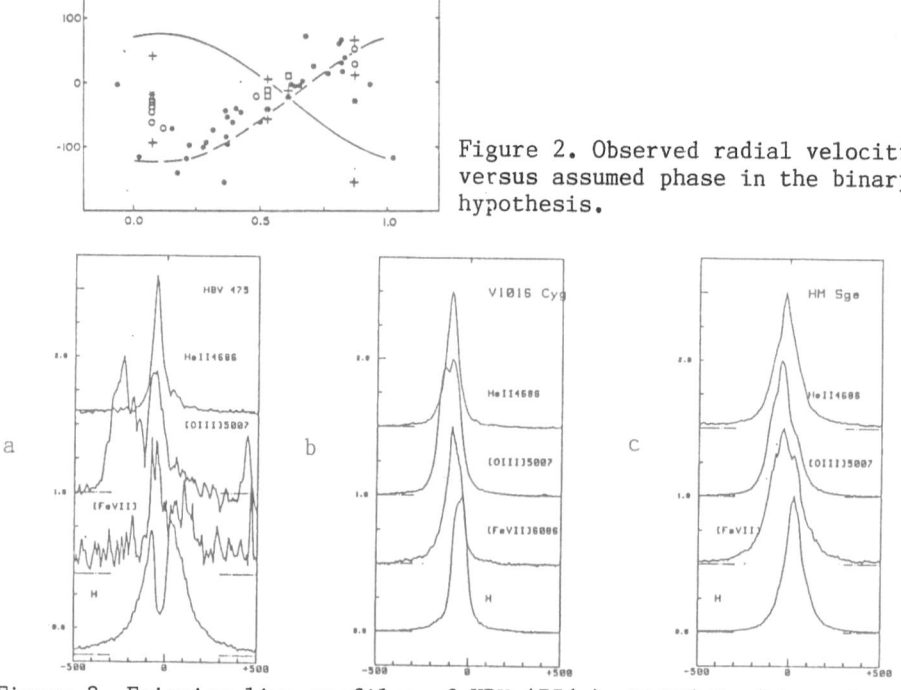

Figure 2. Observed radial velocities versus assumed phase in the binary hypothesis.

Figure 3. Emission line profiles of HBV 475(a), V1016 Cyg(b), and HM Sge(c). Abscissae are velocities in the heliocentric system.

SESSION 4. CONTRIBUTION FOR OTHER INDIVIDUAL OBJECTS

"Fortunately there are no neutrinos for us"

Roberto Viotti

PHOTOMETRIC AND SPECTROSCOPIC VARIATIONS OF THE SYMBIOTIC STAR EG ANDROMEDAE

A. Skopal[1], D. Chochol[1], A. Vittone[2], A. Mammano[3]
1 Astronomical Institute, Slovak Academy of Sciences,
  059 60 Tatranská Lomnica, Czechoslovakia
2 Osservatorio Astronomico Capodimonte,
  Via Moiariello 16, 80131 Napoli, Italy
3 Universita di Messina,Via C. Battisti 90,Messina, Italy

EG And is a symbiotic binary system. The cool component is an M3 III star, the hot one is a subdwarf with temperature 60 250 K and luminosity 1.45 $L_\odot$ (Boyarchuk, 1985). The eclipsing nature of the system suggested from UV spectroscopy by Oliversen et al. (1985) was confirmed photometrically by Chochol et al. (1987). The circular spectroscopic orbit of cool component determined by Oliversen et al. (1985) supposing the orbital period 470 days leads to f(m) = 3.2×10^{-2} $M_\odot$ and detached configuration.

The UBV photometry of EG And was obtained in 1985-87 by 0.6 m telescope at Skalnaté Pleso Observatory. The differential photometry Δm = m(EG And) – m(HD 3914) in U colour is presented in Fig. 1. Our observations cover two recent minima, which occured in 1985 and 1986-87. The eclipse is most pronounced in U colour. The details about UBV photoelectric observations of EG And are in the paper of Chochol and Skopal (1987), where orbital period 480 days was suggested. The phase in Fig. 1 was computed using the period 481.7 days, which we obtained from spectroscopic observations. Therefore a new ephemeris for minima is:

$$JD_{min} = 2\ 446\ 336.7 + 481.7 \text{ days x E} . \qquad (1)$$

As it is possible to see on Fig. 1, the descending branch of the minimum is steeper than the ascending one, that could be caused by an elliptical orbit. This conclusion has been confirmed from spectroscopic elements.

The spectroscopic observations of EG And have been carried out at Asiago Astrophysical Observatory in 1966-69, 1972, 1978, 1983. Altogether 20 spectra were obtained by 1.22 m telescope with a prism spectrograph with dispersion of 6 nm/mm at H_γ. Two spectra with a dispersion 1.8 nm/mm were obtained by 2m telescope at Ondrejov Observatory in 1976, two spectra with the same dispersion were obtained by 0.9m telescope at Torun Observatory in 1986. We measured radial velocities of neutral metals to obtain information about the motion of cool component. The data are presented in Fig. 2, where also the

J. Mikolajewska et al. (eds.), The Symbiotic Phenomenon, 289–290.

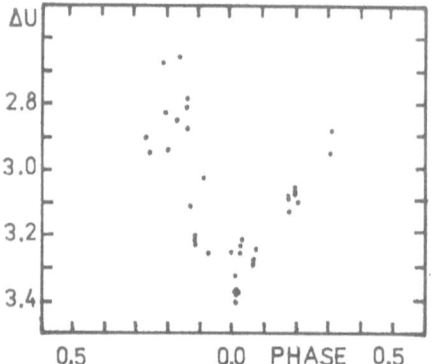

Figure 1. Differential U
photometry of EG And.

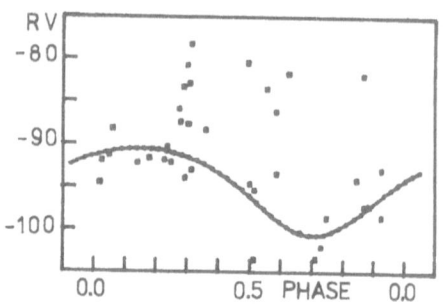

Figure 2. Phase diagram of
the radial velocities of neutral metals and spectroscopic
orbit of EG And.

radial velocities of cool component published by Oliversen et al.
(1985) and Garcia (1986) are included. At first we computed the phases of radial velocities with the period 480 days. We obtained a phase distribution of radial velocities close to this one on Fig. 2. It
was obvious that radial velocities exceeding -87 km/s can not reflect
the orbital motion, so we removed them. Orbital elements were computed using the code SPEL (Horn, 1987):

$P = (481.7\pm2.1)$ days $\qquad$ T(periastr.) = 2 442 324 JD

$e = 0.17\pm0.12$ $\qquad$ T(min.) = 2 442 432 JD

$\omega = 169°\pm43°$

$\gamma = -94.7\pm0.4$ km/s $\qquad$ $f(m) = 6.18\times10^{-3}$ $M_\odot$

$K_1 = 5.1\pm0.6$ km/s $\qquad$ $a_1 \sin i = 47.4$ $R_\odot$

Possible masses of components are e.g. 0.5 $M_\odot$ for a hot component and
4 $M_\odot$ for a cool one (i = 90°). The phases in Fig.2 were computed according to the ephemeris (1). All excluded radial velocities (except
the RV=-81.6 in the phase 0.86, published by Garcia (1986)) were measured by us and occurred in the phases 0.29-0.63, when the cool component was behind the hot one. A nebulosity around the hot component
can influence the absorption lines of neutral metals. It is interesting to note that the strong emission lines of hydrogen were present
in all excluded spectra.

The mass function determined by us is much smaller than the value determined by Oliversen et al. (1985),which leads to detached
configuration. If our elements are real, the cool component easily
overflows the Roche lobe.

REFERENCES

Boyarchuk, A.A.: 1985, In: Proc. ESA Workshop: Recent Results on Ca-
    taclysmic Variables, ESA SP - 236, 97.
Chochol, D., Skopal, A.: 1987, Astrofiz. Issled. BAN (in press).
Chochol, D., Skopal, A., Vittone, A., Mammano, A.: 1987, Astrophys.
    Space Sci. 131, 755.
Garcia, M.R.: 1986, Astron. J. 91, 1400.
Horn, J.: 1987, private communication.
Oliversen, N.A., et al.: 1985, Astrophys. J. 295, 620.

The Search for the Elusive Companion of EG Andromedae

Joseph E. Pesce and Robert E. Stencel
Center for Astrophysics and Space Astronomy
University of Colorado
Boulder, CO 80309-0391 USA
and
Nancy A. Oliversen
Computer Sciences Corporation
Greenbelt, Maryland USA

Abstract: We report observations at opposite quadratures of the interacting symbiotic binary EG Andromedae (HD 4174, Period = 470^d). Correcting for absolute motion at the system, it appears that many of the nebular lines arise from material that moves with the red giant star. The He II feature appears to track the hot component. It may be possible to use this feature in other, similar systems in order to "pin-down" the mass ratio.

The optical spectrum of EG Andromedae was first noted by Wilson (1950) to consist of a high velocity (v_r = -96 km s^{-1}: Oliversen *et al*, 1985) M2 giant upon which were superposed the optical emission lines of O III] and the Balmer series. The ultraviolet spectrum exhibits a far UV continuum and emission lines of C IV, He II, O III] and other species. The companion star is thought to be similar to the central star of a planetary nebula (Kenyon 1983). Oliversen *et al* (1985) have used the spectroscopic motion of the red giant features and an estimate of the mass of the hot object (0.7 M⊙) to approximate the mass ratio of about 3.5.

Oliversen *et al* report that the red giant optical absorption features undergo an estimated 15 km s^{-1} total velocity excursion between orbital quadratures. With a mass ratio of 3.5, we predict that the hot object should exhibit 53 km s^{-1} in spectral line displacement. The motion of such spectral features should be easily measured given the ±3 km s^{-1} precision possible with high-resolution IUE data (Ayres 1985).

Two high-dispersion spectra of EG And were obtained with the *International Ultraviolet Explorer* satellite SWP camera (1150-2000 Å) (IUE: Boggess *et al.* 1978). The first (SWP 23692) was obtained at phase 0.73 (on the Oliversen *et al.* ephemeris), near maximum red giant orbital redshift. The second (SWP 26268) was approximately one half period later at phase 0.40 (maximum red giant blueshift). Both were 345 minute exposures taken on 15/16 August 1984, and 27 June 1985, respectively. After each stellar observation,

J. Mikolajewska et al. (eds.), The Symbiotic Phenomenon, 291–292.
© *1988 by Kluwer Academic Publishers.*

a two minute "wavcal" Pt-Ne emission line spectrum was obtained to provide individual wavelength calibration.

Measurements of the central wavelengths and fluxes of the emission lines were obtained by fitting a least-squares Gaussian profile to the unsaturated portions of the profiles of unblended emission lines. The formal measurement error is typically $\pm 0.015 \text{Å}$ for well exposed emission lines, except He II ($\pm 0.060 \text{Å}$) and O I] ($\pm 0.018 \text{Å}$).

Comparisons of the two corrected stellar spectra were made and the shift in wavelength between the two phases was thus obtained. The greatest velocity shift is for O IV] which corresponds to $+15$ km s^{-1}. Surprisingly, this is the same as the expected velocity sign *and* amplitude of the red giant primary. Rather than finding the -53 km s^{-1} wavelength shift between quadratures for material near the hot companion, we found the majority of the UV emission lines seem to arise from material near the primary (v $= +15$ km s^{-1}). In particular, all of the intercombination lines show a $+7$ to $+15$ km s^{-1} displacement between quadratures. The resonance lines of Si IV and C IV, however, show a much smaller shift, $+1$ to $+6$ km s^{-1}. He II is the only feature possibly showing a systematic shift with the companion star's velocity sign (-2.2 km s^{-1}) but at an amplitude many times less than expected (additional details in Pesce *et al.*, 1987).

We propose the following picture: The intercombination lines arise from a large fraction of the red giant's outer atmosphere that is radiatively heated by the companion star and caused to expand so that part of it overflows the tidal radius and forms a stream directed at the companion. The resonance lines arise from this stream, but from material on the red giant side of the system's inner Lagrangian point, L_1. The He II feature may be associated with the stream and/or accretion on the companion star's side of L_1. He II can possibly be used to track the hot component in symbiotic stars. Future studies of EG And and other systems showing He II emission are planned in order to prove this hypothesis.

Associating the resonance and intercombination lines with the red giant atmosphere supports the analysis by Friedjung, Stencel and Viotti (1983) that the red giant wind plays a major role in the spectrum formation in symbiotic binaries. Also, we are unable to verify the mass ratio of 3.5 deduced by Oliversen *et al* (1985), because the only feature that can be possibly associated with the motion of the hot component, He II, is mutilated by complex absorption patterns possibly associated with a stream and accretion.

REFERENCES

Ayres, T. *et al.* 1985 in *The Future of UV Astronomy*, ed., J. Mead, R. Chapman & Y. Kondo, NASA Conf. Publ. 2349, p.468.

Boggess, A., *et al* 1978 *Nature* **275**, 1.

Friedjung, M., Stencel, R.E., and Viotti, R. 1983 *A&A* **126**, 407.

Kenyon, S.J. 1983, Ph.D.thesis, University of Illinois.

Oliversen, N.A., Anderson, C.M., Stencel, R.E., and Slovak, M.H. 1985, *Ap.J.* **295**, 620.

Pesce, J.E., Stencel, R.E., and Oliversen, N.A. 1987, *Pub.A.S.P.* in press (Nov.).

Wilson, R.E. 1950, *Pub.A.S.P.* **62**, 14.

THE SYMBIOTIC STAR UV AURIGAE

Parag Seal
Indian Institute of Astrophysics
Bangalore 560034
INDIA

1. INTRODUCTION

The symbiotic variable UV Aur was observed by Merrill (1941) and Sanford (1944, 1949). UV Aur is classified as C8 ep carbon variable (Kukarkin et al. 1969) with a hot white dwarf and surrounding nebula. Infrared photometric observations have been done by Allen (1982), Noguchi et al. (1981) and Kenyon and Gallagher (1983) in the near infrared region and by Woolf (1973) at 2.2 micron to 20 micron region. UV Aur has also been observed by IRAS and IUE Satellite. Polarimetry of UV Aur has been studied by Khudyakova (1985) in the optical region.

2. OBSERVATION

The observations of UV Aur were carried out with the 1 m telescope at kavalur during 1986-87 using the universal Astronomical Grating Spectrograph. Two spectra were obtained in the region 4200 A° to 5000 A° and 6000 A° to 7000 A° at a dispersion of 86 A°/mm. Later another spectra (5200 A° to 8000 A°) was also obtained at a dispersion of 50 A°/mm. Spectrophotometry was also done from 4000 A° to 6000 A° region.

3. ANALYSIS

The spectra of UV Aur 5200 A° to 8000 A° region is shown in the Figure. The Balmer lines are the strongest throughout the spectra. The low excitation lines of [Fe II] are present along with the high excitation lines of [Fe VII]. Also absorption bands for Na I D_1 and D_2, TiO and O_2 are present.

From the spectrophotometric data, relative intensities of emission lines are derived from the magnitudes which are normalised to λ = 5000 A°. From the ratio of hydrogen line relative intensities the reddening and colour excess are derived as A_v = 0.78±0.10,

J. Mikolajewska et al. (eds.), The Symbiotic Phenomenon, 293–294.
© *1988 by Kluwer Academic Publishers.*

294

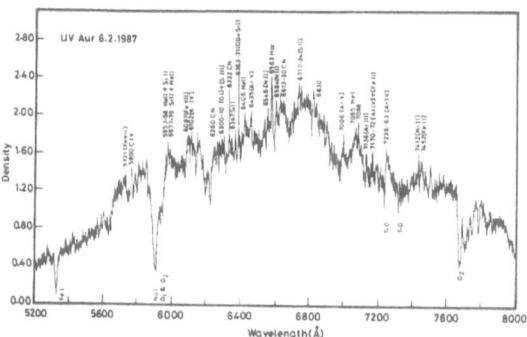

Density tracing of the spectrum of UV Aur

E_{B-V} = 0.25±0.03.

From the relative intensities of H-beta, He I 4471 and He II 4686 lines the temperature of the hot component is derived as T_h = 58.3x10^3 K. From the intensity ratio of [O III] λ 5007 + λ 4959 to λ 4363 as 1.88±0.50, the electron density of the nebular component is derived as 10^8 cm^{-3}, assuming the electron temperature as usual T_e = 17000 K.

From the infrared data (Kenyon et al. 1983) bolometric luminosity and effective temperature are calculated as M_{bol}= -5.2 and T_{eff} = 3200 K. The luminosity is of the order of L ≃ 10^4 L$_\odot$. From IRAS observation, the flux ratio F_ν (25 micron)/F_ν (12 micron) = 0.32 is derived.

References

Allen,D.A. : 1984, Proc.ASA. 5, 369.
Boyarchuk,A.A., Gershberg,R.E. and Pronik,V.I. 1963, Isvestia, Crime 29, 291.
Iijima,T.: 1981, Photometric & Spectroscopic Binary System, NATO Advanced Study Institute, ed. E.B.Carling and Z.Kopal, (Dordrecht: Reidel), p.517.
Kenyon,S.J. and Gallagher,J.S. : 1983, Astron.J. 88, 666.
Khudyakova,T.N.: 1985, Sov. Astron. Lett. 11, 262.
Kukarkin,B.V., Kholopov,P.N., Efremov,Yu,N., et al. : 1969, General Catalog of Variable Stars 1, Nanuka, Moscow.
Merrill,P.W.: 1941, Astrop.J. 93, 40.
Noguchi,K., Kawara,K., Kobayashi,Y., Okuda,H. and Sato,S. ; 1981, Pub. Astron. Soc. Jap. 33, 373.
Sanford,R.F.: 1944, Pub. Astron. Soc. Pac. 56, 122.
Sanford,R.F.:1949, Pub. Astron. Soc. Pac. 61, 261.
Woolf,N.J.: 1973, Astrop. J. 185, 229.

THE THIRD GALACTIC CARBON SYMBIOTIC

Regina E. Schulte-Ladbeck[1,*], D. Jack MacConnell[2,+] and Nelson Zarate[3]

[1]Washburn Observatory, University of Wisconsin-Madison, 475 North Charter Street, Madison, Wisconsin 53706
[2]Astronomy Programs, Computer Sciences Corporation, Space Telescope Science Institute, Homewood Campus, Baltimore, Maryland 21218
[3]Space Telescope Science Institute, Homewood Campus, Baltimore, Maryland 21218

ABSTRACT. Of the more than 100 symbiotic stars now known in our Galaxy, only two have previously been reported to contain carbon stars as their cool components. We here present observations of a third such object, which we wish to call **Weaver's star** since Weaver (1972) first identified it as a symbiotic.

Weaver's star has the equatorial coordinates (1950) $\alpha = 12^h48^m21^s.9$, $\delta = -64°43'48"$ and the galactic coordinates $l = 303°.0$, $b = -2°.2$. These coordinates put the object in the Coalsack region. The spectrum was reported by Weaver to contain a very strong H_α emission line, weaker H_β, and progressively weaker series members until H9, prominent [OIII] $\lambda\lambda$ 4363Å, 4959Å and 5007Å and a He II λ 4686Å line almost as strong as H_β. The continuum was noted as being medium strong at H_α and comparatively fainter in the blue region of the spectrum. However, no absorption bands are mentioned. The photographic magnitudes as observed by Weaver between May 19 and May 29, 1971 did not show significant variability. The values given are U = 14.6, B = 14.3, V = 13.8, and I_{ph} = 12.0, with an internal p.e. of 0^m.07. Weaver's star is included in the various Allen catalogs (e.g. Allen 1984). Allen lists the star under the name *SS 38* (Sanduleak and Stephenson 1973) as a D-type of variable K magnitude (5.7< K < 6.5).

In Figure 1 we present a Reticon spectrum of Weaver's star along with a spectrum of the bright and well known galactic carbon symbiotic UV Aur. Both spectra cover the same far red wavelength region from 6700Å< λ < 9600Å. The characteristic absorption bands of CN are strong in the spectrum of UV Aur and can be easily recognized in the spectrum of Weaver's star, albeit at a lower signal to noise level. We attempted to classify Weaver's star on Richer's system (Richer 1971) by comparing it with UV Aur which Richer assigns to C9 II at minimum light. His criteria are that the absence of the Ca II triplet ($\lambda\lambda$ 8498Å, 8542Å and 8662Å) indicates a type later than C7, whereas a weak K I doublet ($\lambda\lambda$ 7665Å, 7699Å) favours a type close to C9; both criteria apply to our spectra of Weaver's star.

---

*Based on observations collected at the European Southern Observatory, La Silla, Chile.
+Visiting Astronomer, Cerro Tololo Interamerican Observatory, National Optical Astronomy Observatories, operated by the Association of Universities for Research in Astronomy, Inc., under contract with the U.S. National Science Foundation.

J. Mikolajewska et al. (eds.), The Symbiotic Phenomenon, 295–296.
© 1988 by Kluwer Academic Publishers.

296

We have subsequently taken a number of red and blue spectra of Weaver's star during the past year. A large number of strong emission lines is present, as is typical for D-type symbiotics. The ratio of Hα to Hβ is large, 16:1, indicating a large amount of extinction near or on the path to the star. Weaver's star is also listed by Allen as a radio source. We have also found a detection of this object in the IRAS point source catalog.

The discovery of a new galactic carbon symbiotic has important implications for the evolution of symbiotic binaries as well as for the statistics of carbon stars. Weaver's star is the first galactic *D-type* carbon symbiotic and, showing variable K magnitudes, it is perhaps the first galactic symbiotic carbon Mira. Other carbon symbiotics have been located in the Large and Small Magellanic Clouds and the Draco dwarf spheroidal galaxy, respectively. A more detailed discussion will be given elsewhere.

ACKNOWLEDGEMENTS. R.E.S.-L. was supported by the Deutsche Forschungsgemeinschaft. D.J.M. thanks W.P.Bidelman and C.B.Stephenson for calling to his attention the positional coincidence of Weaver's star with that of a C star in an unpublished list of D.J.M.

REFERENCES
Allen, D.A.: 1984, *Proc.A.S.A.* 5, 369
Richer, H.B.: 1971, *Astrophys.J.* 167, 521
Sanduleak, N. and Stephenson, C.B.: 1973, *Astrophys.J.* 185, 899
Weaver, W.B.: 1972, *Publ.astron.Soc.Pacific* 84, 854

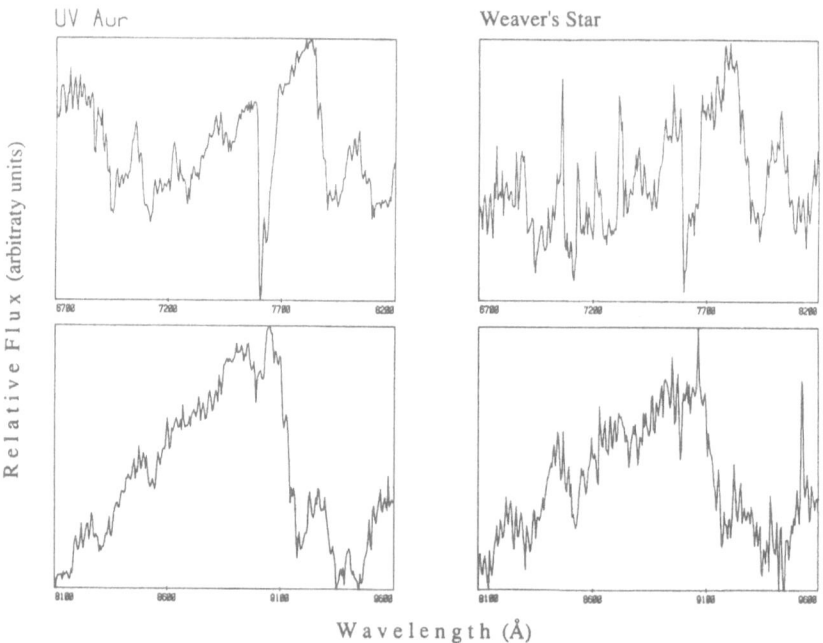

Figure 1. The far red spectrum of the well known galactic carbon symbiotic UV Aur shows the strong CN bands that are typical for carbon stars. The same bands are now identified in Weaver's symbiotic star. Together with UV Aur and UKS Ce-1, Weaver's star is the third carbon symbiotic known in our Galaxy.

BI CRUCIS

A. Altamore[1], C. Rossi[1], R. Viotti[2]

1. Istituto Astronomico, Universita' La Sapienza, Roma, Italy
2. Isituto Astrofisica Spaziale, Frascati, Italy

BI Crucis is a 12 mag star whose optical spectrum is characterized by a red continuum and variable emission line spectrum (Allen 1974, Henize and Carlson 1980, Whitelock et al. 1983). In order to investigate its symbiotic character on 18 February 1983 we have obtained at the 1.5m ESO telescope a 59 Å/mm spectrogram of the 5700-6900 Å region. BI Cru displayed a very rich emission line spectrum with very strong Hα and prominent HeI (5876 and 6678Å) lines. Several FeII lines are also present which appear optically thick (Figure 1). A few absorption features (NaI, 6269-84) of interstellar origin are present. However, we find no trace of TiO bands (or of neutral atoms) in spite of Allen's (1974) finding, but in agreement with Whitelock et al. (1983). Allen (priv. comm.) remarks that in his spectrum there are slight 'waves' in the continuum that looked like TiO absorptions. Thus the symbiotic nature of BI Cru is mostly based on its long term IR variability (T≃280 d, Whitelock et al. 1983). They also found the first overtone vibration rotation band of CO at 2.3 μm in emission. The CO emission band was recently resolved by McGregor et al. (1987). This is the first observation of CO in emission in a symbiotic object. The red continuum is more probably a highly reddened hot continuum. We note that a weak continuum is present in the LWR IUE image taken in March 1981 (Fig.2). This spectrum also shows a few emission lines of MgII and FeII. BI Cru is also a strong IRAS source. Following the model of Kenyon et al. (1986) for D-type symbiotics, the cool component of BI Cru could be reddened by circumstellar dust. A high resolution ESO CAT/CES red spectrum shows Hα doubled by a central absorption extending from -38 to -290 km/s with respect to the center of the emission line which suggests the presence of intermediate velocity winds like in other symbiotic stars.

J. Mikolajewska et al. (eds.), The Symbiotic Phenomenon, 297–298.

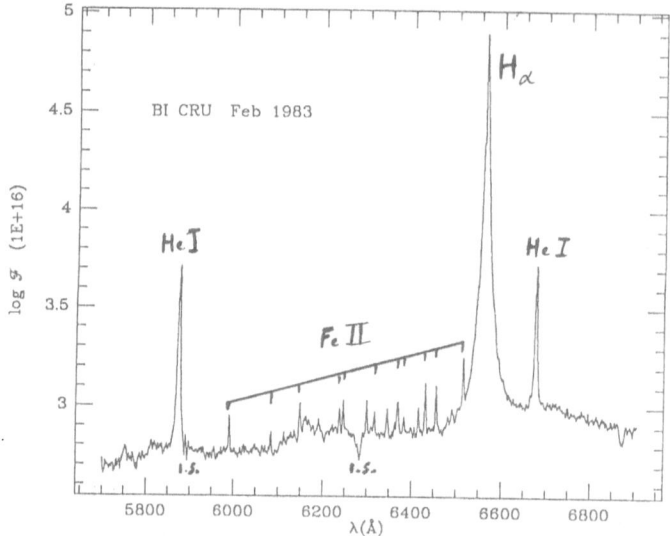

Figure 1. The low resolution spectrum of BI Cru in Feb 1983.

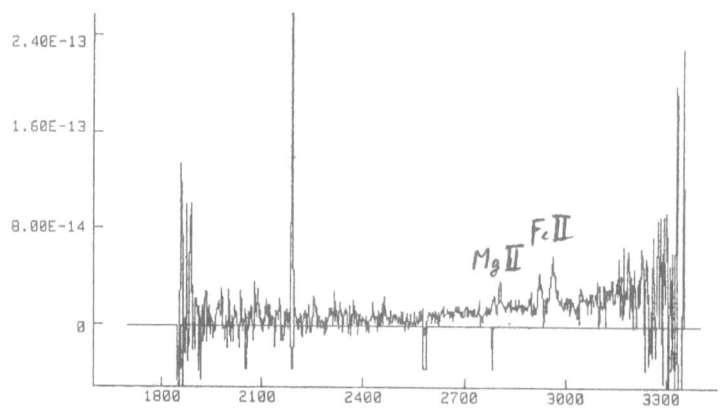

Figure 2. The low resolution UV spectrum in March 1981.

REFERENCES

Allen, D.A.: 1974, Inf. Bull. Var. Stars No. 911.
Henize, K.G., Carlson, E.D.: 1980, P.A.S.P. 92, 479.
Kenoyn, S.J. et al.: 1986, Astron. J. 92, 1118.
McGregor,P.J., Hyland,A.R., Hillier,D.J.: 1987, in press.
Whitelock, P.A. et al.: 1983, M.N.R.A.S. 205, 1207.

EFFECTS OF ECCENTRIC ORBIT OF BF CYGNI ON *IUE* AND OPTICAL SPECTRA

J. Mikolajewska & M. Mikolajewski
Institute of Astronomy, Nicolaus Copernicus University
Chopina 12/18, PL–87100 Torun, Poland

A recent analysis of all available photometric data has resulted in a new ephemeris for BF Cyg: $MIN=JD$ 2415058 + 756.8E. Simultaneously, optical spectra collected in 1979–1986 showed periodic changes of all emission lines (Mikolajewska 1987; Mikolajewska & Iijima 1987). It is interesting that the forbidden lines of [OIII] varied in antiphase to the permitted emission lines and optical brightness.

IUE spectra taken in 1979–1981 showed a strong hot continuum and high ionization resonance emission lines of NV, SiIV CIV, intercombination lines of NIV], NIII], SiIII], CIII] and OIII] as well as HeII emission. The observed $\lambda 2200\text{Å}$ interstellar absorption band suggests $E(B-V)\approx 0.3$. Taking into account the interstellar reddening distribution in the vicinity of BF Cyg (Lucke 1978; Mikolajewska & Mikolajewski 1980), the observed extinction implies a distance $d \lesssim 1.5$kpc. This distance is in good agreement with the observed low value of the systemic radial velocity (~15 km/s, Fig.1) of BF Cyg and the standard galactic rotation law. The standard extinction curve with $E(B-V)=0.3$ (Seaton 1979) was used for reddening correction of the spectra. The UV continuum of BF Cyg can be interpreted as a combination of a hot subdwarf ($T_{eff}\approx 60000$K, $L\approx 2500 L_\odot$ for $d=1.5$kpc) and hydrogen $bf+ff$ emission ($T_e\approx 10000$K). The emission measure of the nebular ($bf+ff$) continuum varied from $\sim 4\times 10^{59}\text{cm}^{-3}$ at maximum to $\sim 10^{59}\text{cm}^{-3}$ at minimum. Assuming cosmic abundance of Si/C the observed SiIII]/CIII] line ratio implies $n_e\approx 2\times 10^{10}\text{cm}^{-3}$ at the photometric maximum and $n_e\approx 3\times 10^9\text{cm}^{-3}$ at the minimum. We assume that these values are representative of the region where the bulk of the HI Balmer and intercombination line emission is produced.

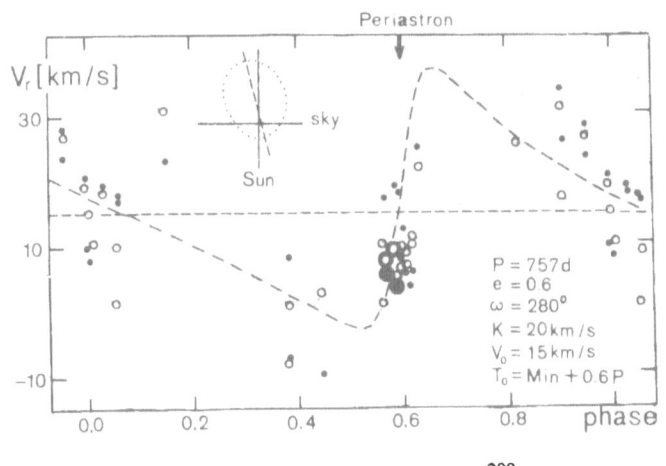

Figure 1

J. Mikolajewska et al. (eds.), The Symbiotic Phenomenon, 299–300.

The ratio of NV(1240)/NIV(1720) gives T_e(NV)$\cong$11500K close to the value derived for the Balmer emission region from the UV continuum fit.

The radio flux at 4.9GHz is equal to 2.06mJy (Seaquist *et al.* 1984) and corresponds to a radio emission measure $n_r^2 V_r \gtrsim 1.6 \times 10^{57}$cm^{-3} (for d=1.5kpc, $T_e \cong$10000K). Seaquist et al. also estimated the effective radius of the "radio photosphere" to be of order 10^{15}cm. Comparison of the radio emission measure with that of the Balmer emission lines and the $bf+ff$ continuum suggests the electron density varies as $1/r^2$ in the nebula. The observed spectral index, $\alpha \cong$1, may suggest that the ionized region is highly asymmetric. We propose that the periodic variations of the emission line fluxes are due to the effects of an elliptical orbital motion of the hot subdwarf ionizing part of the wind from the M giant. The recombination time τ_{rec} is ~1 hour in the region where the bulk of the HI, HeI and

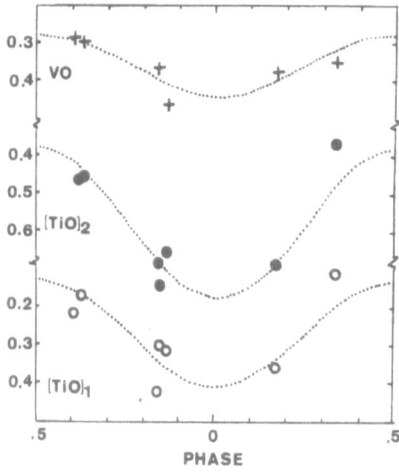

Figure 2

intercombination line emission is formed, so the emission region moves with the ionizing star. Fig.1 presents the composite radial velocity curve of HI – *dots* and HeI – *open circles* (see Mikolajewska 1987 for references); the large marks correspond to averaged data. We suggest an eccentric orbit ($e \cong$ 0.6), with the spectroscopic conjunction coinciding with the photometric minimum. It is remarkable that maximum of HI intensity occurs near periastron (also maximum of the electron density from SiIII]/CIII]), while maximum of [OIII] is at apastron (minimum of n_e). For the [OIII] lines, an additional effect due to (probably) changing electron temperature (by about 10–20% higher at apastron) is possible. The radial velocity curve may be affected by a different wind expansion velocity along the orbit of the ionizing component.

The presence of the hot subdwarf in BF Cyg may cause a substantial reflection effect. The effect may be very strong, because the illuminated hemisphere of the giant is visible when the distance between the components is lowest (periastron). Fig.2 shows the behaviour of [TiO]$_1$, [TiO]$_2$ and [VO] indices of the M giant spectrum given by Kenyon & Fernandez–Castro (1987), as a function of our revised photometric (orbital) phase. They are consistent with M2III ($T_{eff} \cong$3750K at maximum and M4III ($T_{eff} \cong$3550K) at minimum, and require $L_{hot} \cong$2500$L_\odot$ (assuming separation between the components of about 200–300$R_\odot$ at periastron), consistent with the *IUE* data.

REFERENCES:

Kenyon, S.J. & Fernandez–Castro, T., 1987, *Astron. J.*, **93**, 938
Lucke, P.B., 1978, *Astron. Astrophys.*, **64**, 367
Mikolajewska, J., 1987, *Astrophys. Space Sci.*, **131**, 713
Mikolajewska, J. & Iijima, T., 1987, *Acta Astr., in press*
Mikolajewska, J. & Mikolajewski, M., 1980, *Acta Astr.*, **30**, 347
Seaquist, E.R., Taylor,A.R. & Button,S., 1984, *Astrophys. J.*, **284**, 202
Seaton, M.J., 1979, *Mon. Not. R. astr. Soc.*, **187**, 73P.

ULTRAVIOLET TO NEAR INFRARED OBSERVATIONS OF BF Cyg

A. Cassatella, R. Gonzalez-Riestra
IUE Observatory, ESA, P.O. Box 54065, Madrid

T. Fernandez-Castro
Planetario de Madrid, Madrid, Spain

J. Fuensalida
Instituto de Astrofisica de Canarias, La Laguna

A. Gimenez
Instituto de Astrofisica de Andalucia, Granada

In this paper we provide preliminary results of multifrequency observations of BF Cyg carried out in July 1986. The ultraviolet spectra were obtained on July 26, 1986 using the IUE satellite. The optical observations were made at the Observatorio del Roque de los Muchachos (La Palma, Canary Islands) in July 1986 during the night 13/14 using the Isaac Newton 2.5m telescope with the Intermediate Dispersion Spectrograph (IDS, 500 mm camera) and the Image Photon Counting System (IPCS). The infrared observations were made during the night 13/14 of July, 1986, at the Observatorio del Teide (Tenerife, Canary Islands) using the Carlos Sanchez 1.5m telescope and an infrared single-channel photometer with an InSb detector.

In figure 1, the combined ultraviolet and optical energy distribution of BF Cyg on July 26, 1986, is plotted from 1150 to 6750 A. In figure 2, the same information is given in logarithmic scale after correction for reddening assuming a colour excess of E(B-V) = 0.35. A preliminary analysis of the observed spectrum allows one to identificy three well-defined components in BF Cyg (see Fig. 2):

a) A hot ionizing component, with Teff > 60000 K can be deduced from the strength of the HeII 1640 A line. This source, which dominates the spectrum below about 2000 A, is most probably a hot subdwarf.

b) A ionized nebula surrounding the hot source or the whole system, responsible for the conspicuous Balmer recombination continuum in emission easily visible in the range 2600 to 3600 A.

c) A cool giant with an effective temperature of about 3000 K which dominates the spectrum in the near infrared.

301

J. Mikolajewska et al. (eds.), The Symbiotic Phenomenon, 301–302.
© 1988 by Kluwer Academic Publishers.

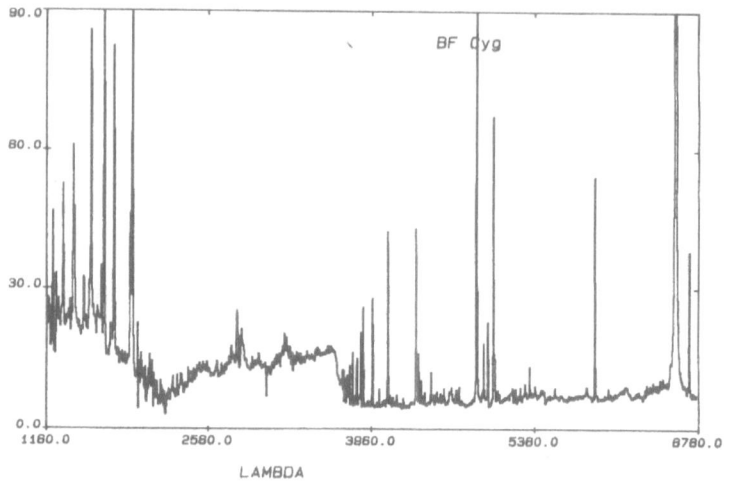

Figure 1. Oberved energy distribution of BF Cyg in the ultraviolet and optical regions.

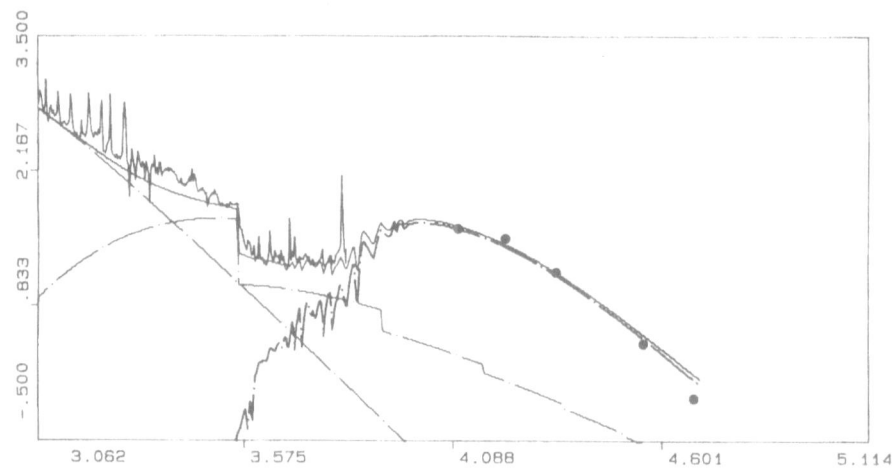

Figure 2. Preliminary model of the energy distribution of BF Cyg. Fluxes were corrected for reddening using E(B-V) = 0.35. The energy distribution is consistent with the presence of three components: a) a hot source represented by a black body with T = 80000 K; b) hydrogen f-b and f-f continuum emission with T_{el} =20000 K; c) a cool component, represented with a standard M5III star (shortward of 10000 A) and with a 3000 K black-body (at longer wavelengths).

ULTRAVIOLET VARIABILITY OF AX PERSEI

Joanna Mikolajewska
Institute of Astronomy, Nicolaus Copernicus University,
Chopina 12/18, PL-87100 Torun, Poland

Recent photometric and spectroscopic observations suggest AX Per might be an eclipsing binary with a period of about 682 days (Kenyon 1986 and references therein). The analysis of optical spectra taken during 1979-1986 has shown periodic minima in all observed permitted lines and a lack of any periodicity in the forbidden lines (Mikolajewska 1987; Mikolajewska & Iijima 1987). A comparison of the available radial velocities with these intensity variations shows that the behaviour of emission lines is consistent with the eclipse interpretation, however the minima (especially in HI and HeI) are too broad to be consistent with eclipses even by a Roche lobe filling red giant. In the following, the UV behaviour of AX Per is analysed using IUE spectra collected during the period 1979-1984.

As in other symbiotic stars, numerous emission lines belonging to many different ionization stages can be identified in the IUE spectra, from OI, MgII and FeII to HeII, NV and MgV. The emission lines appear narrow at high resolution.

Fig.1 displays the composite light curves of the ultraviolet emission lines (intergrated flux), ultraviolet continuum (monochromatic flux), and the SiIII] 1982/CIII] 1908 ratio as a function of orbital phase. The data corresponding to different orbital cycles are marked with different characters. The continuum and all emission lines except OIII 3133 (the Bowen fluoresced line) show minima at the same time as do the optical emission lines and UBV magnitudes. The deepest minimum is observed in the continuum ($\lambda 3000$ Å) and OIII] 1660-6, the depth is comparable with that observed in the optical Balmer lines. In addition, hydrogen free-bound recombinations are responsible for most of the long wavelenth IUE continuum, because the 3000Å flux is correlated with the Balmer line intensities. Unfortunately, the continuum fluxes in the shortwavelength IUE range were unmeasurable in the spectra taken during minima, so any conclusion about the amplitude and shape of the minimum in this spectral range cannot be derived. However, the total IUE flux ($\lambda\lambda 1200-3200$Å) at minimum was at least about 5 times lower than that observed at maximum. Another interesting feature of the variability is that minima in the intercombination lines and in HeII 1640 seem to be narrower than those in the Balmer lines and continuum. The OIII 3133 line, which is probably fluorescently pumped by HeII 304 photons, does not show any variation, suggesting the presence of extended material surrounding AX Persei. The lack of eclipses in optical forbidden lines also indicates the presence of such circumstellar matter.

J. Mikolajewska et al. (eds.), The Symbiotic Phenomenon, 303–304.

304

In order to get some knowledge of the physical conditions in the nebula and their changes with time, the electron temperature, the electron density as well as the degree of ionization were analysed. The measured line ratios were corrected for reddening using the average interstellar extinction curve by Seaton (1979) with E(B-V)=0.27 (Kenyon 1986).

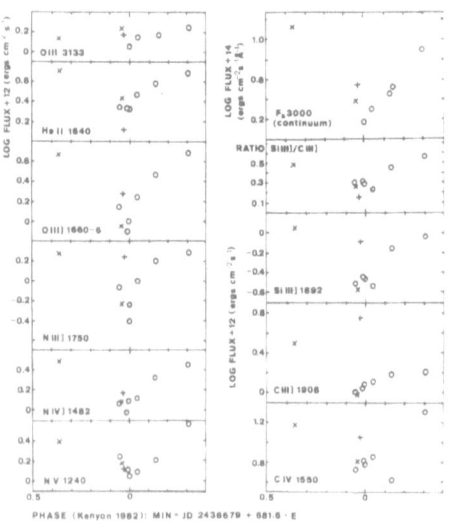

Figure 1

The NV 1240/NIV 1718, NIV 1486/NIII 2064 and CIV 1550/CIII 2297 ratios of fluxes in lines produced by collisional excitation to those produced by dielectronic recombination (Nussbaumer & Storey 1984) give the following values of electron temperature: T_e(NV) $\simeq$ 13000 K, T_e(NIV) $\simeq$ 11000 K, and T_e(CIV) $\simeq$ 11000 K. These temperatures suggest a photoionized region. No variation with time was found. The observed flux ratio NV 1240/CIV 1550 indicated $T_{ion} \simeq$ 120000 K and remained almost constant during the period studied, within the measurement errors, suggesting that the ionization temperature did not vary significantly.

Fig.1 also presents the phase dependence of the electron density sensitive ratio of SiIII] 1892/CIII] 1908 (Nussbaumer & Stencel 1987). The corresponding values of density are 2×10^{10} cm^{-3} at maximum and 4×10^9 cm^{-3} at minimum (cosmic relative abundance of Si/C was assumed). The [NeIV] 1601/2423 line ratio remained constant within the measurement errors and implied $n_e \simeq 10^8$ cm^{-3} for $T_e \simeq$ 13000 K (Nussbaumer 1982) close to the value derived from the optical forbidden lines.

More detailed results of this study will be published elsewhere.

I thank T. Valente from Trieste Astronomical Observatory for help in reduction of the IUE data.

REFERENCES:

Kenyon, S.J., 1982, *Publ. Astr. Soc. Pac.*, **94**, 165.
Kenyon, S.J., 1986, *The Symbiotic Stars*, Cambridge Univ. Press.
Mikolajewska, J., 1987, *Astrophys. Space Sci.*, **131**, 713.
Mikolajewska, J. & Iijima, T., 1987, *Acta Astr.*, **37**, in press.
Nussbaumer, H., 1982, in *The Nature of Symbiotic Stars*, M. Friedjung and R. Viotti eds., *D.Reidel*, p.85.
Nussbaumer, H. & Stencel, R.E., 1987, in *The Scientific Accomplishments of the IUE*, ed. Y. Kondo, *D. Reidel*, p. 203.
Nussbaumer, H. & Storey, P.J., 1984, *Astr. Astrophys. Suppl.* **56**, 293.
Seaton, M.J., 1979, *Mon. Not. R. astr. Soc.*, **187**, 73P.

IUE AND OPTICAL OBSERVATIONS OF HE 2-104

Julie H. Lutz
Program in Astronomy
Washington State University
Pullman, WA 99164-2930
USA

ABSTRACT. He 2-104 has been classified as both a planetary
nebula and as a symbiotic star. Optical and ultraviolet
spectra and CCD images have been obtained in order to learn
more about the evolutionary state of this object. The
spectra of the object indicate that it could be either a
symbiotic star or a very high excitation planetary nebula.
The CCD images show a nebulosity with a radius of about 4
seconds of arc and faint bipolar outer structures extending
out approximately 25 seconds of arc from the center of the
nebula.

1. THE OPTICAL AND ULTRAVIOLET SPECTRA

Spectroscopic observations with a resolution of about 10 A
were obtained of He 2-104 with the SIT Vidicon on the 1.5 m
telescope at CTIO in April 1984 and May 1985. The average
emission line fluxes are given in Table 1. To within the
errors of observation and calibration (about 25%) the
emission line fluxes did not vary in the two epochs of
observation.

The spectrum shows a very steep Balmer decrement and
the strong [O III] 4363 A emission (comparable in strength
to H-gamma) that are characteristic of high density
objects. [Fe VI] is observed in the spectrum, as it is in
some symbiotic stars and a few highly ionized planetary
nebulae (e.g. NGC 7027, M 1-1, He 2-111).

Ultraviolet observations were obtained with the
International Ultraviolet Explorer satellite on April 24
and 25, 1985. Large aperture exposures were made with both
the SWP (160 minutes) and LWP (60 minutes) cameras. The
ultraviolet line fluxes are given in Table 1. The emission
lines are, once again, typical of either a planetary nebula
or a symbiotic star, except that Mg II and Si III] emission

J. Mikolajewska et al. (eds.), The Symbiotic Phenomenon, 305–306.

lines are seen more commonly in symbiotic stars. There is
no evidence for continuum on the IUE spectra, even at the
shortest wavelengths.

CCD IMAGES

CCD images of He 2-104 were obtained with [0 III] and H-
alpha narrow band filters by using a TI chip on the 0.9 m
telescope at CTIO in May 1986. The images show a bright
nebula with a radius of about 4 seconds of arc and faint
bipolar outer structures extending about 25 seconds of arc
out from each side of the center of the nebula. Long
exposures of the faint bipolar loops reveal faint, clumpy
structures that are reminiscent of Herbig-Haro objects.

CONCLUSIONS

The nebulosity associated with He 2-104 would seem to put
the object into the classification of planetary nebula.
However, a 400 day Mira pulsation has been observed in the
infrared (Whitelock, private communication). This means
that there are two possibilities for He 2-104. It could be
one of the rare symbiotics (like R Aqr) that show
nebulosity. It could be a planetary nebula with a binary
nucleus where the cooler component happens to be in the
Mira state. Further studies of this unique object are
necessary.

Table 1

Wavelength A	Flux*	Wavelength A	Flux*	Wavelength A	Flux*
1240	sat.	3868	11.9	4861	13.6
1550	6.3	3889	4.6	4959	16.0
1640	4.5	3968	4.6	5007	43.2
1750	2.4	4072	0.8	5146	0.9
1890	1.2	4101	3.7	5176	1.5
1909	4.8	4340	6.8	5411	0.3
2732	0.5	4363	7.7	5755	0.8
2800	1.9	4471	0.8	5876	5.5
2837	0.8	4640	1.9	6300	2.0
3123	2.9	4686	3.3	6563	155.0
		4716	1.2	6584	12.7

$*10^{-10}$ erg cm^{-2} s^{-1}

SUMMARY OF DISCUSSION ON INDIVIDUAL OBJECTS

Discussion chairman Scott Kenyon
Discussion recorded by David Allen

1. HIGH VELOCITY WINDS FROM THE HOT COMPONENT

A number of contributors had suggested the existence of high-velocity winds which presumably originates in the hot components. One possible piece of evidence was broad wings to Lyα. Kenyon opened the discussion by asking whether these wings indeed arose in a wind or merely reflected scattering processes. The absence of broad wings on other lines, Kwok noted, does not disprove a wind. Viotti did, however, caution that Lyα will be self-absorbed and have scattering wings, so that it is not the best choice. Furthermore, as Selvelli added, even isolated M giants show broad wings to Lyα (Johanssen & Jordan, 1986, *Mon. Not. R. astr. Soc.*, **210**, 239) with a redward shift of about 1 Å, as seen in symbiotic stars. Schmid pointed out that Lyα could be scattered in a high-velocity neutral region illuminated by the ionized gas.

Turning to the anomalous C IV doublet ratios, Allen noted that a wind that was sufficiently optically thick to create a photosphere could explain the correlations seen in RX Pup and Z And between theλ1548/λ1550 ratio and the continuum brightness. Michalitsianos felt that there was a sound theoretical understanding of the effect of the wind on the doublet ratio, giving it the potential to be a good diagnostic of the wind. However, very few stars are bright enough to be studied with IUE.

2. BIPOLAR STRUCTURES

Kenyon steered the discussion to the radio morphologies. The CH Cygni radio jet suggested collimated outflow at high velocities. Seaquist reiterated that the data require expansion rather than the progressive ionization of pre-existing features, and asked whether a disk could produce the collimation. Slovak, noting the absence of a strong magnetic field to serve as a collimator, drew parallels with the T Tauri stars. In the symbiotic novae however, as Kenyon pointed out, little or no disk is likely to form from the wind: why do the radio observations still suggest bipolarity? Chochol proposed the formation of an excretion disk around the hot star, but this was not widely accepted as physically viable due to problems with conservation of angular momentum.

Rao conjectured that magnetic fields too weak to detect could be relevant, particularly in the presence of differential rotation. Webbink performed a back-of-envelope calculation to infer that fields of order 1 gauss were adequate. Livio

J. Mikolajewska et al. (eds.), The Symbiotic Phenomenon, 307–308.
© *1988 by Kluwer Academic Publishers.*

noted that fewer than 20% of planetary nebulae are spherical whereas there is little evidence that novae are bipolar (Friedjung challenged the latter claim), so that the cool star may be responsible. The SiO masers in VX Sgr have a bipolar distribution, Stencel added, thanks to a surface field of a few tens of Gauss.

The possibility was aired that the morphology is more apparent than real. Laminar sheets seen edge-on can appear bipolar, Slovak noted, but Friedjung reckoned an unacceptable proportion would have to be seen in this way. Both Viotti and Allen questioned whether we are merely recording two components, perhaps associated with the two stars. Viotti showed a radio map of α Sco in which this was exactly the situation.

3. ORBITAL PARAMETERS

Liebowitz raised a specific problem with the quoted mass ratio in AG Peg, which did not permit acceptable masses for the stars. Slovak felt that this was a case of underestimating the errors in the original determination by Cowley & Stencel (1973, *Astrophys. J.*, **184**, 687), and Webbink added that the velocity curves had not been corrected for reflection effects.

Turning to eccentricities, Liebowitz noted that Garcia's orbital data suggested small values, in contrast to non-interacting binaries of comparable orbital periods. Nussbaumer added that an attempt to explain V1329 Cyg using a large eccentricity had failed to work out. Webbink expected very small eccentricities among the symbiotics because the circularisation timescale is shorter than the evolution timescale for the stars in question. A thorn in the side of this argument, he admitted, is the existence of the barium stars (G-K giant plus probable white dwarf) which have $\bar{e} \sim 0.2$, and so have managed to interact and retain finite eccentricities.

SESSION 5. SYMBIOTIC PHENOMENA AND STELLAR EVOLUTION. CONCLUSIONS

"There are some dangers in reasoning"

Michael Friedjung

THE FORMATION AND EVOLUTION OF SYMBIOTIC STARS

R. F. Webbink
Department of Astronomy, University of Illinois
1011 W. Springfield Ave., Urbana, IL 61801, U.S.A.

ABSTRACT. The evolutionary origins of symbiotic stars containing (i) disk-accreting main sequence stars, (ii) wind-fed, shell-burning white dwarfs, and (iii) disk-accreting neutron stars are described. Of particular interest are those white dwarf systems which have orbital periods too short to have escaped tidal mass transfer prior to becoming symbiotics. We show here that, under suitable circumstances, low-mass, long period binaries may undergo quasi-conservative mass transfer, rather than evolving through common envelope evolution to the cataclysmic variable state, thus accounting for the existence of these systems. Approximate expressions are given for the lifetimes, and relative efficiencies (mass accreted/mass of donor) for different modes of interaction among symbiotic binary systems.

1. INTRODUCTION

It is now apparent that symbiotic stars constitute a very heterogeneous class of objects. At least three types of symbiotic stars may be identified according to the energy source for their hot components: (i) disk-accreting main sequence stars (e.g., CI Cyg: Kenyon, et al. 1982; Mikołajewska and Mikołajewski 1983; Mikołajewska 1985); (ii) wind-fed, shell-burning white dwarfs with extended envelopes (e.g., AG Peg: Gallagher, et al. 1979; Keyes and Plavec 1980); and (iii) disk-accreting neutron stars (V2116 Oph = GX 1+4: Cutler, Dennis, and Dolan 1986). Those of type (ii) can be further subdivided into (iia) systems of relatively short orbital period (P < 4 yr), in which the donor star is a normal late-type giant or bright giant (S-type symbiotics, as in the example, AG Peg, cited above); and (iib) systems of indeterminant, but undoubtedly long orbital period (P ≥ 15 yr), which generally show strong near-infrared flux excesses, and whose cool components are frequently long-period variables in their own right (D-type symbiotics: see Whitelock 1987). In turn, the existence of symbiotic stars of type (iia), which have orbital periods too short to escape tidal mass transfer in future, implies the existence of yet another class (iia'), in which the burning white dwarf is fed by disk accretion from a Roche-

311

J. Mikolajewska et al. (eds.), The Symbiotic Phenomenon, 311–321.
© *1988 by Kluwer Academic Publishers.*

lobe-filling cool companion. There is as yet no well-established
example of a type (iia') system (cf. Kenyon and Webbink 1984), but it
should be noted that nuclear burning on a typical white dwarf produces
~30 times as much energy per gram as does accretion, making the
presence of an accretion disk problematical in the presence of a
burning white dwarf. The status of the yellow (D'-type) symbiotic
stars (Glass and Webster 1973; Allen 1982) within this classification
scheme is not yet clear.

2. ORIGINS OF SYMBIOTIC STARS

It is evident that binary systems may arrive at a symbiotic state
by a number of evolutionary paths. Some of these are quite straight-
forward. For example, practically any binary with a low-mass
secondary, and a sufficiently large initial orbital separation to
permit the more massive primary to reach the giant branch, is a
candidate for a type (i) system (disk-accreting main sequence star), in
the preceding classification scheme. Given a larger orbital separation
yet, the more massive primary of such a system might evolve to a white
dwarf stage without ever undergoing tidal mass loss. Such a system
could become a type (iib) symbiotic as the secondary ascends the giant
or asymptotic giant branch (Tutukov and Yungel'son 1976).

Other symbiotic stars (types iia, iia', and iii) must be the
products of earlier phases of binary interaction, however. They have
orbital periods too short to have escaped such a fate as the
progenitors of their hot components evolved to the degenerate state.
This raises some interesting problems in understanding their
evolutionary history, because systems of similar total mass and angular
momentum have also been identified as the progenitors of cataclysmic
variables (Paczyński 1976; Ritter 1976; Webbink 1976). Those
progenitors must have passed through a phase of common envelope
evolution (Paczyński 1976; Meyer and Meyer-Hofmeister 1979) which
removed much of the mass, and most of the initial angular momentum of
the initial binary, leaving remnant systems with orbital periods of the
order of hours or days, rather than months or years. How is it
possible that the symbiotic stars escaped this fate?

The critical question for the occurrence of common envelope
evolution, as it is now understood, is the stability of the binary
against dynamical time scale mass transfer. If the mass ratio, M_1/M_2
(donor/accretor), at the onset of mass transfer exceeds some critical
value, q_{crit}, dynamical instability of the donor star develops
(Paczyński, Ziołkowski, and Żytkow 1969), leading to common envelope
evolution; otherwise, mass transfer proceeds quasi-conservatively,
i.e., in a manner which is at least approximately conservative of the
total mass and angular momentum of the binary. In turn, q_{crit} depends
on the structure of the donor star. Values of q_{crit} are not yet known
for detailed models of giant star donors, but the results for condensed
polytropes (Hjellming and Webbink 1987) are well-approximated

(for $m_c \gtrsim 0.2$) by

$$q_{crit} = 0.362 + \frac{1}{3(1 - m_c)} \quad , \tag{1}$$

where m_c is the fraction of the mass of the donor star contained within its degenerate core. This finding opens the possibility that binary systems of sufficiently long period and low mass to develop relatively large core masses before mass transfer may survive with their masses and angular momenta more or less intact (see Figure 1).

Can common envelope evolution itself give rise to symbiotic-like binaries? Adopting a very idealized model for this type of evolution (as described by Webbink 1984), we obtain the distribution of remnant systems illustrated in Figure 2. Among remnant systems with nondegenerate components still massive enough to evolve to the giant branch within the age of the Galaxy, none has an orbital period much exceeding one year, regardless of initial mass ratio. Since the distributions portrayed here optimistically assume perfect efficiency in converting orbital energy of the original binary to unbinding the envelope of the donor star, this remnant orbital period limit is itself undoubtedly an upper limit. Common envelope evolution thus fails to account for the long orbital periods of known symbiotic stars with hot white dwarf components.

Quasi-conservative evolution, on the other hand, succeeds in this regard, as illustrated in Figure 3. Two potential populations of symbiotic stars arise from this type of mass transfer. At larger mass and long orbital period, we find the remnants of intermediate mass binaries (initial primary masses of ~3-12 $M_\odot$) which encountered mass transfer while the initial primary was crossing the Hertzsprung gap (see Webbink 1979). At lower mass, a second population resulting from low mass systems of nearly equal initial masses and relatively long orbital period (as described above) appears. Because of the rapid increase in the initial mass function toward lower masses, it is undoubtedly this second population which should dominate among types (iia) and (iia') symbiotics. The orbital periods predicted for these systems on the assumption of conservation of total mass and angular momentum are somewhat longer than established orbital periods for symbiotics of this type, but this discrepancy disappears altogether if these systems lose one-third to one-half of their initial angular momenta during the course of mass transfer, as Algol binaries (which undergo this same mode of mass transfer) appear to do (Giuricin and Mardirossian 1981). Because the radii of the white dwarf progenitors in these systems are strongly dependent on their core masses, the white dwarf masses in these low-mass symbiotics should be strongly correlated with their binary periods.

The existence of V2116 Oph, a type (iii) system with a disk-accreting neutron star, strongly suggests that at least a few low-mass white dwarf symbiotics are able to avoid common envelope evolution

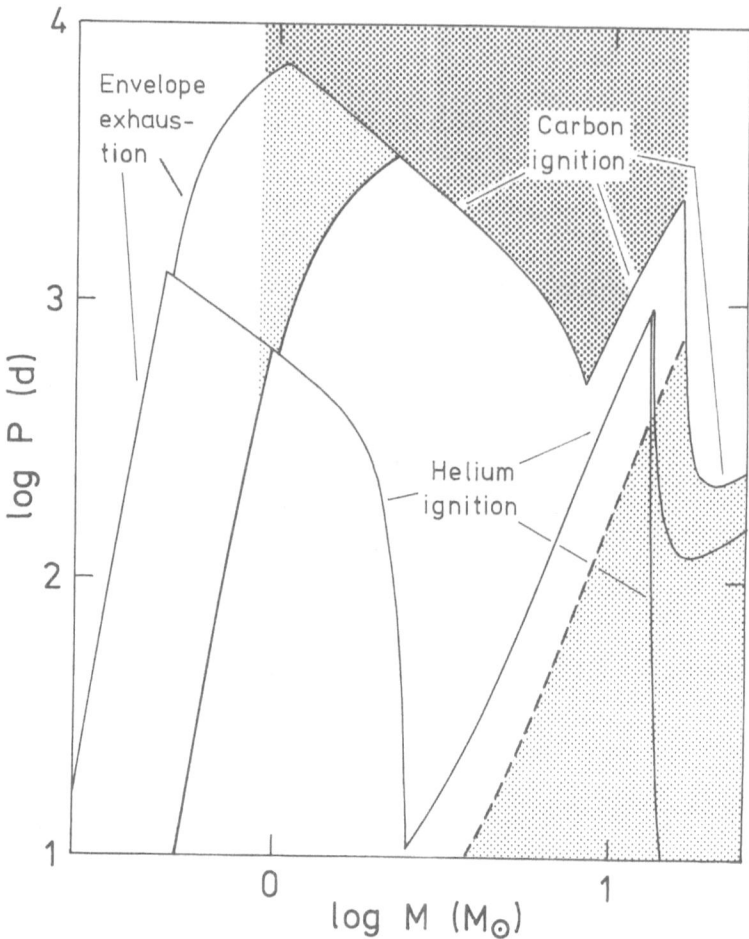

Figure 1. The orbital period-primary mass diagram for the progenitors
of symbiotic stars containing hot white dwarfs. The orbital periods of
binary systems with lobe-filling components at various critical phases
of their evolution are labeled in the diagram. The heavy dashed line
at lower right denotes the base of the giant branch; the heavy solid
line paralleling that labeled "envelope exhaustion" marks the limiting
period above which binaries of unit mass ratio are stable against
dynamical mass transfer. Lightly shaded regions denote systems of unit
mass ratio which undergo quasi-conservative evolution within 1.3×10^{10}
yr. Systems in the unshaded region between them undergo common
envelope evolution. The heavily shaded region at long period denotes
long-period systems which may contain late-type giants, but interact
only via stellar winds.

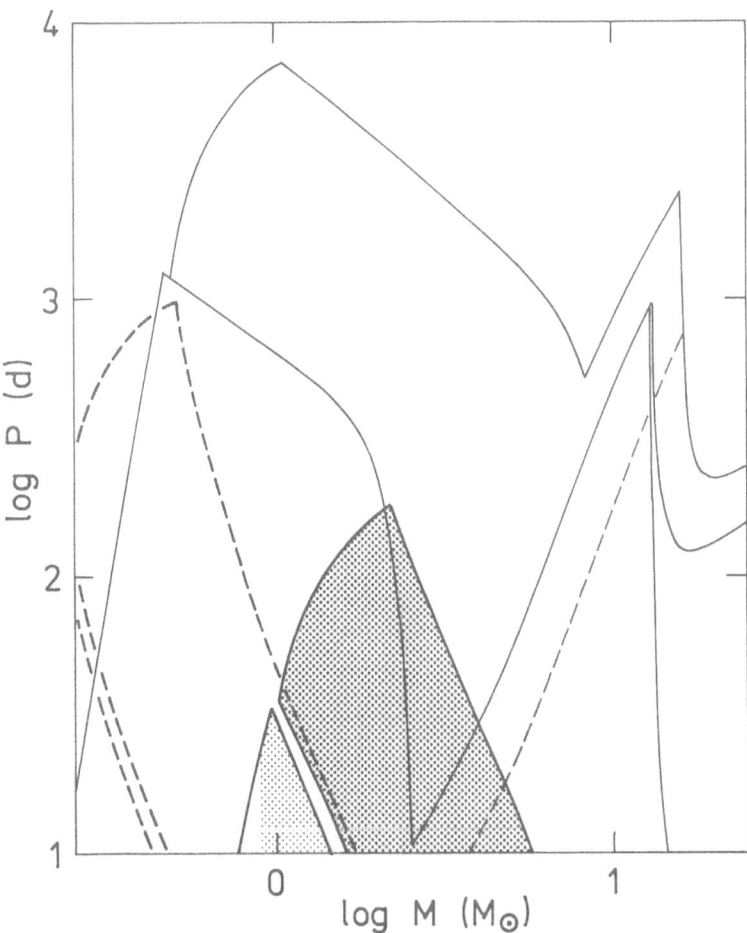

Figure 2. The period-mass diagram for remnants of common envelope evolution. Various critical periods are indicated as in Figure 1. The mass in this case is that of the non-degenerate component. Enclosed within the heavy solid lines are remnants of systems with unit initial mass ratios. The shaded regions mark systems which reach the second phase of mass transfer within 1.3×10^{10} yr: heavy shading, those with CO white dwarfs; light shading, those with He white dwarfs. The heavy dashed lines enclose the corresponding regions for binaries with initial mass ratios $M_1/M_2 = 2.5$.

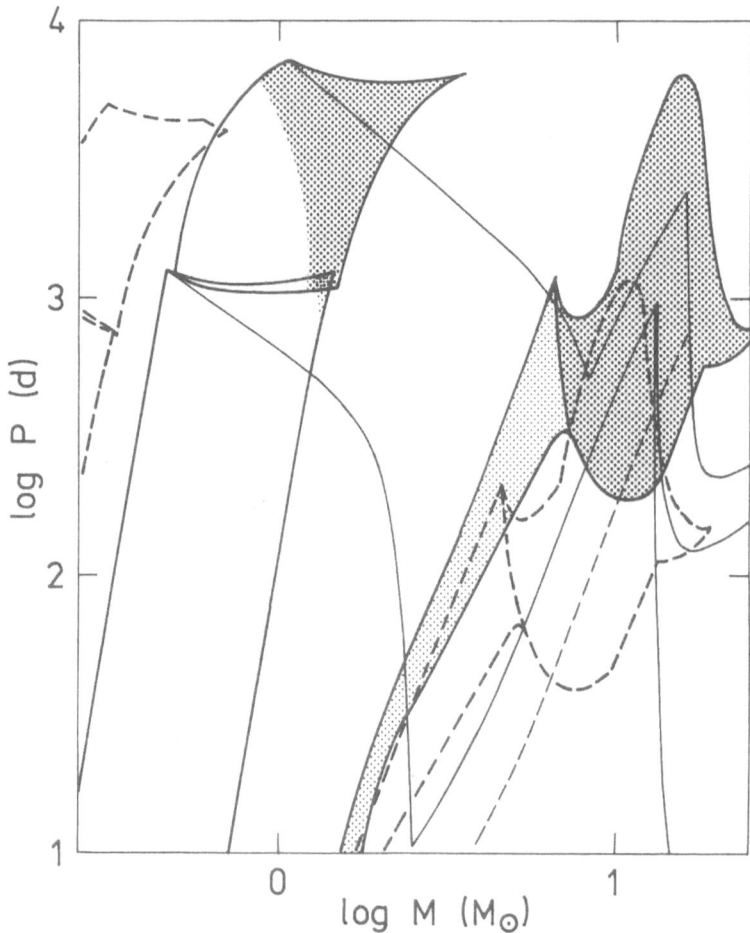

Figure 3.   The period-mass diagram for remnants of conservative mass transfer.   As in Figure 2, the mass is that of the non-degenerate component, the heavy solid lines enclose remnants of systems with unit initial mass ratios, the heavy dashed lines those of systems with initial mass ratios of 2.5.   Once again, the shaded regions mark systems which reach the second phase of mass transfer within 1.3×10^{10} yr: heavy shading, those with CO white dwarfs; light shading, those with He white dwarfs.

during their second phases of mass transfer as well. Taam and van den Heuvel (1986) argue that the high magnetic moment deduced for the X-ray pulsing neutron star in this system indicates that it is a very young neutron star, and therefore probably a product of accretion-induced collapse of a white dwarf. The progenitor of V2116 Oph is therefore itself likely to have been a type (iia'), disk-accreting white dwarf symbiotic.

3. LIFETIME OF THE SYMBIOTIC STATE

All of the varieties of symbiotic stars enumerated above involve interaction between the two components of a binary system, with the accretion of mass by the hot component via either stellar winds (iia, iib) or tidal mass transfer (i, iia', iii). The duration of the symbiotic phase should therefore reflect, at least crudely, the duration of this accretion phase. Moreover, the fraction of this time during which the hot component is sufficiently luminous to produce a recognizably symbiotic spectrum must depend as well on the amount of matter accreted by that component during the interaction lifetime. Both of these factors depend primarily on the properties of the donor star and the orbital parameters of the binary, and only very weakly on the nature of the accreting hot component.

Order of magnitude expressions are given in Table 1 for the durations (Δt) and relative efficiencies ($\Delta M_2 / M_1$ = fraction of the initial mass of the donor, star 1, accreted by the hot component, star 2) of each of these modes of interaction. In these expressions, P is the orbital period, τ_R the e-folding time scale for the growth of the donor star in radius due to nuclear evolution, τ_W the e-folding time scale for the growth in wind mass loss rate, m_c the fraction of the mass of the donor star contained within its degenerate core, $R_{1,max}$ the maximum radius achieved by the donor star, and A the orbital separation. In the case of wind accretion, the Bondi-Hoyle approximation has been adopted, with the wind velocity from the giant assumed equal to its escape velocity, and the contribution of orbital motion to the relative velocity of the accretor neglected (see Tutukov and Yungel'son 1976). In the case of tidal mass transfer, analytic solutions for the time-development of the mass transfer rate are given by Webbink and Iben (1987) for the early phases of both dynamical time

TABLE 1. MASS TRANSFER MODES

Type	Δt	$\Delta M_2 / M_1$
Pre-dynamical	$(\tau_R^2 \, P)^{1/3}$	$(P/\tau_R)^{1/3}$
Quasi-conservative	τ_R	$1 - m_c$
Wind	$(\tau_W^{-1} + \tau_R^{-1})^{-1}$	$\frac{1}{4} \, (M_2/M_1)^2 \, (R_{1,max}/A)^2$

scale and quasi-conservative mass transfer. Long term evolution in the quasi-conservative case is described by Webbink, Rappaport, and Savonije (1983).

If we adopt a simple parametric model of giant or asymptotic giant branch evolution, the expressions given in Table 1 can be rewritten in terms of the initial mass of the donor star, M_1, and the ratio, Π, of the orbital period, P, to the limiting period for tidal interaction, P_{RG} or P_{AG}. The resultant expressions, for donors of relatively low initial mass ($M_1 < 2.5\ M_\odot$), are listed in Table 2. Stellar winds have here been treated in Reimers' (1975) approximation, insofar as the wind accretion models and evaluation of P_{AG} are concerned; they have been neglected in evaluating the expressions for tidal mass transfer and P_{RG}.

TABLE 2. SYMBIOTIC STAR LIFETIMES

	Δt (yr)	$\Delta M_2 / M_1$
(a) *Red Giant Branch Donor* (critical period $P_{RG} = 640^d\ M_1^{-0.89}$)		
Pre-dynamical	$6 \times 10^4\ M_1^{-0.33}\ \Pi_{RG}^{-0.12}$	$0.005\ M_1^{-0.28}\ \Pi_{RG}^{0.56}$
Quasi-conservative	$1.1 \times 10^7\ M_1^{-0.05}\ \Pi_{RG}^{-0.68}$	$1.-0.46\ M_1^{-1.08}\ \Pi_{RG}^{0.14}$
Wind ($P/P_{RG} \equiv \Pi_{RG} < 1$)	$1.9 \times 10^6\ M_1^{-0.05}\ \Pi_{RG}^{-0.68}$	$0.010\ M_1^{-2.25}\ \Pi_{RG}^{0.99}$
Wind ($\Pi_{RG} > 1$)	$1.9 \times 10^6\ M_1^{-0.05}$	$0.010\ M_1^{-2.25}\ \Pi_{RG}^{-1.33}$
(b) *Asymptotic Giant Branch Donor* (critical period $P_{AG} = 2200^d\ M_1^{0.74}$)		
Pre-dynamical	$3 \times 10^4\ \Pi_{AG}^{0.33}$	$0.015\ \Pi_{AG}^{0.33}$
Quasi-conservative	2×10^6	$1.-0.62\ M_1^{0.59}\ \Pi_{AG}^{0.24}$
Wind ($P/P_{AG} \equiv \Pi_{AG} < 1$)	4×10^5	$0.04\ (1.-0.58\ M_1^{-0.69})\ \Pi_{AG}^{1.66}$
Wind ($\Pi_{AG} > 1$)	4×10^5	$0.04\ (1.-0.58\ M_1^{-0.69})\ \Pi_{AG}^{-1.33}$

N.B.: All masses are in solar units.

It is evident from Table 2 that the existence of even a small minority of symbiotic stars undergoing quasi-conservative Roche lobe overflow could dominate symbiotic star statistics. Such systems have not only significantly longer lifetimes than other modes of interaction, but also much higher efficiencies in accreting the envelopes of their companion stars. The absence of examples of type (iia') symbiotics suggests that conditions for this type of mass transfer rarely arise among those symbiotic stars containing hot white dwarfs which have previously undergone mass transfer. We should therefore expect that white dwarf (type ii) systems are primarily wind-fed, as observed, but with an appreciable minority of short-period systems in pre-dynamical Roche lobe overflow, and with a distinct preference for nearly-lobe-filling configurations even among wind-fed systems. Among main sequence star accretors (type i systems) on the other hand, quasi-conservative mass transfer may well occur in long-period, low-mass binaries of nearly equal initial masses (see Figure 1). If such systems transfer mass in bursts (and they achieve peak accretion rates in excess of $\sim 10^{-6}$ $M_\odot yr^{-1}$ — see Kenyon and Webbink 1984), they may constitute the majority of type (i) systems.

The remaining factor governing the detectability of symbiotic systems is of course the nature of the hot component. Hydrogen-burning white dwarfs typically liberate $\sim$3000 times as much energy per gram of accreted material as do non-burning, accreting main sequence stars. They are thus able to produce detectable nebular emission at much lower accretion rates (such as those occurring via stellar winds) than main sequence accretors, which require Roche lobe overflow (cf. Kenyon and Webbink 1984). Neutron stars release $\sim$30 times as much energy per gram of accreted material as hydrogen-burning white dwarfs, but their characteristic spectral energy distribution is so hard that they are no more efficient at producing hydrogen-ionizing photons than disk-accreting cold white dwarfs. Thus, they probably also require Roche lobe overflow to produce symbiotic spectra.

4. CONCLUSIONS

The above considerations regarding the origin and stability of symbiotic binaries lead to the following broad conclusions:

1. Long-period symbiotics (P > 15 yr) are powered by wind-accreting white dwarfs (Tutukov and Yungel'son 1976), having evolved to their present state without tidal interaction.

2. Symbiotics powered by accretion onto main sequence stars are necessarily shorter in orbital period, as they require Roche lobe overflow, and may occur over a wide range of initial conditions. However, those systems with orbital periods of $\sim$2-10 yr with initial mass ratios near unity may be very long-lived in the symbiotic state.

3. Short-period (P $\leq$ 15 yr) symbiotics containing hot white dwarf components are products of quasi-conservative mass transfer. Two populations may occur: (a) relatively massive systems ($M_{cool} \sim$ 4-10 $M_\odot$, P $\sim$ 100-1000 days); and (b) low-mass systems ($M_{cool} \sim$ 1-2 $M_\odot$, P $\sim$ 200-5000 days). Some low-mass systems containing massive white dwarfs may avoid common envelope evolution, surviving as symbiotic stars for a very long time ($\sim$2 x 10^6 yr) in a Roche lobe-filling state;

but such systems are evidently relatively rare. Except for these
systems, those in which interaction occurs via stellar winds should
dominate statistically, but it remains problematical why observed
systems of this type so often underfill their Roche lobes by a large
margin (see Kenyon and Gallagher 1983; Kenyon and Fernandez-Castro
1987; and discussions therein). The white dwarf masses among low-mass
systems should be correlated with their orbital periods.

This research was supported by National Science Foundation grants AST
83-17916 and AST 86-16992. I also thank Dr. W. Dziembowski, Director,
and the staff of the N. Copernicus Astronomical Center, Warsaw, where a
portion of this work was completed, for their gracious hospitality.

REFERENCES

Allen, D.A. 1982, in *IAU Colloquium No. 70, The Nature of Symbiotic
    Stars*, ed. M. Friedjung and R. Viotti (Dordrecht: Reidel), p. 27.
Cutler, E.P., Dennis, B.R., and Dolan, J.F. 1986, *Astrophys. J.*, 300,
    551.
Gallagher, J.S., Holm, A.V., Anderson, C.M., and Webbink, R.F. 1979,
    Astrophys. J., 229, 994.
Giuricin, G., and Mardirossian, F. 1981, *Astrophys. J. Suppl.*, 46, 1.
Glass, I.S., and Webster, B.L. 1973, *Monthly Notices R. Astr. Soc.*,
    165, 77.
Hjellming, M.S., and Webbink, R.F. 1987, *Astrophys. J.*, 318, 794.
Kenyon, S.J., and Fernandez-Castro, T. 1987, *Astr. J.*, 93, 938.
Kenyon, S.J., and Gallagher, J.S. 1983, *Astr. J.*, 88, 666.
Kenyon, S.J., and Webbink, R.F. 1984, *Astrophys. J.*, 279, 252.
Kenyon, S.J., Webbink, R.F., Gallagher, J.S., and Truran, J.W. 1982,
    Astr. Astrophys., 106, 109.
Keyes, C.D., and Plavec, M.J. 1980, in *IAU Symposium No. 88, Close
    Binary Stars: Observations and Interpretation*, ed. M. J. Plavec,
    D. M. Popper, and R. K. Ulrich (Dordrecht: Reidel), p. 365.
Meyer, F., and Meyer-Hofmeister, E. 1979, *Astr. Astrophys.*, 78, 167.
Mikołajewska, J. 1985, *Acta Astr.*, 35, 65.
Mikołajewska, J., and Mikołajewski, M. 1983, *Acta Astr.*, 33, 403.
Paczyński, B. 1976, in *IAU Symposium No. 73, Structure and Evolution
    of Close Binary Systems*, ed. P. Eggleton, S. Mitton, and J.
    Whelan (Dordrecht: Reidel), p. 75.
Paczyński, B., Ziołkowski, J., and Żytkow, A. 1969, in *Mass Loss from
    Stars*, ed. M. Hack (Dordrecht: Reidel), p. 237.
Reimers, D. 1975, *Mem. Soc. R. Sci. Liége*, 6ᵉ ser., 8, 369.
Ritter, H. 1976, *Monthly Notices R. Astr. Soc.*, 175, 279.
Taam, R.E., and van den Heuvel, E.P.J. 1986, *Astrophys. J.*, 305, 235.
Tutukov, A.V., and Yungel'son, L.R. 1976, *Astrofiz.*, 12, 521 (English
    transl.: 1977, *Astrophys.*, 12, 342).
Webbink, R.F. 1976, *Astrophys. J.*, 209, 829.

Webbink, R.F. 1979, in *IAU Colloquium No. 53, White Dwarfs and Variable Degenerate Stars*, ed. H. M. Van Horn and V. Weidemann (Rochester: U. Rochester Press), p. 426.

Webbink, R.F. 1984, *Astrophys. J.*, **277**, 355.

Webbink, R.F., and Iben, I., Jr. 1987, in *IAU Colloquium No. 95, The Second Conference on Faint Blue Stars*, ed. A.G.D. Philip, D.S. Hayes, and J. Liebert (Schenectady: L. Davis Press), in press.

Webbink, R.F., Rappaport, S., and Savonije, G.J. 1983, *Astrophys. J.*, **270**, 678.

Whitelock, P. 1987, *Publ. Astr. Soc. Pacific*, **99**, 573.

RECURRENT NOVAE

Mario Livio
Dept. of Astronomy, Univ. of Illinois, Urbana, IL 61801
and Dept. of Physics, Technion, Haifa 32000, Israel

ABSTRACT. An examination of the existing observational material on
all the objects classified as recurrent novae, reveals that only
T Pyx, U Sco, T CrB and RS Oph deserve such a classification. We
review the properties of outbursts powered by thermonuclear runaways
and accretion events and determine their observational consequences.
We conclude that the outbursts of T CrB and RS Oph are very probably
caused by accretion events from a giant companion onto a main
sequence star. The outbursts of T Pyx and U Sco may be powered by
thermonuclear runaways on the surface of white dwarfs with masses
close to the Chandrasekhar limit. However, some questions, in
particular regarding the composition of the accreted material in
U Sco remain to be answered.

1. INTRODUCTION

Recurrent novae (RN) are intermediate between classical novae (CN)
and dwarf novae (DN) in terms of their recurrence timescales ($\sim$
10-80 yr) and magnitude ranges in outburst ($\sim$ 7-11 mag). Before we
start discussing these systems, we would like to adopt the definition
proposed by Webbink, Livio, Truran and Orio (1987, hereafter WLTO) in
order to avoid ambiguity with CN and DN:
    RN are systems which exhibited two or more distinct outbursts,
reaching absolute magnitudes at maximum comparable with those of CN,
and in which the ejection of a discrete shell in outburst, at
velocities $V_{ej} \gtrsim 200$ km s^{-1}, has been observed.
    Based on this definition, WLTO were able to exclude a number of
systems, traditionally classified as RN from this group. These
include:
WZ Sge
    Now recognized as a DN of the SU UMa type (Patterson et al. 1981,
    Vogt 1981).
VY Aqr
    Had extremely frequent eruptions (e.g. in 1907, 1929, 1934, 1939,
    1940, 1942, 1958, 1962, 1964, 1965, 1966, 1967, 1973, 1983,

J. Mikolajewska et al. (eds.), The Symbiotic Phenomenon, 323–334.

1986, 1987, (see IAUC 4414 for the last one). Also, its
spectrum at minimum strongly resembles that of a DN. This
system is thus a DN.

V1195 Oph
The small amplitude of the outburst, its length, and the light
curve (Plaut 1968) are fairly convincing that this system is a DN.

V616 Mon
This object is really a soft x-ray transient, possibly containing
a black hole (McClintock and Remillard 1986).

RZ Leo
This system had confirmed outbursts in 1918 and 1984. Based on
the small amplitude of the eruption, the abrupt decline to minimum
and the spectrum during the decline (Cristiani et al. 1985), WLTO
concluded that this system is a DN, possibly similar to WZ Sge.

V 1017 Sgr
An examination of the light curves of this symbiotic system by
WLTO, revealed the fact that it is very probably a CN (which had a
nova-type outburst in 1919), which has also undergone DN eruptions
(in 1901 and 1973). This makes the system similar to V446 Her,
Q Cyg, V 3890 Sgr, Nova Vul (1979), WY Sge, GK Per and possibly BV
Cen (see Livio 1987 and references therein). The fact that the
system is a cataclysmic variable, coupled with the BVRI colors at
maximum, (which are consistent with the presence of an accretion
disk, Vidal and Rodgers 1974), suggest that the G5 IIIp star in
this system (Kraft 1964) fills its Roche lobe. Attempts for a
spectroscopic orbit determination are thus strongly encouraged.
The orbital period can be expected to be in the range 2-20 days.

V529 Ori
A study of the history of this object (Ashworth 1981) demonstrated
that there is only one reliable observation of the system. It is
therefore not a RN.

The above discussion leaves only four systems which fall under
our definition of RNe: T Pyx, U Sco, T CrB and RS Oph. Before
examining these systems in some detail we would like to discuss some
aspects of the physics of thermonuclear runaways (TNR) and accretion
events. This will help us to determine the applicability of
theoretical models to specific RN outbursts.

2. OUTBURSTS POWERED BY THERMONUCLEAR RUNAWAYS

The success of TNRs in explaining CN outbursts suggests the
possibility that a similar mechanism might be operating in RN. The
basic problem that has to be solved in this context is that of the
recurrence timescale, τ_{rec}, which is much shorter in RN than in CN.
The most important physical parameters determining τ_{rec} are the mass
of the white dwarf M_{WD}, the accretion rate $\dot{M}$ and to a lesser extent
the white dwarf's luminosity L_{WD} and the heavy element contents of
the accreted material, Z. Ignoring for the moment the effects which
may be introduced by variations in L_{WD} and in Z, TNRs occur when the
pressure at the base of the accreted envelope reaches a critical

value (Fujimoto 1982, MacDonald 1983, Truran and Livio 1986) of the
order of $P_{crit} \simeq 2 \times 10^{19}$ dyne cm^{-2}. The pressure at the white
dwarf-envelope interface is given by

$$P_{base} \approx \frac{GM_{WD} \, \Delta M_{acc}}{4\pi R_{WD}^4} \, , \tag{1}$$

where ΔM_{acc} is the mass of the accreted material. We thus obtain a
recurrence timescale of

$$\tau_{rec} \simeq 1.2 \times 10^4 \, (\frac{\dot{M}}{10^{-8} M_{\odot} yr^{-1}})^{-1} (\frac{M_{WD}}{M_{\odot}})^{-1} (\frac{R_{WD}}{6 \times 10^8 cm})^4 \text{ yr.} \tag{2}$$

Using a mass-radius relation for the white dwarf, we can, therefore,
obtain the recurrence time as a function of the white dwarf mass (for
a given accretion rate), this is shown in Fig. 1. Two things become
immediately apparent from the figure: (1) <u>accretion rates in excess
of 10^{-8} M$_{\odot}$/yr are required, in order to obtain the recurrence times
observed in RNe</u>, (2) for accretion rates typical to cataclysmic
variables (e.g. Patterson 1984), <u>the mass of the white dwarf must be
larger than 1.3 M$_{\odot}$</u>.
    In fact, the situation is further complicated by the fact that
for accretion rates above a certain critical value $\dot{M}_{weak}$ (which
depends on the white dwarf mass), TNRs are considerably weakened
because of the strong compressional heating which results in ignition
under only mildly degenerate conditions (Kutter and Sparks 1980,
Prialnik et al. 1982). For $\dot{M} > \dot{M}_{weak}$, no mass ejection is obtained
and thus, such outbursts do not satisfy our definition of RNe. The
exact value of $\dot{M}_{weak}$ depends somewhat on L_{WD} and Z. In Fig. 1 we
have plotted recurrence times as a function of M_{WD}, based on average
values of $\dot{M}_{weak}$ (the dashed curve) taken from numerical results.
From this plot we see that in order to obtain recurrence times
shorter than 50 years, <u>the mass of the white dwarf must be close to
the Chandrasekhar limit</u> (and $\dot{M} \gtrsim 1.7 \times 10^{-8}$ M$_{\odot}$/yr).
    The above constraints on the white dwarf mass and the accretion
rate have important consequences for the luminosity obtained from TNR
powered models. The maximum luminosity obtained from shell burning
is always comparable to that given by Paczynski's (1971) core mass-
luminosity relation. Since Paczynski's luminosity is very close to
the Eddington limit for very massive white dwarfs (the two
luminosities cross at $M_{WD} \sim 1.39$ M$_{\odot}$), we can predict that the maximum
luminosity of a TNR powered RN should be

$$L_{max} \gtrsim L_{EDD} \simeq 5.2 \times 10^4 \, L_{\odot} \, (\frac{M_{WD}}{1.38 \, M_{\odot}}) \tag{3}$$

At quiescence the bolometric luminosity of the system must exceed the accretion luminosity, thus

$$L_{BOL} \gtrsim \frac{GM_{WD}\dot{M}}{R_{WD}} \approx 270 \ L_\odot \ (\frac{M_{WD}}{1.38 \ M_\odot}) \ \frac{\dot{M}}{(1.7x10^{-8}M_\odot yr^{-1})} \ (\frac{R_{WD}}{1.9x10^8 cm})^{-1} . (4)$$

Furthermore, if accretion takes place via a disk, then we can use the calculations of Warner (1987) and of WLTO to obtain the absolute visual magnitude M_v. We find that TNR powered RNe, must occupy at quiescence the region curve the upper curve in Fig. 2.

The exact numerical values can change somewhat, depending for example on the white dwarf luminosity. A large number of numerical simulations of TNRs on the surface of a 1.38 $M_\odot$ white dwarf, have shown that the dependence of the recurrence timescale on L_{WD} can be fitted approximately by the relation $\tau_{rec} \sim L_{WD}^{-0.28}$ (Livio, Hayes and Truran 1987).

2. OUTBURSTS POWERED BY ACCRETION

Accretion powered outbursts can occur in principle in (at least) three different ways. (i) An instability associated with the mass losing component, causing bursts of mass transfer. (ii) An instability in the accretion disk. (iii) A time dependent accretion rate, modulated by orbital eccentricity. Unfortunately all of these mechanisms suffer from large uncertainties, thus making their predictive power rather uncertain.

Dynamical instability of the red component has been particularly emphasized by Bath (1969, 1972, 1975) as a mechanism for dwarf nova eruptions. Recent two-dimensional numerical calculations by Edwards and Pringle (1987) have shown that short bursts of mass transfer are indeed obtained this way, for a polytropic equation of state with n = 3/2. While it is not clear yet from these calculations, whether the amount of mass transferred by low mass main sequence stars is sufficient to power DN eruptions, lobe-filling giants are found to transfer ~ 10^{-4} $M_\odot$ (see discussion of T CrB below). These models are not sufficiently developed to predict the recurrence time of the outbursts. All that can be said at this point is that τ_{rec} should be longer than the timescale for the envelope to regain its thermal equilibrium, thus (for parameters appropriate for T CrB)

$$\tau_{rec} \gtrsim \frac{R_{gas} \ T \ \Delta M_t}{\Delta L} \approx 6.3 \ (\frac{T}{10^5 K}) \ (\frac{\Delta M_t}{6x10^{-3} \ M_\odot}) \ (\frac{\Delta L}{0.1L})^{-1} yr \qquad (5)$$

where ΔM_t is the mass that has been affected by the instability and ΔL is the luminosity deficit obtained in the mass transfer process.

Disk instability models produce a limit cycle behavior in which the disk undergoes transitions between a low and high viscosity states (e.g. Meyer and Meyer-Hofmeister 1984, Smak 1984, Lin, Papaloizou and Faulkner 1985, Cannizzo, Wheeler and Polidan 1986 and references therein). These transitions are a consequence of the double-valued nature of the viscosity - surface density functional relation. Matter in the disk accumulates, until the surface density Σ at some radius exceeds a certain critical value, at which point that annulus becomes thermally unstable, subsequently dragging the entire disk to the high state. An important property shared by the models of all the different groups working on disk instabilities, is that above a certain critical accretion rate, $\dot{M}_{crit}$, the disk lies on the stable (high-temperature) branch of the T_{eff}-Σ curve. Thus, eruptions caused by a disk instability can be expected only if the accretion rate (onto a 1 $M_\odot$ white dwarf) satisfies (e.g. Shafter, Wheeler and Cannizzo 1986)

$$\dot{M} \lesssim \dot{M}_{crit} \approx 3 \times 10^{-9} \, P_4^{1.8} \, M_\odot/yr \qquad (6)$$

where P_4 is the orbital period (in units of 4 hours). If this condition is translated to an absolute visual magnitude at quiescence, it requires that RNe powered by a disk instability, should occupy at quiescence only the region below the lower curve in Fig. 2.

Unfortunately, the recurrence intervals that can be obtained in the disk instability model are very uncertain, because they depend quite sensitively on the unknown viscosity parameter in the cold state α_c. Similarly, the outburst duration is largely determined by α_H, the viscosity parameter in the hot state. In general, the recurrence interval is given by

$$\tau_{rec} \simeq \frac{\Delta M_{outburst}}{\dot{M}_s} \qquad (7)$$

where $\Delta M_{outburst}$ is the mass removed during the outburst and $\dot{M}_s$ is the mass transfer rate from the secondary. A very rough estimate for τ_{rec} can be obtained by taking the viscous drift timescale in the annulus at which the instability starts. This gives (Lin and Shields 1986, Cannizzo, Shafter and Wheeler 1987),

$$\tau_{rec} \simeq 8.8 \times 10^7 \, [F(q)]^{0.61} \, (\frac{M_{WD}}{M_\odot})^{0.67} \, P_4^{0.41} \, (\frac{\alpha_c}{0.1})^{-1.1} sec \qquad (8)$$

where $F(q)$ is a function of the mass ratio $q = M_s/M_{WD}$ given by (Eggleton 1983)

$$F(q) = \frac{0.49q^{2/3} (1+q)^{1/3}}{0.6q^{2/3} + \ln(1+q^{1/3})} \qquad (9)$$

and we have assumed that the instability starts at the outer disk edge. The timescale found in actual calculations by Cannizzo, Wheeler and Polidan (1986), is shorter than the one given by eq. (8) by a factor of about 30, due to the fact that only a small fraction of the disk's mass is accreted during an outburst. Their results show that the recurrence time is proportional to $\alpha_c^{-1.23}$.

The decay timescale of the outburst is given in the disk instability model by (Smak 1984, Cannizzo, Shafter and Wheeler 1987)

$$\tau_d \simeq 38.3 \ [F(q)]^{0.45} \ (\frac{M_{WD}}{M_\odot})^{0.67} \ P_4^{0.3} \ (\frac{\alpha_H}{0.1})^{-0.8} (\frac{\delta r/r}{0.1}) \ \text{days} \quad (10)$$

where $\delta r/r$ is the fractional width of the cooling front which transfers the disk back into the cold state (typically of order 0.1). From eq. (8) and the following discussion we see that for disk instabilities to produce outbursts on recurrence times typical to RNe, the viscosity parameter in the cold state must be very low, $\alpha_c \lesssim 10^{-3}$.

Finally, an orbital eccentricity (if it exists) cannot be the only cause for the outbursts of the four RNe, since for two of them (T CrB and RS Oph) the orbital parameters are known. Also, the outbursts are not periodic. Eccentricity may play a certain minor role in the outbursts of T CrB (see WLTO).

3. INDIVIDUAL SYSTEMS

We shall now review some of the properties of the four RNe and examine them in relation to possible outburst models. An extensive discussion can be found in WLTO, here we shall add some recent developments.

T Pyx

This relatively regular RN has undergone outbursts in 1890, 1902, 1920, 1944 and 1966 (thus it can experience an outburst any day now!). Its corrected colors are $(B-V)_o = -0.26$, $(U-B)_o = -1.25$ and $(V-R)_o = -0.11$ (WLTO and references therein). Distance estimates based on the equivalent width of the interstellar calcium K line (Catchpole 1969) and the fact that the slow outburst development indicates a luminosity at maximum not much exceeding the Eddington value, give: $1050\text{pc} \lesssim D \lesssim 4500$ pc. The absolute visual magnitude at minimum is thus $(M_v)_{min} = 2.4 \pm 1.6$, which, when combined with the colors leads to the conclusion that the system cannot contain a red giant (making the system different from T CrB and RS Oph, to be discussed below). This limits the orbital period of the system to $P_{orb} \lesssim$ days (rather than hundreds of days), and makes it similar to cataclysmic variables. We can compare some of the details of the observations at quiescence and at outburst with the predictions of TNR models and accretion events. The main points of agreement with

TNR models are: (1) The light curve development and the appearance of several velocity systems are very similar to slow novae, (2) on the rise to maximum the broadband colors become redder, characteristic of an expanding photosphere, (3) the mass of the ejected shell $\sim 10^{-6}$ $M_\odot$ is consistent with the requirements of TNR models. The main difficulty for a TNR model is presented by the colors at minimum, which are much bluer than expected from an accretion disk. WLTO suggested that nuclear burning continues in T Pyx even at minimum, producing a bolometric luminosity $L_{BOL} \sim 3 \times 10^3$ $L_\odot$ accompanied by a strong reflection effect from the secondary. It is not yet entirely clear whether the outburst characteristics can be reproduced under such conditions, this problem is currently investigated (Livio, Hayes and Truran 1987).

On the other hand, it is quite clear that an accretion event model, similar to DN eruptions, encounters severe difficulties: (1) In DN outbursts the broadband colors become bluer, in particular near the peak. (2) In DNe the rise to maximum spans only a few orbital periods in contrast to the slow rise of T Pyx. (3) The dominance of a hot continuum source at minimum indicates that accretion at a high rate (which would make the disk stable) is ongoing.

We therefore conclude that the outbursts of T Pyx are probably powered by TNRs, although some unanswered questions still remain.

U Sco

The observed properties of this system led WLTO to conclude that this system consists of a white dwarf accreting through a hot accretion disk from a G3-6 III-IV companion. The success of Starrfield, Sparks and Truran (1985) to reproduce the gross outburst characteristics, by means of a TNR resulting from accretion of hydrogen rich material, at $\dot{M} \simeq 1.7 \times 10^{-8}$ $M_\odot/yr$, onto a 1.38 $M_\odot$ white dwarf, has further led WLTO to the suggestion that U Sco is probably powered by a TNR. The main difficulty encountered by disk instability models has been in the fact that the hot disk, implied by the observations of U Sco at quiescence, would be expected to lie on the stable branch.

Some recent developments, however cause us to re-open the question of the outburst mechanism of U Sco. (1) U Sco was detected in outburst again in May 1987 (IAUC 4395, 4396, 4397, 4399, 4405), only eight years after its previous outburst. This poses serious difficulties to TNR models, since it requires an extremely high accretion rate (see discussion in section 2). (2) The spectrum of the system at quiescence shows no hydrogen and is dominated by a helium emission line spectrum (Hanes 1985). In outburst, optical and UV observations also reveal a strong depletion of hydrogen with respect to helium, He/H $\sim$ 8:1 by mass (Barlow et al. 1981, Williams et al. 1981). In a recent work, Truran et al. (1988) have attempted to produce TNRs for different values of H/He ratios. They found that visually bright outbursts are not obtained for H/He $\leq$ 1 (by mass). This is simply a consequence of the fact that the ratio of energy generated per gram in hydrogen burning, to the binding energy per gram is

$$\frac{\epsilon_{nuc}}{\epsilon_{bind}} \simeq 6.5 \, X_H \, (\frac{M_{WD}}{1.38 \, M_\odot})^{-1} \, (\frac{R_{WD}}{1.9 \times 10^8 \, cm}) \, , \tag{11}$$

where X_H is the hydrogen mass fraction. Thus, for too low values of X_H, the photosphere does not expand to $\sim 10^{12}$ cm, which is essential in order for the outburst energy to come out in the visual.

The outburst mechanism of U Sco is thus not clear yet. Possible solutions to the outlined difficulties are:

(1) The outburst is caused by a TNR, but for a yet unknown reason, the hydrogen in the transferred material is not detected (after all hydrogen is detected in outburst).

(2) The outburst is caused by a disk instability, but the transferred material is very rich in helium. In such a case, due to the higher ionization temperature, even the hot disk implied by observations can become unstable (Smak 1983). This solution poses some difficulties from an evolutionary point of view (WLTO).

(3) The outburst is caused by a TNR, but the physics of the TNR development has to be modified, to allow the transport of most of the energy generated (e.g. by more efficient convection), to be deposited into only a small fraction of the envelope.

T CrB and RS Oph

These two systems are very similar, as can be seen from Table 1 and represent the short period end of S-type symbiotics. Their properties have been extensively discussed by Webbink (1976), Livio, Truran and Webbink (1986), WLTO and Garcia (1987). The models proposed by Webbink and by Livio, Truran and Webbink for the outbursts of T CrB and RS Oph are essentially identical and involve the episodical transfer of a chunk of matter of $\Delta M \sim 10^{-4}$ - 10^{-3} $M_\odot$ from the red giant onto an accreting main sequence star. The differences between the two systems, according to the model of Livio et al. (1986), arise simply because of the difference in the evolutionary status of the accreting star. Due to the higher effective accretion rate in the RS Oph system and the larger total accreted mass, the main sequence star accretor is assumed to be in a bloated configuration ($R \sim 12 \, R_\odot$). Consequently, the stream of matter from the giant impacts the stellar surface directly, resulting in one outburst. In T CrB the stream passes around the accreting star, either skimming its surface or colliding with itself and circularizing (thus producing the first maximum). In either case, a disk is formed subsequently and the decay of this disk produces the second maximum. A recent calculation by Edwards (1987, private communication) has shown that for a giant of the type found in T CrB, mass of the order of 10^{-4} $M_\odot$ will indeed be transferred in a mass transfer instability.

The main difference between the outbursts of T CrB and RS Oph and those of other symbiotic systems also believed to be powered by accretion events, is in the very dynamic, shock-type outbursts of RS Oph and T CrB (see Livio, Webbink and Truran 1986). This is very probably a consequence of the shorter orbital period, since the luminosity of the principal outburst can be expected to scale like L_0 ~ $\Delta E_{kin}/\tau_{dyn}$ ~ $P_{orb}^{-5/3}$ (the available kinetic energy is inversely proportional to the binary separation).

4. CONCLUSIONS

An examination of the observational material suggests that only four systems: U Sco, T Pyx, T CrB and RS Oph deserve a classification as recurrent novae.

The outbursts of T CrB and RS Oph are probably caused by accretion events initiated by bursts of mass transfer from giant companions onto main sequence stars. Thermonuclear runaways on the surface of very massive white dwarfs appear as a likely cause for the outbursts of U Sco and T Pyx, but a number of points have to be clarified, in particular the composition of the accreted material in U Sco, before final conclusions can be drawn.

As a final remark we would like to point out that <u>the outbursts of RNe are expected to exhibit a smaller range than CNe</u>. This is a consequence of the fact that either the presence of a giant (required for the accretion model) or the presence of a very massive white dwarf accreting at a high rate (required for the TNR models) result in a brighter system at minimum.

ACKNOWLEDGEMENT

This work has been supported in part by NSF grant AST 86-11500 at the University of Illinois. I would like to thank the LOC and in particular J. Mikolajewska for their support.

References

Ashworth, W. B. Jr. 1981, <u>Quart. J.R.A.S.</u>, 22, 22.
Barlow, M. J., Brodie, J. P., Brunt, C. C., Hanes, D. A., Hill, P. W., Mayo, S. K., Pringle, J. E., Ward, M. J., Watson, M. G., Whelan, J. A. J. and Willis, A. J. 1981, <u>M.N.R.A.S.</u>, 195, 61.
Bath, G. T. 1969, <u>Ap. J.</u>, 158, 571.
Bath, G. T. 1972, <u>Ap. J.</u>, 173, 121.
Bath, G. T. 1975, <u>M.N.R.A.S.</u>, 171, 311.
Cannizzo, J. K., Wheeler, J. C. and Polidan, R. S. 1986, <u>Ap. J.</u>, 301, 634.
Cannizzo, J. K., Shafter, A. W. and Wheeler, J. C. 1987, preprint.
Catchpole, R. M. 1969, <u>M.N.R.A.S.</u>, 142, 119.
Cristiani, S., Duerbeck, H. W. and Seitter, W. C. 1985, IAUC 4027.
Edwards, D. A. and Pringle, J. E. 1987, preprint.
Eggleton, P. P. 1983, <u>Ap. J.</u>, 268, 368.
Fujimoto, M. Y. 1982, <u>Ap. J.</u>, 257, 767.
Garcia, M. R. 1986, <u>Astron. J.</u>, 91, 1400.
Hanes, D. A. 1985, <u>M.N.R.A.S.</u>, 213, 443.

Kraft, R. P. 1964, Ap. J., 139, 457.

Kutter, G. S. and Sparks, W. M. 1980, Ap. J., 239, 988.

Lin, D. N. C., Papaloizou, J. and Faulkner, J. 1985, M.N.R.A.S., 212, 105.

Lin, D. N. C. and Shields, G. A. 1986, Ap. J., 305, 28.

Livio, M. 1987, Comments on Ap., in press.

Livio, M., Truran, J. W. and Webbink, R. F. 1986, Ap. J., 308, 736.

Livio, M., Hayes, J. and Truran, J. W. 1987, in preparation.

MacDonald, J. 1983, Ap. J., 273, 289.

McClintock, J. E. and Remillard, R. A. 1986, Ap. J., 308, 110.

Meyer, F. and Meyer-Hofmeister, E. 1984, Aston. Ap., 132, 143.

Paczynski, B. 1971, Acta. Astron., 21, 417.

Patterson, J. 1984, Ap. J. Suppl., 54, 443.

Patterson, J., McGraw, J. T., Coleman, L. and Africano, J. L. 1981, Ap. J., 248, 1067.

Plaut, L. 1968, Bull. Astr. Inst. Netherlands Suppl., 3, 1.

Prialnik, D., Livio, M., Shaviv, G. and Kovetz, A. 1982, Ap. J., 257, 312.

Shafter, A. W., Wheeler, J. C. and Cannizzo, J. K. 1986, Ap. J., 305, 261.

Smak, J. 1983, Acta Astron., 33, 333.

Smak, J. 1984, Acta Astron., 34, 161.

Starrfield, S., Sparks, W. M. and Truran, J. W. 1985, Ap. J., 291, 136.

Truran, J. W. and Livio, M. 1986, Ap. J., 308, 721.

Truran, J. W., Livio, M., Hayes, J., Starrfield, S. and Sparks, W. M. 1988, Ap. J., in press.

Vidal, N. V. and Rodgers, A. W. 1974, P.A.S.P., 86, 26.

Vogt, N. 1981, Ap. J., 252, 653.

Warner, B. 1987, M.N.R.A.S., in press.

Webbink, R. F. 1976, Nature, 262, 271.

Webbink, R. F., Livio, M., Truran, J. W. and Orio, M. 1987, Ap. J., 314, 653.

Williams, R. E., Sparks, W. M., Gallagher, J. W., Ney, E. P., Starrfield, S. G. and Truran, J. W. 1981, Ap. J., 251, 221.

NOTE ADDED IN PROOFS:

Spectral observations obtained by Sekiguchi et al. (1987, private communication) of the 1987 outburst of U Sco generally support the TNR model for the outburst. However, a similar overabundance of helium over hydrogen as seen in the previous outburst (see text) has been observed.

Observations of T CrB during quiescence (Selvelli, Cassatella and Gilmozzi, private communication) revealed a UV luminosity of the order of 2×10^{35} erg s^{-1}. This would require an accretion rate of the order of 10^{-6} $M_\odot$/yr onto a main sequence star. These authors regard this, and the absence of the CIV line in their high resolution spectrum as evidence for the presence of a white dwarf.

TABLE 1

A comparison of properties of T CrB and RS Oph

Property	T CrB	RS Oph
Orbital Period	227.5 days	230 days
Companion	M3 III giant	M0-M2 III giant
Ejection Velocity at Maximum	~ 5000 km/sec	~ 3500 km/sec
Evolution of Light Curve	very rapid	very rapid
Spectrum During Late Decline	High excitation forbidden emission line spectrum	High excitation forbidden emission line spectrum
τ_{rec}	80 years	22 years
Secondary Maximum	Present	Absent
$M_{hot} \sin^3 i$	$1.80 \pm 0.20\ M_{\odot}$	Unknown

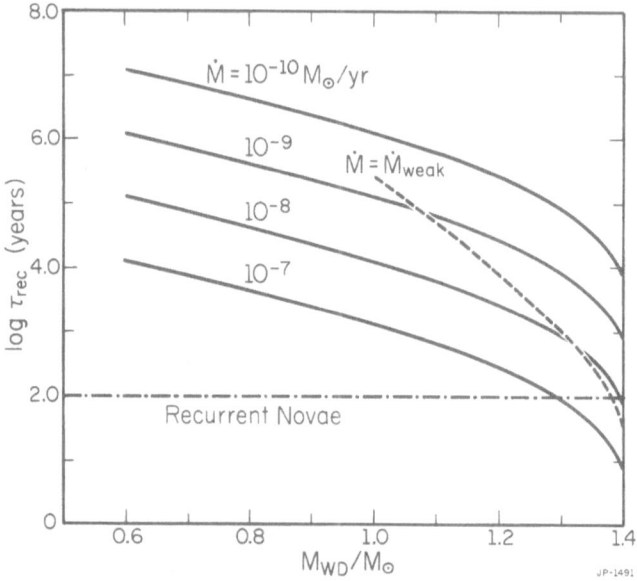

Fig. 1: The recurrence time as a function of the white dwarf mass. The dashed line gives the recurrence time for the maximum accretion rate capable of producing a strong TNR.

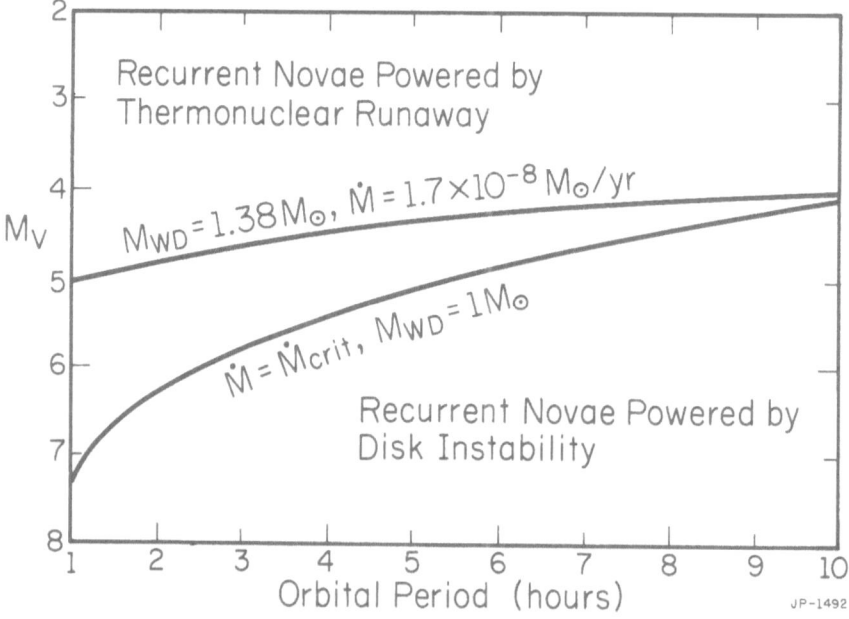

Fig. 2: The visual magnitude of the accretion disk at quiescence as a function of the orbital period. Recurrent novae powered by TNRs or disk instability must occupy the marked regions.

UNRAVELLING THE MULTIPLE COMPONENT RADIO EMISSION OF RS OPH IN OUTBURST

A.R. Taylor[1], M.F. Bode[2], R.J. Davis[1] and R.W. Porcas[3]
1. NRAL, Jodrell Bank, Nr. Macclesfield, SK11 9DL, UK
2. School of Physics and Astronomy, Lancashire Polytechnic, PR1 2TQ, UK.
3. MPIfR, 5300 Bonn 1, FRG.

ABSTRACT VLBI observations of RS Ophiuchi during its 1985 outburst enable us to start to disentangle the multiple component radio emission. It appears that the emission at low frequencies was predominantly non-thermal, whereas at high frequencies the emission is most likely thermal. It is then possible to probe the physical properties of the remnant.

1. INTRODUCTION

RS Ophiuchi is a recurrent nova which has undergone 5 recorded outbursts, the latest of which was on 1985 January 26th ($t=t_0$). Radio observations showed a rapid rise in flux to a peak around 1 month from outburst (Hjellming et al 1986). Radio spectra showed the presence of at least two components, one at low frequencies with spectral index typically -0.08, and another at high frequencies with index of order $+0.5$.

2. OBSERVATIONS AND RESULTS

Single baseline (Jodrell-Effelsberg) observations at 4.9GHz were carried out on 1985 March 8th (day 40). A good fit to the limited data can be obtained with a 30mJy source of angular size 8m.a.s.. As the total 4.9GHz flux on day 40 was 60mJy, this suggests a possible core-halo structure. Far more information on the source geometry was derived from 1.7GHz observations on April 13th (day 77) using data from the Jodrell, Effelsberg and Westerbork telescopes. We have re-analysed the visibility data, but essentially derive a best fit source geometry very similar to that given in Porcas et al (1987). This consists of a dominant central source and two symmetrically placed weaker sources, forming a linear structure of total extent around 200m.a.s. at approximately position angle 90°.

3. DISCUSSION AND CONCLUSIONS

The brightness temperature of the radio central source on day 77 is 6×10^{6}K. This implies brightness temperatures on day 55, when the first EXOSAT data were obtained (Mason et al 1987), in excess of 10^{7}K. If the radio and X-ray emission arise from the same mass of gas, such high thermal temperatures would give rise to around 100

335

J. Mikolajewska et al. (eds.), The Symbiotic Phenomenon, 335–336.

336

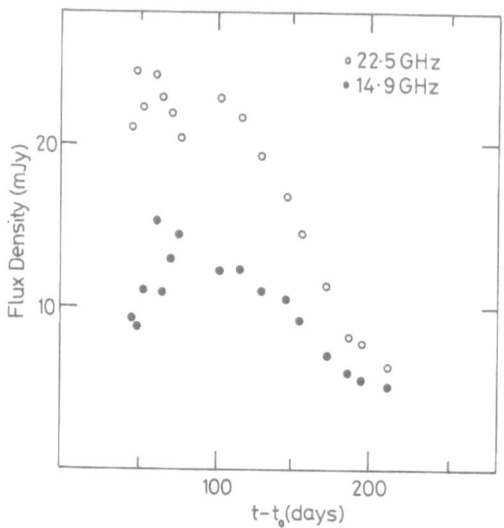

Fig. 1. Thermal radio light curves (see text for details).

times the X-ray flux observed. Thus the low frequency flux at early times must be non-thermal.

A thermal interpretation of the high frequency emission is permitted as $T_b \sim 10^5 K$ here. Figure 1 shows the light curves which result by subtracting the non-thermal component, assumed to have spectral index -0.08 and to dominate the flux at 1.5GHz, from the high frequency fluxes. The results are now similar to the radio development of classical novae. These light curves also resemble those of the coronal lines (e.g. [FeXI], Snijders 1987). Simple arguments lead to an emitting mass of gas around $2 \times 10^{-6} M_\odot$, with kinetic energy of order 2×10^{43} ergs. The equipartition field strength of 52mG in the central component is similar to that required by Bode and Kahn (1985) to explain anisotropic ejection. The efficiency of conversion of the outburst energy into relativistic electrons and enhanced fields (around 0.1%) is similar to that in young supernova remnants (Reynolds and Chevalier 1984). A full discussion of this work will be given in Davis et al (1987, in preparation).

REFERENCES

Bode, M.F., and Kahn, F.D., 1985, Mon. Not. R. Astr. Soc., _217_, 205.
Hjellming, R.M., et al, 1986, Ap.J. (Lett.), _305_, L71.
Mason, K.O. Cordova, F.A., Bode, M.F., and Barr, P., 1987, in _RS Oph (1985) and the Recurrent Nova Phenomenon_, ed Bode (VNU Sci. Press), p167.
Porcas, R.W., Davis, R.J., and Graham, D.A., 1987, in _RS Oph (1985) and the Recurrent Nova Phenomenon_, ed Bode (VNU Sci. Press), p203.
Reynolds, S., and Chevalier, R.A., 1984, Ap.J. (Letts), _281_, L33.
Snijders, M.A.J, 1987, in _RS Oph (1985) and the Recurrent Nova Phenomenon_, ed. Bode (VNU Sci. Press), p51.

THE 1987 OUTBURST OF THE RECURRENT NOVA U SCO

K. Sekiguchi[1], M.W. Feast[1], P.A. Whitelock[1], M.D. Overbeek[2],
W. Wargau[3] and J. Spencer Jones[1].

1. S.A. Astronomical Observatory, P.O. Box 9, Observatory 7935, S.A.
2. Box 212, Edenvale 1610, Transvaal, South Africa.
3. University of South Africa, P.O. Box 392, Pretoria 0001, S.A.

ABSTRACT. Spectral observations obtained soon after the 1987 brightening of U Sco support a thermonuclear runaway model for outbursts of this object. Spectra later in the decline are, however, more characteristic of a hot accretion disc. These observations are reconciled in a model where the low-mass high-velocity shell ejected from the surface of the white dwarf collides with the accretion disc causing it to brighten.

1 INTRODUCTION

The recurrent nova U Sco, has the fastest decline rate of all known novae and one of the shortest documented recurrence times amongst recurrent novae. Its outbursts have been modelled as thermonuclear runaways (TNRs) on a white dwarf with a mass near the Chandrasekhar limit (Starrfield et al. 1985, Webbink et al. 1987). It is therefore seen as a likely precursor of a type I supernova. We have obtained observations in May 1987 during the fifth recorded outburst. Of particular interest is an optical spectrum obtained closer to visual maximum than any previous spectra.

2 THE 1987 OUTBURST

Overbeek, who regularly monitors U Sco, found it to be at $m_v = 10.8$ on 1987 May 16.09. It appears that maximum light must have occurred around May 12/13 when U Sco was unobservable due to its close proximity to the moon. This highlights the possibility of missing completely such a rapidly evolving outburst. Subsequent photometry and spectroscopy suggest that this was in all respects a normal maximum. The recurrence time of previous recorded outbursts of U Sco averaged 39 yr. Theoretical models of TNRs were able to match this time scale only by assuming the mass of the white dwarf was at the Chandrasekhar limit. The last interval of only 8 yr between outbursts obviously presents a challenge to these models.

J. Mikolajewska et al. (eds.), The Symbiotic Phenomenon, 337–338.

338

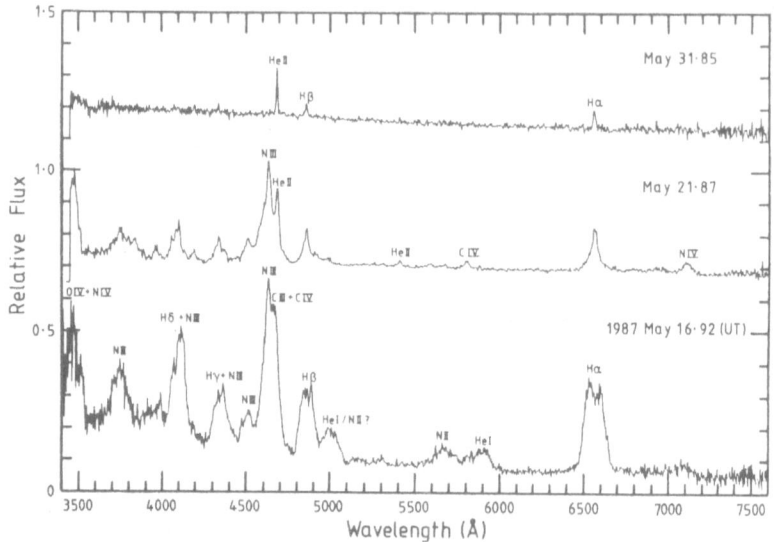

Fig. 1.   Spectra of U Sco obtained approximately 4, 9 and 16 day after
the estimated time of visual maximum.

3   SPECTRA

The spectra shown in Figure 1 were obtained at SAAO at approximately 4,
9 and 16 day after the estimated time of maximum light.   They can be
compared with spectra obtained during the 1979 outburst and discussed by
Barlow et al. (1981).
    The earliest spectrum shows the broad (FWZI ~ 10000 km s^{-1}) flat-
topped emission lines indicative of an optically thin shell expanding at
constant velocity.   Presumably the result of a low-mass shell ejected
at high velocity (~5000 km s^{-1}) following the TNR.   In contrast the
last spectrum shows only narrow (FWZI ~ 1500 km s^{-1}) emission lines of
HI and HeII.   It is similar to that seen in U Sco during quiescence
(Hanes 1985) which has been interpreted as originating from an accretion
disc (Webbink et al. 1987).   We therefore suggest that although the
outburst on U Sco originates as a TNR its later development is due to
brightening of the disc caused by the impact of TNR ejecta.
    A paper presenting the detailed observations and discussion
summarized here has been submitted to Monthly Notices of the Royal
Astronomical Society.

REFERENCES

Barlow, M.J. et al., 1981.   MNRAS, **195**, 61.
Hanes, D.A., 1985.   MNRAS, **213**, 443.
Starrfield, S., Sparks, W.M. & Truran, J.W., 1985.   Ap. J., **291**, 136.
Webbink, R.F. et al., 1987.   Ap.J., **314**, 653.

What can we learn from ζ Aur binary systems?

K.-P. Schröder
Hamburger Sternwarte, Universität Hamburg
Gojenbergsweg 112
2050 Hamburg 80
West Germany

Abstract. ζ Aur binaries are wide eclipsing systems (P ≃ 1...10 yrs), containing a late supergiant primary (G to M type) and an early dwarf companion (AO V ... B3 V). Some recent studies of wind phenomena (mass loss rates), chromospheric properties and wind acceleration (of the supergiant) as well as wind accretion phenomena introduced by the gravitational interaction with the companion are presented, summarizing previous work of the Hamburg 'binary wind team'(mainly D. Reimers, R. Baade, A. Che-Bohnenstengel, K. Hempe, K.-P. Schröder).

I) Introduction

In recent years it has been recognized that mass loss in red giants is decisive for the final evolution and fate of low and intermediate mass stars. IUE observations of ζ Aur systems offered a new possibility to obtain accurate mass loss rates.(Reimers, 1987).
Unlike common symbiotic stars, ζ Aur systems are well detached binaries (fig. 1), not resolved optically but spectroscopically. While the giant (G - M type) dominates the visual flux, the hot companion outshines him in the UV, though comparably pointlike. Three aspects are offered to the observer of ζ Aur systems:
First, IUE spectra show a wealth of resonance lines, formed in the supergiant wind upon the B star spectrum. Applying a non-sperical radiative transfer scheme (in the two level approximation, Hempe, 1982) Che et al (1983) where able to fit the line profiles by chosing adequate mass loss, wind velocity and turbulence parameters.
Second, the B star companion interacts locally with the supergiant wind. Complex CIV and SiIV features are observed, originating from a shock front and wind accretion onto the hot companion. Thinking in terms of mass loss determination, these phenomena are an unwanted disturbance of the wind, which is approximated to be continuous and sperically symmetric. Studies were made by A. Che-Bohnenstengel and D. Reimers (1986) as well as by Ahmad et al (1983) and Ahmad (1986).

J. Mikolajewska et al. (eds.), The Symbiotic Phenomenon, 339–346.

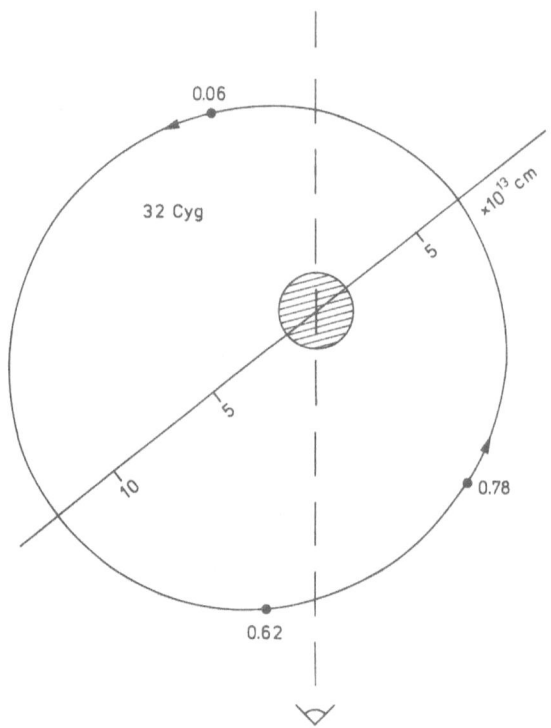

fig. 1: The orbit of the B star relative to the K supergiant
32 Cyg with some observed phases indicated, the B star not
drawn to scale.

Third, during eclipse of the companion by the supergiant chromosphere,
numerous absorption lines occur upon the B star spectrum. By means of
curves of growth, columndensities as a function of height can be determined. Integrating over a simple parameterized density model, Schröder
(1985) was able to derive the density distribution of the chromospheres
of three K supergiants. Assuming pure rayleigh scattering, these kind
of density models are also able to explain the wavelength dependence of
the eclipse light curves (Schröder, 1986). Fe I /Fe II ionization ratios
give us information about the electron densities of these chromospheres.
The steep density gradient (steeper than $\sim r^{-2}$) proves the acceleration
of the wind and we are able to fix the location of the so far not well
understood wind acceleration region in the upper chromosphere.
Thus, ζ Aur stars are the only stars besides the sun, where winds and
chromospheres an be studied in spatial (height) resolution, the B star
serving as an ideal probing light sourge.
   There is only one disadvantage with ζ Aur systems: there are only very
few of them because of their special character. Very well known are
ζ Aur, 31 Cyg and 32 Cyg (see Wright, 1970). New discoveries are 22 Vul
(Parsons and Ake, 1983) and HR6902 (Griffin & Griffin, 1986). A related
but more complex problem is VV Cep. δSge undergoes chromospheric eclipses
only.(Reimers and Schröder, 1983).

II) The supergiant wind

The wind is visible at all phases in P Cyg type profiles (during total eclipse of b star: pure emission lines) of ions like Fe II, Si II, S II, Mg II, C II, Al II and O I . These lines are formed by scattering of B star photons in the wind of the red giant. A few lines like Fe II UV mult. 9 (at 1270 Å) are seen in pure absorption due to the branching ratios of the upper levels which favour reemission as Fe II UV mult. 191 photons (Hempe and Reimers, 1982; Baade, 1986).

Theoretical modelling of wind line profiles and of their phase dependency has yielded accurate mass loss rates and wind velocities for a number of systems (Table 1). It has turned out, that a good mass loss determination requires both phases with the B star in front (showing wind material at about terminal velocity only, which yields wind turbulence v_{tur} from the width of the profiles) and phases with the B star behind the red supergiant (which yields the windvelocity v_w from the profiles, whose widths are about $2v_w + v_{tur}$ then). Typically, $v_w \simeq 2v_{tur}$ is obtained. Further details can be found in Che, Hempe and Reimers (1983). It turned out that it was possible to match the circumstellar line profiles at all phases with <u>one</u> set of parameters v_w, v_{tur} and – within a factor of 2 – one mass loss rate $\dot{M}$ (c.f. fig. 2). That means, at least in the orbital plane the envelope asymmetries (in density) are within a factor of 2 on a scale of several K giant radii. Table 1 gives a summary of the final parameters chosen by the work of Che et al, 1983.

Table 1. Summary of wind and turbulent velocities, and mass loss rates for ζ Aur, 32 Cyg and 31 Cyg

	v_{wind} {km/s}	$v_{turbul.}$ {km/s}	v_w {km/s}	v_{tur} {km/s}	$\dot{M}$ {$M_\odot$/yr}
ζ Aur	20...45	20...40	40	30	$0.63 \ 10^{-8}$
32 Cyg	30...60	15...30	60	25	$2.80 \ 10^{-8}$
31 Cyg	30...80	10...30	80	20	$\geq 1.0 \ 10^{-8}$

From the observed population of excited Fe II levels from 32 Cygni at distances of ≥ 5 K giant radii, Che–Bohnenstengel (1984) found $T_e = 4800$ K for $n_e/n_H \simeq 0.01$ and $T_e \simeq 10^4$ K for smaller electron densities. The LTE value would be 4200 K.

In 31 Cyg, the more extended strömgren sphere (the companion is the earliest of these three systems) complicates the quantitative interpretation of the wind line profiles in terms of mass loss.

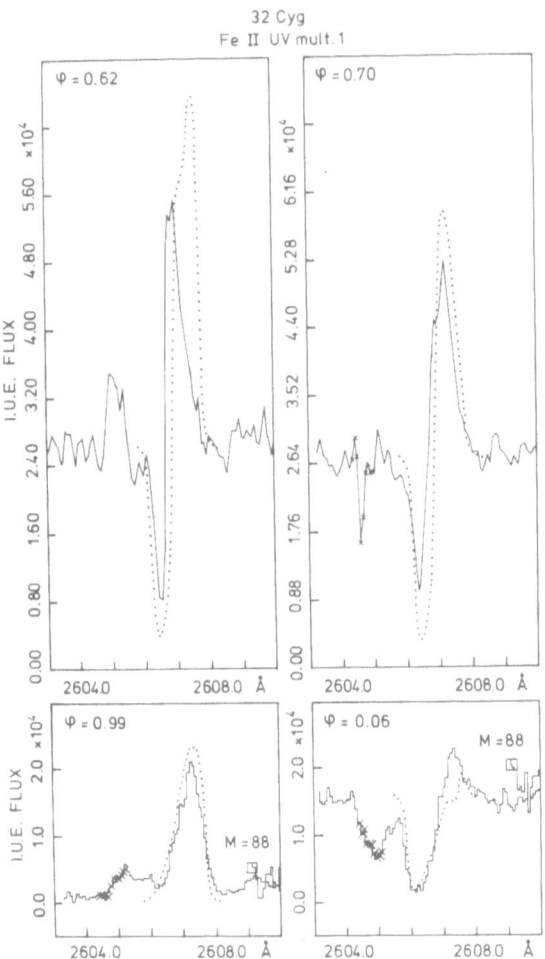

fig.2: Comparison of theoretical (.....) Fe Ⅱ UV mult. 1 wind
lines with the observation for 32 Cygni at various phases.

III) Shock front and accretion phenomena

There are several broad, complex high excitation lines observed in
N^{4+}, C^{3+}, Si^{3+}, Al^{2+} and Fe^{2+} ions, which cannot be explained by resonance scattering of B star photons in the wind, since the wind tempera-
ture is far too low to produce sufficient of such ions. The special
phase dependence of the emission and absorption features also excludes
an origin in a hot transition region of the red supergiant. Che-Bohnenstengel and Reimers (1986) studied the cIV and SiIV resonance doublets
in detail and proposed the following configuration (fig.3):
Formed by the supersonoc motion of the B star in the wind of the supergiant, there is a shock cone. The B star is located near to its apex.

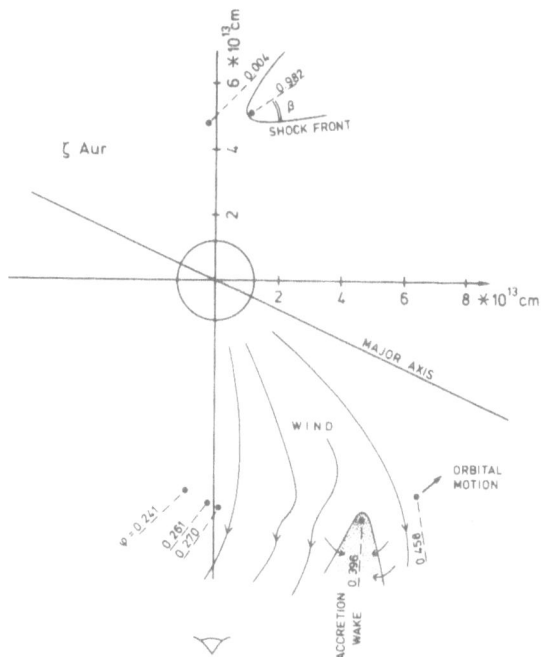

fig. 3: A roughly to scale presentation of shock front and
accretion phenomena of ζ Aur at various observed phases. The
disturbances in the wind at phase $\phi = 0.396$ are shown qualitatively. Broken lines indicate the aberration of the shock
cone axis (arctan v_{orbit}/v_{wind}) at each phase.

The emission region has an extension comparable to the supergiant diameter, since the (asymmetric) emission remains visible during total
eclipse of the companion, but its intensity has decreased considerable.
In the back of the cone, there is a clumpy accretion wake, causing
complex absorption features (observed in CIV and SiIV)shortly after
primary eclipse (with the cone then opened to the observer, semi angles
beeing about 25°...45°). Comparing the emission measures of CIV and
SiIV, Che-Bohnenstengel and Reimers (1986) estimated the temperature in
the shock cone region to be about 50000...80000 K. The density is enhanced there by about 3 orders of magnitude (compared to the wind), n_e
is of the order of $10^8...10^9/cm^3$ ($10^7....10^8/cm^3$ in the accretion wake).
    Applying the Theory of Livio and Warner (1984), they found ζ Aur and
δ Sge beeing candidates for an accretion disk, formed inside of the
shock cone. Actually, only these two systems have broad <u>symmetric</u> emission features, observed in the SiIV and CIV lines (except during ζ Aur
eclipse since the ζ Aur accretion disk is small enough to be totally
eclipsed by the giant) and at δ Sge Fe II UV mult. 78 lines and others,
emitted from the edge of the disk.

IV) Chromospheric eclipses

The extended chromosphere – where the wind already starts to expand –
could be studied by means of curves of growth, applied first by O.C.
Wilson, H.G. Groth, K.O. Wright and others in the 1950's. Today, we
benefit from much better atomic data and the IUE data are a major advance
in several aspects: the comparably pointlike b star provides a smoth
continuum (whereas at ≥ 400 nm the K giant contributes to the flux with
his complex spectrum). On it, one can observe numerous absorption lines
of Fe II, Ti II, V II, Fe I and more, up to heights h' (projected binary
separation) of partly more than one supergiantradius above the photosphere. The resultant height depending columndensities N(h') can well be
reproduced by a numerical integration of a simple density model (along
the relevant lines of sight), as shown by fig. 4. The density $\rho(h)$ is
represented by a power law of the form

$$\rho(h) \sim r^{-2} * h^{-a}$$

with 'a' of the order of 2.5 ... 3.5 (Schröder, 1985).

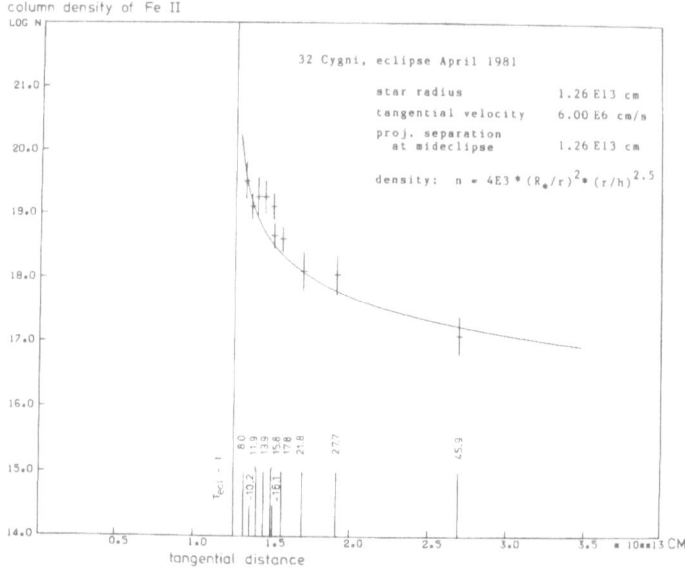

fig. 4: Observed columndensities obtained by means of curves
of growth of Fe II UV lines from 32 cyg chromospheric eclipse
versus tangential distance, compared to a track of theoretical
columndensities, calculated by numerical integration over the
density model specified in the plot.

In addition to absorption lines, the wavelength dependence of eclipse
continuum light curves can be reproduced by such density models for the

lower chromosphere. The parameters are slightly different but the density fits to the absorption line density in the overlapping height range. (Schröder, 1986). When continuum fluxes from IUE high resolution spectra are obtained (from line free sections only), pure rayleigh scattering turns out to explain the (line absorption cleaned) continuum opacity very well (fig. 5).

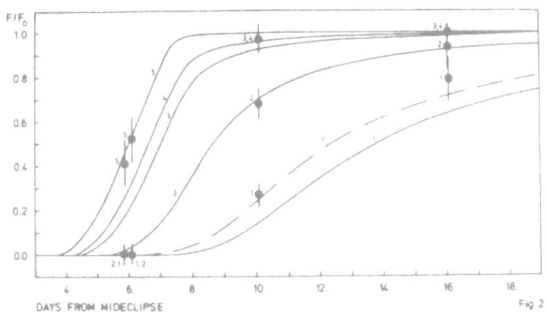

DAYS FROM MIDECLIPSE   Fig 2

fig. 5: Measured   32 Cyg normalized continuum fluxes (cleaned from line absorption) are indicated by dots at 1350Å (1), 1513Å (2), 1783Å (3), 1960Å (4) and 2992Å (5). Solid lines: fluxes from the best fitting model of a rayleigh scattering chromosphere at these wavelengths.

Observation of the iron ionization ratio Fe I /FeII $\simeq 10^{-3.5}$ and iron ionization equilibrium calculation yield electron densities of $n_e \lesssim 10^{-4}...10^{-2}$, increasing with height. While the metals are mainly in the first ionization stage (due to the radiation field of the B star, radiative ionization dominates), hydrogen has to be regarded as beeing nearly neutral (the strömgrenspheres of the B stars do not reach down into the chromospheres). Making a simple hydrogen ionization calculation, Schröder (1986) found Te $\lesssim$ 8000 K ... 11000 K, slightly increasing with height (h $\lesssim$ 0.5 R$_*$). Therefor , no dust formation is possible and radiative pressure on dust grains can be ruled out as a wind acceleration mechanism in these K supergiants.

Applying the equation of continuity to $\rho(h)$ and knowing the mass loss rate $\dot{M}$, v(r) can be derived. For 32 Cyg and 31 Cyg, the observed velocity is consistent with the thus derived terminal velocity. The density gradient is showing actually the acceleration of the wind. The ζ Aur density gradient is steeper than it should be from acceleration only. At that 1979 eclipse, we certainly observed a local mass loss deficiency.

On scales less than about one giant radius, the chromospheric matter is not as homogeneous as we, for simplicity, assume it to be. At egress of 1981 eclipse of 32 Cyg, Schröder (1983) found a very compact cloud of a diameter of about 1/6 giant radius close to the limb of the giant. It was about ten times denser than the surrounding matter and by rayleigh scattering, it produced a second dip in the shorter wavelength light curves. From old Ca II K observations, additional line absorptions at radial velocities up to ±100 km/s are known near eclipse (Wright, 1970).

These timedepending features are intrinsic problems, when studying wind acceleration. But in prinziple, ζ Aur systems for the first time enable us to determine location and fundamental physical parameters of the wind acceleration region, giving important constrains to every theory on wind acceleration mechanisms.
In conclusion, by their special character ζ Aur systems teach us a lot about the fundamental parameters of red supergiant winds and chromospheres. The interactions of the companion with the wind are restricted to the local environment and do not dominate the processes in the circumstellar matter further.

Acknowledgement: The Deutsche Forschungsgemeinschaft supported the binary project with several grants to the author and his collegues.

References:

Ahmad, I.A., Chapman, R.D., Kondo, Y,1983, Astrophys. J. 126, L5–L7
Ahmad,I.A., 1986, Astrophys. J. 301, 275
Baade, R., 1986, Astron Astrophys. 154, 145
Che, A., Hempe, K., Reimers, D.,1983, Astron. Astrophys. 126, 225
Che-Bohnenstengel, A.,1984, Astron Astrophys. 138, 333
Che-Bohnenstengel, A., Reimers, D., 1986, Astron. Astrophys. 156, 172
Griffin, R., Griffin, R., 1986, J.Astrophys.Astr. 7, 195
Hempe, K.,Reimers, D., 1982, Astron. Astrophys. 107, 36
Hempe, K., 1982, Astron. Astrophys. 115, 133
Livio,M., Warner, B., 1984, The Observatory 104, 152
Parsons, S.B., Ake, T.B., 1983, Inf. Bull. variable Stars No. 2334
Reimers, D., Schröder, K.-P., 1983, Astron.Astrophys. 124, 241
Reimers, D., 1987, Proc. IAU Symp. 122, 307
Schröder, K.-P., 1983, Astron. Astrophys. 124, L16
Schröder, K.-P., 1985, Astron. Astrophys. 147, 103
Schröder, K.-P., 1986, Astron. Astrophys. 170, 70
Wright, K.O., 1970, Vistas in Astronomy 12, 147

Summary of Final Discussion

Robert E. Stencel
Center for Astrophysics and Space Astronomy
University of Colorado
Boulder, CO 80309-0391 USA

The last session of the Colloquium was Chaired by Harry Nussbaumer and opened with statements from members of a panel comprised of Livio, Mikolajewska, Luud, Viotti, Magalhaes, Slovak, Kwok and Whitelock. Viotti began by comparing the available data for symbiotics historically and since IAU Colloquium 70 in 1981. He emphasized new UV, X-ray, infrared and radio observations. He also sketched the recent history of wavelength specific publications. Whitelock stated that it now seems clear all D-types probably contain Miras and that S-types do not evolve into D-types. Luud stated that his group intends to continue its near infrared observations with the 2 meter at Tartu, and that it is important to confirm conclusions based on IRAS data. He mentioned that masses for the Miras are needed given the discrepancy between observed and evolutionary masses.

Sun Kwok then presented his synthesis of the meeting in Table 1. This proved both helpful and provocative.

Webbink reminded the participants that it is important to compare symbiotics with related objects, like the Barium binaries. Yungleson added Bq or B[ ] stars to Webbink's suggestion (cf. Wackerling, in Mem. R.A.S. volume 73). Stencel emphasized the discontinuity in mass loss rates when dust formation begins (can increase the rate by a factor of 100 or more). This lead Stencel to further suggest that maybe the S-types are all first ascent red giants, the D'-types blue loop objects (He core burning main sequence) and the D-types all AGB stars. Whitelock objected, saying it probably is not that simple.

Mario Livio called for measured abundances of ejecta as a way of telling about the nature of the hot object. He complained about the lack of time for polarimetry reports at the meeting, indicating that polarimetry can provide important evidence about disks. Magalhaes answered, agreeing that IR polarimetry particularly could give useful diagnostics. He added that emission lines produced by fluoresence should be polarized and that planned space astronomy experiments like Astro's WUPPE and the Hubble Space Telescope's FOS should detect symbiotics. Schwarz and Viotti both talked about the importance of high resolution and high signal-to-noise spectroscopy to identify which features belong to the hot component.

Slovak wondered whether sporadic mass transfer could create disk instabilities. Cas-

J. Mikolajewska et al. (eds.), The Symbiotic Phenomenon, 347–348.
© *1988 by Kluwer Academic Publishers.*

Table 1: Cool Components in Symbiotic Binary Stars

Mira on AGB	M Giant	K or Earlier
M3-M10	M0-M3	
large dM/dt	low dM/dt	very low dM/dt
dusty	little dust	no dust
D-type	S-type	S'-type
nebular emission from		no radio emission

em.
lines → * O

$n_e \sim 10^6$ radio

em. lines

* $n_e \sim 10^{10}$

ionized M star wind		neb em from accretion
	Outburst Energy	
thermonuclear		gravitational
rapid accretion	low accretion rate	rapid accretion
H-burning steady	H-burning sporadic	
	Outburst Interval	
long ($> 10^2$ yr)	short (1 yr)	short (1 yr)
	White Dwarf Wind	
strong	weak	none

satella restated that the UV behaviour of these systems during outburst remains a puzzle. Vogel suggested the outburst could also be triggered by an increase in mass loss rate leading to a substantial continuum brightening, according to the ionization model. Seaquist responded, saying you might then expect the radio source to fade then as the ionization volume decreases. Webbink pointed out that the original nova model fits symbiotics nicely. The nova wind matches the orbital speed of the companion in energetics and strongly implies a key role. Solf asked why symbiotic optical and radio outflows however do not look like nova shells.

Nussbaumer exhorted everyone to publish their data, including negative results. Slovak asked whether the theorists could predict mass loss rates from symbiotic star P-Cygni profiles, and provide an interpretation for the high velocity line wings. Luud hoped more accurate distances could be obtained. Stencel suggested everyone use the meeting participant list for distribution of preprints and observation coordination.

CONCLUDING REMARKS

M. Friedjung
Institut d'Astrophysique
98 bis, Boulevard Arago, 75014 Paris, France

ABSTRACT. After a brief description of the difficulties and pitfalls of finding a correct interpretation of the symbiotic phenomenon, I emphasize what I consider to be the highlights of the meeting. Directions for future research are indicated.

In this talk I shall try to stimulate the later general discussion, and try to be a little provocative. Since the first IAU colloquium on this subject at the Haute Provence Observatory, there has been much progress in the physical understanding of symbiotic stars, so we decided to treat subjects in a different order. We wanted to consider physical processes early on during the meeting, and then see how observations fitted ! If both the theory of physical processes and the observations are good enough, it should not matter where we start, as the same reality is always studied. If our concepts are sufficiently broad and rigorous and our perceptions sufficiently numerous and accurate, we may expect to be able to reach reality. If not we may hope at least to obtain a useful way of representing in our minds what we know or think we know.

One question one can ask is : is there one symbiotic phenomenon ? Also what is it or what are they ? David Allen spoke to us about a primitive mammal, which he compared with a symbiotic star. This mammal is I believe only found in Australia; many symbiotic stars have been discovered in Australia by David Allen, but we cannot conclude that they are primitive ! Such reasoning is not valid; we must be somewhat more rigorous. We must look for contradictions between the observations and accepted ideas, and test the latter. In addition, nothing must be considered impossible, so we must not be chained to physical prejudices!

Almost all who work in the field of the symbiotic phenomenon consider symbiotic stars binary. A poll conducted by David Allen at the start of our meeting showed only one colleague who supported single star models. Since then I have received a telex from another who also supports such models. They are in a very small minority. This was not the situation 30 years ago. At that time Gauzit gave what he considered good reasons for believing the symbiotic star AX Per single. In the light of present day knowledge we can say that he was not aware of the

349

J. Mikolajewska et al. (eds.), The Symbiotic Phenomenon, 349–354.
© 1988 by Kluwer Academic Publishers.

complexities of binaries. The work of Boyarchuk in the optical about 20 years ago turned the tide towards binary models; one can perhaps state that ultraviolet observations with IUE played the main role in finally killing single star models. In this connections I would like to say how sorry I am that Professor Boyarchuk was unable to participate in this meeting.

Binary processes are however dependent on many different parameters; indeed one can wonder whether there are not too many parameters to derive definite conclusions. Symbiotic stars are often compared with cataclysmic binaries, whose general nature "we know". The latter are believed to consist of a white dwarf accreting from a Roche lobe filling companion, usually not very far from the main sequence. The white dwarf is thought to be surrounded by an accretion disk when the magnetic field is unimportant; when the field is large accretion is thought to take place via an accretion column. If you do not believe this type of model for cataclysmic binaries you will have a lot of trouble getting your papers accepted, while even if they are accepted after being read by a soft hearted referee, they will not be read by many people ! The question is how relevant are such ideas for symbiotic binaries. The latter appear to always have a cool giant mass donnor star which does not need to fill its Roche lobe. A cool giant can have a strong wind, especially if it is a Mira variable; while the main sequence, white dwarf or possibly neutron star mass gainer believed to be present, may accrete from this wind or by Roche lobe overflow. This new "orthodoxy" which is the framework in which most of us make our interpretations is less simple than that for cataclysmic binaries. Red giants and their winds are less well understood than stars near the main sequence, and all sorts of other physical processes can occur.

It is very easy to fall into traps or to become confused, even if the basic model just described is accepted as true. Some years ago I supported a model in which the symbiotic phenomenon was due to increased solar type activity of the cool giant, associated with a higher rotational velocity than for normal cool giants, because of tidal locking of the rotational and orbital periods. A region similar to the solar transition region might then produce the high ionization emission lines observed, while small variations in the wind from the cool giant could cause large changes in the accretion rate to the compact component, and hence in the nature of any accretion disk. This model was proposed because early IUE observations of Z And suggested that the hot continuum was not hot enough to produce the highest ionization lines in photoionized regions, while some lines at least were formed in a region where high temperature radiation was diluted, that is far from that where the hot continuum was formed, while a certain form of reasoning suggested that this region was thin. The fact that high ionization resonance emission lines of CI Cyg unlike other lines of this star were little or not eclipsed also seemed to support the model. However it now appears that enough high energy radiation is generally present for photoionization. Mikolojewska showed that the high ionization emission lines of CI Cyg had radial velocity variations probably in phase with those of the hot component, and first results on the widths of absorption lines of the cool component of CI Cyg obtained by me in collaboration with several French colleagues, suggest that the rotation of CI Cyg may not be tidally locked to its orbital period. The model may also have other problems. However even if effects of increased activity of the cool giant are less important than I thought, the possibility of their presence should not be forgotten in future interpretations. In any case the wind from the cool component seems often to be dominant for emission line formation, event if it is photoionized by radiation from the hot component.

We can now ask whether our meeting has succeeded. Our theoretical models are still very incomplete, so we had to talk a lot about observations on the first day, though the programme was arranged in such a way that we were building up towards theoretical models. The second day was devoted to theory, but when we tried to compare theory with observations of individual objects on the third day, something funny happened. The individual stars did not always seem to like our theories very much, so our models clearly need much improvement. Talks on related objects on the fourth day may however help us in this.

If I try to see what were the highlights of the meeting, the work on orbits seems particularly important. The work on the radial velocity variations of the absorption lines of the cool component has very much strengthened the binary interpretation since 1981; at that time the lack of reliable orbits was a weak point of such an interpretation. I was very impressed by the work on determining orbits from the reflection effect, even for such crazy objects as symbiotic stars where other effects can be strong, one can get reasonable results which even agree with those obtained from radial velocities. Physically, changes in the amount of accretion due to variations of the mass transfer rate associated with an eccentric orbit cannot be important in the amount of radiation emitted by a symbiotic binary for which the method works; either the luminosity directly due to the accretion is small, or the orbit is circular.

The cool component of a symbiotic binary is more "normal', and better understood than the hot one; the talks on the cool component helped us to better understand these components, which are a good starting point in the study of any symbiotic binary. The differences between binaries containing a Mira variable and those containing another type of cool giant were made very clear for us and should be the basis for future classification. Classification into S and D type symbiotics can have traps, as shown by the study of IRAS observations. Symbiotic stars may also make an important contribution to the study of the formation and destruction of circumstellar dust.

At the end of the first day several talks were given on observational methods which can give information about the geometry of symbiotic stars and their nebulae. Radio and optical observations show large deviations from spherical symmetry for the nebulae; jets and bipolar structure is common. Indeed even if an object appears spherically symmetric, it was pointed out that we may be seeing a bipolar structure nearly along its axis ! Polarization methods need perhaps to be further developed before all their possibilities can be realized.

The theoretical talks were on relatively simple physical models that is on photoionization of the cool star's wind, colliding winds, accretion disks including their instabilities and possible formation following accretion from a wind, and on thermonuclear processes resulting from accretion on to a white dwarf. All these and probably quite many other effects need to be taken into account in interpretation; each effect does not occur in isolation from the others. A subject where progress has in particular been made in recent years is that of disk formation following accretion from a wind.

The posters were presented towards the end of the second day. I was particularly struck by a poster indicating that the CNO abundancies of symbiotic stars are similar to those of normal M stars (see review by H. Nussbaumer in this volume), one on classification from emission line ratios (winner of the poster prize of IAU colloquium n° 103) and one showing that the visible surface of the active component of PU Vul is getting smaller and hotter. This selection is somewhat preliminary and personnal; reading the final texts in the proceedings of this meeting may show others to have been at least equally important.

As far as the presentations on individual stars are concerned, the fact that the luminosity of the hot component of Z And cannot be due to accretion from a wind is striking, as are the differences in outburst behaviour of Z And and AG Dra observed in the ultraviolet. In the former the very hot source previously seen disappeared (or perhaps rather cooled) as might have been expected from already known optical behaviour, while in the latter it is not clear whether the temperature changed or not. CI Cyg is important to study because of its eclipses, while the interpretation of CH Cyg is extremely uncertain. It is still not clear what is the best model for the excitation of the jet of R Aqr. I am rather surprised about the interpretation of the emission line profiles of RX Pup as produced by rings; similar stationary features are seen in the line profiles of classical novae and interpreted as due to ejection deviating from spherical symmetry, that is ejection of polar caps and equatorial rings. Why cannot such an interpretation be used for RX Pup, especially as it was active some years ago, perhaps somewhat resembling symbiotic novae whose properties were also presented ?

Among the subjects discussed on the last day let me note the interest of ξ Aur/VV Cep systems, which show certain similar phenomena to those of symbiotic stars, but in a less violent way. The detailed interpretation of such phenomena in ξ Aur/VV Cep systems may help us to see what to look for in symbiotic binaries. As far as recurrent novae are concerned, I must admit that I am somewhat sceptical about detailed interpretations. This is because even for classical novae, the measured accretion rate after the end of the explosion is too high compared with theoretical prediction, and it is not certain whether this problem can be solved by the "hibernation" model. Finally progress has been made in understanding the evolution of symbiotic and related stars.

On several occasions classical novae were mentioned and compared with symbiotic novae. Let me, as someone who has worked on novae for many years, point out some differences. Novae are cataclysmic binaries. There is very good evidence that after the initial explosion an optically thick wind is generated, which continues for a considerable time. This wind appears to be probably accelerated by radiation pressure and there are fairly good reasons for supposing the luminosity well above the Eddington limit in this stage. At such a time it is difficult to place the exploded star in the HR diagram as its surface is not seen. Physically this situation is attractively explained if the cool component is considered as then revolving inside the expanded white dwarf; viscous dissipation can produce a super Eddington luminosity, and perhaps also prevent the white dwarf moving further to the right (to cool effective temperatures) in the HR diagram. On the other hand there is as far as I am aware no compelling reason for believing that symbiotic novae and symbiotic stars in general have optically thick winds. The orbital separation appears to be much larger, so one component should never revolve inside an expanded other component, so never producing a super Eddington luminosity in this way. This is a basic diffrerence in the model, which should not be forgotten.

Let me finally turn to what the future holds for us. On the observational side we need more and better orbital data. It is particularly important to determine the orbital eccentricity, in order to see whether in some cases mass transfer variations and hence variations in the accretion rate, can be produced by varying orbital separation. This has a major bearing on certain models for symbiotic stars.

The cool companion needs to be better studied by high resolution spectroscopy in the infrared, and particularly by Fourier transform spectroscopy (FTS). By such methods we can find out to what extent the cool companion is really "normal", obtain better luminosities and determine the abundances of various elements in its atmosphere. In addition parts of the nebular envelope in front of the cool component can absorb line radiation; this geometry is different from that encountered at shorter wavelengths, and can give us new information. For instance a poster presented here by Bensammar and others including me describes FTS observations of CI Cyg in eclipse. A P Cygni absorption feature of the HeI 10830 Å line with a terminal velocity of the order of 150 km s^{-1} was observed, which was not seen during earlier observations at phase 0.5. It is in my opinion premature to interprete this...

Eclipsing symbiotic stars such as CI Cyg and as pointed out the under studied AR Pav, need to be examined in much more detail. The differing eclipses of different spectral features can give us much geometrical information. High spatial resolution methods can be expected to progress. FeII emission lines when seen, can be studied by the self absorption curve method developed by me and Muratorio, which also can give geometrical information. CH Cyg is a good example.

It was also pointed out that we must better observe PU Vul. This star, which is probably a symbiotic nova, could make the transition to the "nebular stage " rather quickly, and we should try to observe the transition.

On the theoretical side we need to better understand accretion disks. The classical Kenyon and Webbink work on the radiation emitted by a symbiotic star containing a disk was done assuming that a disk emits as a sum of black bodies; better assumptions need to be made in future. It was also pointed out that a disk emitting radiation by gravitational dissipation would be hard to detect, if it surrounded a white dwarf undergoing a thermonuclear event (though in fact such a disk should reprocess radiation from the white dwarf). The question is, can theory predict how observers should best be able to detect disks. In addition what kinds of disks are possible ? Our Greek American colleagues have proposed a thick accretion disk model for R Aqr, but as pointed out by one participant, it is not clear whether such a disk is physically possible.

Ionization models need to be further developed, and combined with detailed analysis of emission line profiles and fluxes. It is not sufficient to predict "radial velocities" for the emission lines, as already done profiles must be predicted. Future work should not only take account of the wind of the cool component, but also other regions able to emit line radiation, such as regions associated with accretion disks.

Additional problems are associated with the effect of the hot component on the cool one, on its outer layers, its wind, and on any dust that is present. What kind of wind comes from the hot component and/or any disk that may exist ? Colliding wind theories may need to be further extended, while we do not understand yet the physics of jets and bipolar flows, also seen for many other kinds of object. The last is a general astrophysical problem, studies of symbiotic stars may help in its solution.

It is clear that much work remains to be done. Our subject is still far from dead !

OVERHEARD IN TORUN (*)

"Who needs Quasars ?"   (Allen)

"You shouldn't throw away information"   (Magalhaes)

"The theory of the boundary layer is -- er -- OK, I'm not dealing
with the boundary layer"   (Duschl)

"Mira, as you know, does not have a period"   (Livio)

"The star just goes a little bit round the bend"   (Friedjung)

"I will not be surprised if there is a disk and I will not be surprised
if there is not a disk. It's the best way to be as a theoretician"
                                                              (Livio)

"Nature is not malicious, but it has a sense of humour"   (Schwartz)

"R Aquarii is psychotic"   (Michalitsianos)

"Models come and go, but observations are forever"   (Michalitsianos)

"Flip-flop is not going to help you here"   (Livio)

Viotti: "You see, you are the father of the symbiotic novae."
Allen: "Oh dear -- who was the mother ?"

"It's a real curse this FeII ... it's like a 17-years old youngster --
doesn't need anything to be excited"   (Nussbaumer)

"Now things are rather confused, as is normal in our subject"
                                                              (Friedjung)

"Progress would probably be easier without observations"   (Nussbaumer)

(*) Mostly collected by D. Allen. A few more sentences are reproduced
on the title page of each session.

Another sort of symbiotic?

(Reproduced by permission of Australian Geographic).

SUBJECT INDEX

When a subject in this index of subjects corresponding to each word or object in the following object index is quoted on many pages of an article, or it is the main topic of that article, only the title page followed by a '+' is indicated.

Abundance, chemical composition 49, 50, 59, 60, 108, 109, 163, 194, 199, 200, 279+, 299, 324, 329, 330, 331, 347, 352, 353.
Accretion 171+, 262, 279+, 323+, 351, 352, 353.
   disk accretion 6, 7, 18, 21, 56, 62, 71, 73, 81, 85+, 98, 104, 107+, 119+, 127, 137+, 149+, 161, 177, 178, 185, 187+, 193, 203, 215, 216, 223, 227, 233+, 239, 242, 246, 247, 307, 311+, 323+, 337+, 343, 350, 352, 353, 354.
   wind accretion 6, 7, 21, 39, 55, 56, 98, 127+, 149+, 161, 164, 174, 183, 185, 203, 224, 233+, 250, 253, 283, 311+, 342, 343, 350, 351, 352.
   instabilities 37, 38, 137+, 233+, 254, 323+, 347.
Angular momentum 149+, 254, 307, 312, 313.
Asymptotic giant branch 6, 8, 56, 62, 130, 135, 177, 347, 348.
ζ Aur-type binary 61, 62, 339+, 352.
BQ[] star 347.
Barium star 200, 308, 347.
Binary model/binary system/binarity 5, 11+, 23, 27+, 60, 73, 74, 88, 108, 207, 209, 221+, 227, 233+, 247, 253, 254, 274, 279, 283, 303, 311+, 339+, 349, 350.
Bipolar flow/bipolar nebula 9, 60, 81, 85+, 137, 177, 203, 242, 276, 285, 305+, 307, 308, 351, 354.
Boundary layer 107+, 119+, 147, 154, 214.
Bright spot/hot spot 223, 254.
Carbon star 8, 25, 109, 293+, 295+.
Cataclysmic variable 6, 7, 8, 19, 27, 154, 177, 178, 312, 324, 328, 350, 352.
Chandrasekhar limit 325, 337.
Chronosphere/transition region/corona 25, 57+, 73, 177, 190, 203, 339+, 350.
Colliding wind 74, 116, 129+, 246, 250, 351, 354.
Common envelope 161, 312, 313, 315.
Cool (late-type) component
   luminosity class 3+, 23, 27+, 37+, 183, 187, 200, 207, 209, 253, 289, 300, 302, 329, 333.
   spectral type 23, 27+, 37+, 88, 183, 185, 197, 200, 204, 207, 209, 235, 253, 261, 263, 269, 270, 283, 289, 291, 295, 300, 302, 329, 333.
D-type symbiotic star 11+, 37, 47+, 72, 73, 74, 95, 103, 117, 123+, 136, 177, 235, 245, 246, 249, 270, 274, 295+, 297, 311, 347, 348, 351.

OBJECT INDEX

Z And 11, 12, 27+, 37+, 42+, 45, 70, 101, 137, 147, 161, 164, 181+, 262, 307, 350, 352.
EG And 16, 27+, 39, 40, 101, 289+, 291+.
R Aqr 7, 9, 11+, 47+, 66, 69+, 77+, 86, 96, 97, 98, 101+, 103, 116, 125, 152, 215, 234, 235+, 246, 306, 352, 353.
VY Aqr 323, 324.
UV Aur 16, 27+, 65+, 101+, 293+, 295+.
ξ Aur 339+.
V CVn 91.
TX CVn 27+, 152.
R Car 67+.
S Car 67+.
SX Cas 25.
V641 Cas 23+.
BV Cen 324.
V704 Cen 47+.
V835 Cen 47+.
T Cep 67+.
VV Cep 23+, 340.
o Cet 6, 23, 24, 47+, 93, 94, 98, 156, 157, 183, 184.
5 Cet 25+.
R CrB 59.
T CrB 8, 11+, 37+, 107, 109, 323+.
BI Cru 47+, 297+.
Q Cyg 324.
BF Cyg 11, 37+, 43+, 70, 152, 299+, 301+.
CH Cyg 7, 9, 16, 20, 23+, 65+, 70, 71, 77+, 86, 88, 95, 96, 97, 101, 209+, 219+, 221+, 223+, 225+, 227+, 229+, 231+, 233+, 242, 276, 307, 352, 353.
CI Cyg 6, 11+, 27+, 37+, 95, 101+, 115, 137, 145, 147, 161, 164, 178, 187+, 193+, 197, 262, 311, 350, 352, 353.
V1016 Cyg 11+, 47+, 65+, 69+, 77+, 86, 101, 108, 134, 152, 164, 168, 262, 269+, 285+.
V1329 Cyg (=HBV 475) 20, 70, 108, 152, 164, 168, 169, 269+, 285+, 308.
31 Cyg 340, 341, 345.
32 Cyg 339+.
HR Del 85.
AG Dra 19, 20, 27+, 33+, 37, 39, 42+, 45, 70, 101, 152, 155, 156, 164, 177, 185, 199+, 205+, 262, 352.
RY Gem 25.
DQ Her 85.
YY Her 115.
V443 Her 40.
V446 Her 324.
RW Hya 6, 11, 27+, 70, 183.
RZ Leo 324.
BX Mon 152.

363

Hen 1410 70.
Hen 1591 47+.
He2-38 47+.
He2-104 47+, 305+.
He2-106 70.
He2-111 305.
He2-127 47+.
He2-139 47+.
He2-147 47+.
He2-171 47+, 70.
He2-176 70.
He2-390 47+, 70.
HR 6702 193+.
M 1-1 305.
NGC 2392 85.
NGC 7027 305.
Nova Vul (1979) 324.
SS 38 47+, 295+.
SS 96 70, 79, 82.
SS 122 47+, 70.
Th 3-7 70.
UKS Ce-1 296.
W16-312 47+, 70.